Functionalized Graphene Materials and Their Applications

Functionalized Graphene Materials and Their Applications

Linjie Zhi

China University of Petroleum (East China), China

Yingjie Ma

National Center for Nanoscience and Technology, China

Debin Kong

China University of Petroleum (East China), China

Translated by **Sijia Hao**

*Beijing Institute of Aeronautical Materials, China &
Beijing Institute of Graphene Technology Ltd. Co., China*

W🌐World Scientific

NEW JERSEY • LONDON • SINGAPORE • BEIJING • SHANGHAI • TAIPEI • CHENNAI

Published by

World Scientific Publishing Co. Pte. Ltd.

5 Toh Tuck Link, Singapore 596224

USA office: 27 Warren Street, Suite 401-402, Hackensack, NJ 07601

UK office: 57 Shelton Street, Covent Garden, London WC2H 9HE

Library of Congress Control Number: 2025005545

British Library Cataloguing-in-Publication Data
A catalogue record for this book is available from the British Library.

功能化石墨烯材料及应用
Originally published in Chinese by East China University of Science and Technology Press
Copyright © East China University of Science and Technology Press, 2022

FUNCTIONALIZED GRAPHENE MATERIALS AND THEIR APPLICATIONS

ISBN 978-981-98-0607-2 (hardcover)
ISBN 978-981-98-0608-9 (ebook for institutions)
ISBN 978-981-98-0609-6 (ebook for individuals)

For any available supplementary material, please visit
https://www.worldscientific.com/worldscibooks/10.1142/14124#t=suppl

Desk Editors: Murali Appadurai/Rhaimie Wahap

Typeset by Stallion Press
Email: enquiries@stallionpress.com

Preface

This book proposes the systematic classification of and exact definitions for all the products from graphene functionalization, which are illustrated by several representative examples. Additionally, the corresponding preparation strategies for functionalizing graphene are summarized and demonstrated by some typical approaches. All the products from graphene functionalization are defined as functionalized graphene materials, which fall into two categories: functionalized graphene and functionalized graphene composite. Functionalized graphene is the product of modifying graphene by tuning its composition, framework, dimension, and morphology, and functionalized graphene composites are hybrids of graphene (or functionalized graphene) and other materials, including small molecules, polymers, metals, inorganic compounds, and carbon nanotubes (CNTs). Functionalized graphene materials are prepared through two strategies, "top-down" and "bottom-up," each of which has its advantages and shortcomings and includes many corresponding preparation methods. The selection of preparation strategies depends on the application requirements, as different applications require different types of graphene. Both strategies are elucidated with detailed examples through an extensive analysis of the literature. Finally, the major challenges and perspectives of utilizing functionalized graphene materials are discussed.

About the Authors

Linjie Zhi, a professor and doctoral supervisor at China University of Petroleum (East China), is a recipient of the National Science Fund for Distinguished Young Scholars (in 2014), has been selected for the "Hundred Talents Program" by the Chinese Academy of Sciences, and has also been selected as a "Highly Cited Researcher" by Clarivate Analytics for eight consecutive years. He heads the 'Carbon-rich Nanomaterials' Partner Group of the Max Planck Institute for Polymer Research. He serves as a member of the Chinese Society of Particuology, the Chinese Society for Composite Materials, the Energy Society, and the China Innovation Alliance of the Graphene Industry (CGIA). He also serves as an editorial board member of several academic journals, including *Adv. Mater. Tech.* and *New Carbon Mater*. He has successfully undertaken more than 20 projects which lie in key and joint key projects of the Major Research Plan of National Natural Science Foundation of China, international cooperation project of the Ministry of Science and Technology of China, the "863" program, the International Cooperation and Exchange Programs of National Natural Science Foundation of China, as well as National Science Fund for Distinguished Young Scholars. He has published over 260 research papers in international journals, such as *Nat. Commun., Angew. Chem. Int. Ed., J. Am. Chem. Soc., Adv. Mater., and Adv. Funct. Mater.*, which have been cited more than 38,000 times, with an H-factor of 86. He has filed more than 40 national invention patents, and three of these patent packages have achieved technology transfer and implementation.

Yingjie Ma is an associate researcher at the National Center for Nanoscience and Technology. He has engaged in research on the application of organic functional materials and carbon materials in the field of energy storage and energy conversion. He is a young editorial board member of *Acta Physico-Chimica Sinica*. He has presided over three projects, including the general program of the National Natural Science Foundation of China, the general program of the Beijing Natural Science Foundation, and the Zhejiang Provincial Fund project. He has published over 30 academic papers, which have been cited over 2,000 times, with an H-factor of 22. He is also the owner of seven authorized Chinese invention patents.

Debin Kong, a professor and doctoral supervisor at China University of Petroleum (East China), is a Taishan Scholar Young Expert and "Guanghua Scholar." He is a member of the Energy Storage Technology Branch of China Internal Combustion Engine Society, a technical expert of the "Inorganic Non-metallic Materials Industry Center of National New Materials Testing and Evaluation Platform" of the Ministry of Industry and Information Technology, and a guest editor of *Molecules*. He has presided over more than 10 national, provincial, and ministerial projects, such as the name of this project is "InterGovernmental Key Special Project", which is funded by "National Key Research and Development Program" and the Special Project for High-Quality Development of the Ministry of Industry and Information Technology of the People's Republic of China. He is the owner of 6 authorized patents (one Japanese patent) and has co-authored 4 academic monographs, including *Functionalized Graphene Materials and Applications and Graphene Electrochemical Energy Storage Technology*. He has published over 60 academic papers in international journals such as *Nat.Commun., Energ. Environ. Sci.*, and *Adv. Mater.*, and his work has been cited more than 5,000 times with six papers being selected as ESI highly cited papers. His research results have been honored with the "Carbon Journal Prize" by *Carbon*, an authoritative journal in the field of carbon materials.

About the Translator

Sijia HAO has been a senior engineer at the Research Center of Graphene Applications at AECC Beijing Institute of Graphene Technology Co., Ltd. He has led a research group at Beijing Institute of Graphene Technology Ltd. since 2020. Prior to joining BIAM, he received his PhD from the Tokyo Institute of Technology under the supervision of Toshiaki Enoki in 2012 and conducted post-doctoral research at Tsinghua University. He has actively worked on research and development of graphene nano-composites for a variety of applications. He has co-authored and translated several books on graphene and other carbon nanomaterials.

Contents

Chapter 1

Overview

1.1 Introduction of Graphene

Graphene is a two-dimensional (2D) structure consisting of a single-atom thick sheet of carbon crystallized as a honeycomb structure monolayer, which is generally considered to be the fundamental building block for all sp^2 graphitic materials, including 0D fullerenes and 1D carbon nanotubes, stacked into 3D graphite (Fig. 1.1). Graphene is considered to be the thinnest, strongest, lightest, stiffest material ever created, and it is also an excellent thermal and electrical conductor, demonstrating great application prospects in many fields. The term graphene was introduced by chemists Hanns-Peter Boehm, Ralph Setton, and Eberhard Stumpp in 1986 as a combination of the word "graphite," referring to carbon in its ordered crystalline form, and the suffix "ene," referring to polycyclic aromatic hydrocarbons in which the carbon atoms form hexagonal, or six-sided, ring structures. The word graphene, when used without specifying the form (e.g., bilayer graphene, multilayer graphene), usually refers to single-layer graphene.

In 1924, a British crystallographer, John D. Bernal, was the first to propose the lamellar structure of graphite; that is, different carbon atomic layers are stacked in an ABAB manner, and the interplanar distance is 0.3354 nm, but there is no chemical bond between the layers, so the out-of-plane interaction is weak.[1] Since van der Waals forces are the only interactions between adjacent carbon layers to maintain the layered structure of graphite, the atomic layers can easily slide onto each other and, hence, graphite can be used as a lubricant and pencil lead. It is also true

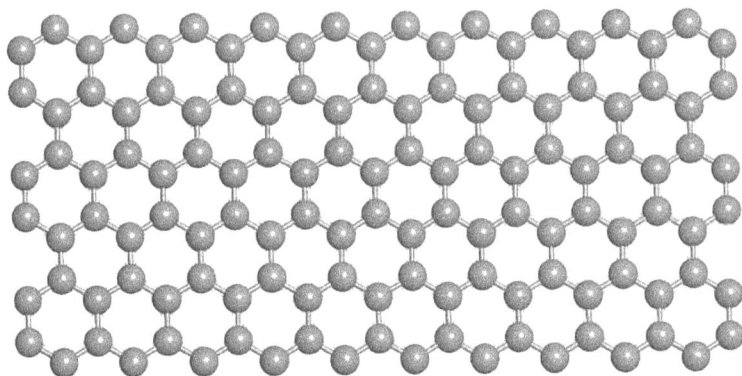

Figure 1.1. Schematic diagram of graphene structure.

that the lack of a chemical bond between adjacent layers leads to weak interlayer bonding, which offers the possibility of fabricating graphene via the mechanical exfoliation method.

Since the lamellar structure of graphite is determined, researchers have devoted themselves to obtaining ultrathin graphite sheets by peeling 3D graphite crystals. In the 1940s, a series of theoretical analyses predicted that monolayer graphite exhibited novel electronic properties.[2] Therefore, research on the peeling of graphite never stopped. In 1962, when Hanns-Peter Boehm and his co-workers used a transmission electron microscope (TEM) to observe graphite flakes in a reduced graphitic oxide solution, the thinnest flakes were observed with an average thickness of 3.7 Å (1 Å = 10^{-10} m), but unfortunately, they only simply described this discovery as follows: "This observation confirms the assumption that the thinnest of the lamellae really consisted of single carbon layers."[3] In 1988, Kyotani *et al.* employed the template method to prepare monolayer graphene between the lamellae of montmorillonite. However, once the template was removed, the sheets would self-assemble to form bulk graphite.[4] In the late 1990s, Ruoff and co-workers tried to isolate thin graphitic flakes on SiO_2 substrates by mechanical rubbing of patterned islands on HOPG (Highly Oriented Pyrolytic Graphite), after which graphite flakes with a thickness of ~200 nm were obtained using the tip of an atomic force microscope (AFM).[5,6] Using a similar method, Kim and co-workers at Columbia University fabricated thin graphite flakes with a thickness of 20–30 nm.[7,8] In addition, several research groups, such as Enoki's,[9,10] have also synthesized very thin films of graphite, but these are multilayer graphite sheets, and there is still a long way to go before obtaining single-atom-thick graphite.

In 2004, Andre Geim and Konstantin Novoselov at Manchester University used mechanical cleavage for the first time to obtain a monolayer graphite, namely, "graphene."[11] According to a definition by the International Union of Pure and Applied Chemistry (IUPAC),[12] graphene is a single layer of graphite structure, describing its nature by analogy to a polycyclic aromatic hydrocarbon of quasi infinite size, but it should be noted that in this definition, the term graphene should be used only when the reactions, structural relations or other properties of individual layers are discussed. Geim defined graphene as a single atomic plane of graphite, which — and this is essential — is sufficiently isolated from its environment to be considered free-standing.[13] Geim emphasized the free-standing nature of graphene, that is, a single-atom-thick sheet of graphite that does not adhere to a foreign substrate is considered to be graphene. Free-standing is an essential feature of graphene. More than 80 years ago, Peierls and Landau argued that strictly 2D crystals were thermodynamically unstable and could not exist in nature.[14,15] Therefore, it was believed that graphene was an integral part of larger 3D structures, usually grown epitaxially on top of monocrystals with matching crystal lattices.[16] Without such a 3D base, 2D materials were presumed not to exist, until 2004, when common wisdom was supported by the experimental discovery of graphene and other free-standing 2D atomic crystals.[11]

Among various 2D materials, graphene has received extensive research attention in the last 2−3 decades due to its fascinating properties. The discovery of graphene provided an immense boost and a new dimension to materials research and nanotechnology. For example, graphene's charge carriers exhibit giant intrinsic mobility that exceeds 200,000 $cm^2/(V \cdot s)$. Single-layer graphene shows extremely high transparency and only absorbs 2.3% of visible light; its Young's modulus is as high as 1.0 TPa, which is the strongest ever measured, and it exhibits mammoth thermal conductivity of up to 5000 $W/(m \cdot K)$. Due to its unique structure and properties, graphene offers an ideal platform for experimental verification of a variety of theoretical analyses that could not be realized earlier and provides a new material for scientific research and application technology development in many fields. Many spectacular phenomena of graphene have been observed, such as the half-integer quantum Hall effect of electrons and holes even at room temperature, extraordinarily high carrier mobility (more than 10 times the mobility of commercial silicon wafers), and ultrahigh sensitivity of single-molecule detection. Further, graphene also exhibits other superior characteristics of an optical, thermal, or mechanical nature. In short, within a very short timescale of less than 20 years, graphene has presented many unique properties and fascinating functions. It thus offers new challenges for materials scientists and great opportunities for further research and development.

1.2 Functionalized Graphene Materials

1.2.1 *Why functionalization*

After more than ten years of development, graphene has already shown promise in the fields of energy storage, catalysis, microelectronics, biomedicine, and the environment and has revealed great potential for wide-ranging applications. It is worth noting that in order to meet the requirements of applications in different fields, graphene is required to be functionalized, providing functionalized graphene with various structures and tunable properties. Functionalization is one of the efficient ways of tailoring the electrical, optical, and mechanical properties of graphene, which will find more applications in diverse fields. In fact, a vast majority of graphene utilized in various fields is functionalized graphene instead of pristine graphene. This issue is discussed in the following.

Firstly, the functionalization of graphene is required to improve its processability. Pristine graphene is a 2D crystal composed of sp^2-hybridized carbon atoms in the form of a hexagonal lattice without any dangling bonds, exhibiting high chemical stability. The inert surface of graphene makes it difficult to interact with other materials (such as solvents). Furthermore, graphene is easy to aggregate due to the strong van der Waals force between the graphene layers, making it difficult to dissolve in water and commonly used organic solvents. This is one of major obstacles for further research and application of graphene. In order to take fully advantage of the fascinating properties of graphene and enable this material to be processed, such as improved dispersibility and processability in solvent and matrix, effective functionalization is needed.

Secondly, the functionalization of graphene is capable of adding new features to graphene materials. While the inherent properties of graphene are not compromised, functionalized graphene exhibits a series of novel properties that not only enrich the scientific knowledge of graphene

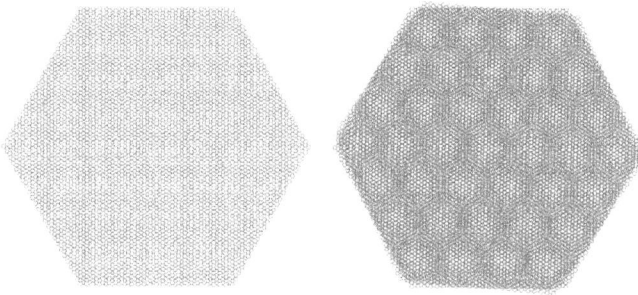

Figure 1.2. Schematic illustration of "magic-angle" graphene superlattice.[19]

materials and exploit the applications of graphene but also provide new material systems for some applications. As shown in Fig. 1.2, 2D topological superlattices consisting of two graphene layers have been found to be excellent platforms for obtaining attractive physical features that were not previously expected.[17,18] When the twist angle of two layers of graphene stack together is cloase to the magic angle of 1.1°, the Fermi velocity vanishes at the Dirac points with the formation of a nearly flat band. Superconductivity with zero resistance is stunningly detected at an extremely low temperature of 1.7 K. Another example is graphene quantum dots, which consist of graphene nanoparticles with a size less than 100 nm, exhibiting pronounced quantum confinement effect.

Last but not least, the functionalization of graphene is done to meet the needs of specific fields. Applications in different fields have different requirements for graphene. Hence, it is necessary to functionalize graphene to meet the required properties of applications. For example, graphene itself possesses zero band gap, doping or striping would be useful for band gap opening. When graphene is used as a catalyst, it is necessary to introduce catalytic sites by doping or etching. When graphene is used as an electrode material for supercapacitors, doping is adopted to improve the affinity of graphene to the electrolyte or to introduce pseudocapacitive active sites. When graphene is used in aqueous or organic solvent systems, it is necessary to graft water-soluble or fat-soluble chains onto graphene in order to achieve sound solubility in the corresponding solvent. When graphene is used as an adsorption material, it needs to be in a porous 3D structure to enhance its adsorption capacity.

However, functionalized graphene has not been systematically classi-fied or strictly defined until now. Systematic classification and exact defi-nition will be beneficial for research and applications of functionalized graphene. Therefore, in order to facilitate the research and application of graphene materials and promote their further development, based on the authors' work in the field of graphene, this book reviews recent research findings on functionalized graphene, proposes a systematic classification of functionalized graphene materials and their corresponding definitions, and also discusses their major challenges and perspectives.

1.2.2 *Functionalized graphene*

In general, via functionalization, the features of graphene such as compo-sition, size, shape, and structure can be modified, consequently tuning its electronic structure, solubility, mechanical and chemical properties, and even endowing it with new properties. For example, heteroatoms such as N, S, P, and B can be incorporated into graphene by doping to change the elemental composition. Etching of graphene via physical or chemical means could modify the carbon framework, such as introducing ordered nanopores or disordered defects, thus tuning the electronic structure of graphene and opening its bandgap or making it catalytically active. The structure of graphene can also be tuned by decorating functional groups through covalent modification. Graphene in different sizes and shapes can be manipulated by means of controlled synthesis or etching. Graphene can be rolled into a 1D tube, stacked into a double-layer or triple-layer struc-ture, or fabricated into a 3D porous structure to achieve the manipulation of graphene micromorphology.

1.2.3 *Functionalized graphene composites*

To meet specific needs, graphene or functionalized graphene has been widely employed as the main constituent of novel composite materials with enhanced and new properties. The successful combination of two or more components can generate composite materials, which show the indi-vidual properties of each component as well as completely new proper-ties. This is currently expanding the fields of application of existing materials and opening new avenues across different scientific fields, enhancing their technological progress. In composite materials, graphene or functionalized graphene plays a specific role as a substrate or a functional component.

As a substrate material, graphene or functionalized graphene is often used to support active materials in composite materials. For instance, in the application of single-atom catalysts, functionalized graphene with heteroatom doping or with defects is employed as a support material to anchor single-atom active centers, making them highly dispersed and inhibiting their aggregation. Another example is that in the application of graphene sensors, functionalized graphene not only functions as the conductive layer but also serves as a recognition site for the substrate support sensor.

Compared with substrate materials, graphene or functionalized graphene is more commonly used as the functional component of functionalized graphene composites. For example, single-layer or few-layer graphene functions as the active layer on a flexible polymer substrate, giving rise to a flexible graphene transparent conductive film. In a transparent conductive film of silver nanowires, graphene oxide is used as a binder to firmly "weld" the silver nanowires together, resulting in a silver nanowire network to enhance its conductivity. In photocatalysts, graphene nanosheets provide highly efficient electron transport channels to enhance the electron transfer efficiency in the catalytic process. In organic solar cells, graphene oxide is used to modify the hole transport layer to reduce its interfacial resistance. In lithium-ion batteries, highly dispersed graphene sheets are used as conductive agents together with silicon carbon active materials to form the anode of the battery. In addition, graphene sheets as active materials are hybridized with polymer substrates to support electromagnetic wave-absorbing materials.

The hybridization of graphene or functionalized graphene with one or more other materials would give rise to a wide variety of functionalized graphene composite materials with rich functions, which is also the most preferred application form of graphene. In a broad sense, functionalized graphene and functionalized graphene composite materials can be termed "functionalized graphene materials." In fact, the vast majority of graphene used in various fields is referred to as functionalized graphene materials.

1.3 Summary

Graphene is an emerging material with great application potential. It has been just around 20 years since graphene was first reported, and despite remarkably rapid progress, only the very tip of the iceberg has been uncovered so far. It should be noted that due to technical reasons or application requirements, pristine graphene is not the mainstream application mode of graphene, and it is often necessary to functionalize graphene or hybridize it with other materials. The following chapters of this book will present the basic concepts, preparation methods, and applications of functionalized graphene materials.

References

1. Bernal, J. D. and Bragg, W. L. (1924). The structure of graphite. *Proc. R. Soc. Lond. Ser. A-Contain. Pap. Math. Phys. Character*, 106(740), 749–773.
2. Wallace, P. R. (1947). The band theory of graphite. *Phys. Rev.*, 71(9), 622–634.
3. Boehm, H. P., Clauss, A., Fischer, G., and Hofmann, U. (1962). Surface properties of extremely thin graphite lamellae. In *Proceedings of the Fifth Conference on Carbon*, pp. 73–80.
4. Kyotani, T., Sonobe, N., and Tomita, A. (1988). Formation of highly orientated graphite from polyacrylonitrile by using a two-dimensional space between montmorillonite lamellae. *Nature*, 331(6154), 331–333.
5. Lu, X., Yu, M., Huang, H., and Ruoff, R. S. (1999). Tailoring graphite with the goal of achieving single sheets. *Nanotechnol.*, 10(3), 269–272.
6. Lu, X., Huang, H., Nemchuk, N., and Ruoff, R. S. (1999). Patterning of highly oriented pyrolytic graphite by oxygen plasma etching. *Appl. Phys. Lett.*, 75(2), 193–195.
7. Zhang, Y., Small, J. P., Amori, M. E. S., and Kim, P. (2005). Electric field modulation of galvanomagnetic properties of mesoscopic graphite. *Phys. Rev. Lett.*, 94(17), 176803.
8. Zhang, Y., Small, J. P., Pontius, W. V., and Kim, P. (2005). Fabrication and electric-field-dependent transport measurements of mesoscopic graphite devices. *Appl. Phys. Lett.*, 86(7), 073104.
9. Affoune, A. M., Prasad, B. L. V., Sato, H., Enoki, T., Kaburagi, Y., and Hishiyama, Y. (2001). Experimental evidence of a single nano-graphene. *Chem. Phys. Lett.*, 348(1), 17–20.
10. Kikuo, H., Yousuke, K., Kazuyuki, T., Jérôme, R., and Toshiaki, E. (2002). Novel electronic wave interference patterns in nanographene sheets. *J. Phys.: Condens. Matter.*, 14(36), L605–L611.
11. Novoselov, K. S., Geim, A. K., Morozov, S. V., Jiang, D., Zhang, Y., Dubonos, S. V., Grigorieva, I. V., and Firsov, A. A. (2004). Electric field effect in atomically thin carbon films. *Sci.*, 306(5696), 666–669.

12. Fitzer, E., Kochling, K.-H., Boehm, H. P., and Marsh, H. (1995). Recommended terminology for the description of carbon as a solid (IUPAC Recommendations 1995). *Pure & Appl. Chem.*, 67(3), 473–506.
13. Geim, A. K. (2009). Graphene: Status and prospects. *Sci.*, 324(5934), 1530–1534.
14. Peierls, R. (1935). Quelques propriétés typiques des corps solides. *Annales de l'institut Henri Poincaré*, 5(3), 177–222.
15. Landau, L. D. (1937). Zur Theorie der phasenumwandlungen II. *Phys. Z. Sowjetunion*, 11(545), 26–35.
16. Venables, J. A. and Spiller, G. D. T. (1983). Nucleation and growth of thin films. In Binh, V. T. (ed.), *Surface Mobilities on Solid Materials: Fundamental Concepts and Applications*, pp. 341–404. Springer US, Boston, MA.
17. Cao, Y., Fatemi, V., Demir, A., Fang, S., Tomarken, S. L., Luo, J. Y., Sanchez-Yamagishi, J. D., Watanabe, K., Taniguchi, T., Kaxiras, E., Ashoori, R. C., and Jarillo-Herrero, P. (2018). Correlated insulator behaviour at half-filling in magic-angle graphene superlattices. *Nature*, 556(7699), 80–84.
18. Cao, Y., Fatemi, V., Fang, S., Watanabe, K., Taniguchi, T., Kaxiras, E., and Jarillo-Herrero, P. (2018). Unconventional superconductivity in magic-angle graphene superlattices. *Nature*, 556(7699), 43–50.
19. Jarillo-Herrero Group. http://jarilloherrero.mit.edu/research/.

Chapter 2

Definition and Classification of Functionalized Graphene Materials

To date, much effort has been devoted to functionalized graphene materials (FGMs). As a result, numerous different products of FGMs have been fabricated. In this respect, functionalization strategies vary according to the type of graphene and the nature of the component used to functionalize it. In the literature, there are numerous scientific articles concerning the methods to functionalize graphene and its derivatives.[1,2] Therefore, it is necessary to clarify the related concepts, main preparation strategies, and systematic classification of FGMs. In this chapter, according to the composition and characteristics of the material system, the classification, definition, and preparation strategies of FGMs are outlined on the basis of the body of literature on graphene (Fig. 2.1).

2.1 Definition of FGMs

Functionalized graphene (FG) refers to the graphene material obtained by tailoring the elemental composition, structure, size, and morphology of graphene through physical or chemical approaches or both. FG features the structure of graphene, that is, sp^2-hybridized carbon atoms in a 2D hexagonal lattice, which retains some of the intrinsic properties of graphene. Meanwhile, FG shows new structural arrangement and new physical and chemical properties. The functionalization of graphene allows the application of graphene materials in specific fields as well as the

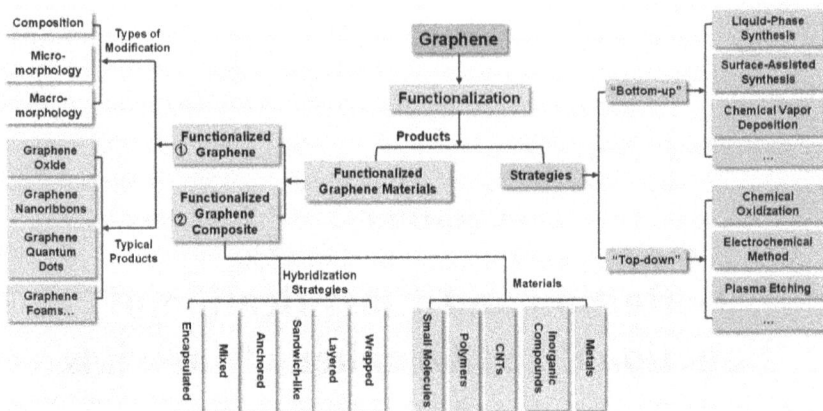

Figure 2.1. Graphene functionalization: Classification, definition, and preparation strategies.

development of new graphene materials, and further reveals the properties of graphene.

Functionalized graphene composites (FGCs) refer to the composite materials fabricated by hybridizing graphene or FG with one or more functional materials, aiming to meet the specific demands of certain applications. In FGCs, graphene or FG usually retains its original composition, structure, physical and chemical properties (under some circumstances, the composition and structure of graphene may be modified by doping), and functions in combination with other functional materials. The utilization of the properties of both graphene and other functional materials allows FGCs to meet the requirements of different applications.

FGMs are collectively referred to as both FG and FGCs. In fact, from the perspective of the material system, almost all graphene materials currently used can be categorized as FGMs (Fig. 2.1).

2.2 Classification of FGMs

2.2.1 *Functionalized graphene*

As mentioned, FGMs can be classified into FG and FGCs, and FG is prepared by modifying one or more aspects of graphene, including the composition, structure, dimension, and morphology.

Quaternary N   Pyrrolic N

Pyridinic N

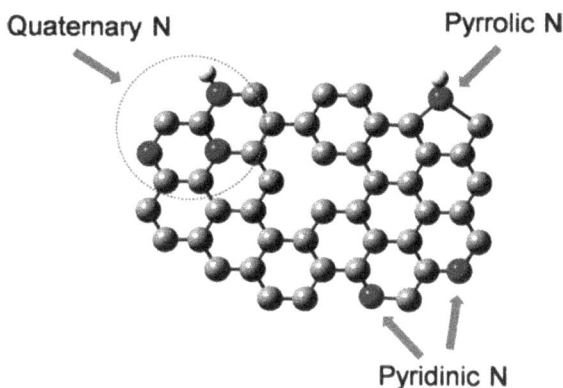

Figure 2.2. Illustration of FG with a modified composition: nitrogen-doped graphene for biosensing.[3]

1. Composition

Heteroatoms such as O, N, S, P, or B are incorporated into graphene by doping to obtain heteroatom-doped FG (Fig. 2.2). Compared to carbon, heteroatoms have different valence electrons and electronegativities. If the carbon atoms of graphene are substituted with these heteroatoms, electrons will be injected into or extracted from graphene, causing a Fermi level lift and consequently producing n-type or p-type doped graphene. After being doped with functionalized groups or heteroatoms, the lattice symmetry of graphene is broken, resulting in a gap between π and π^* bands and converting graphene into a semiconductor. In particular, carbon atoms on graphene near heteroatoms are more reactive. In addition, these heteroatoms and surrounding carbon atoms in doped graphene could serve as catalytic active sites for graphene catalysts or function as redox active sites for graphene electrodes. Therefore, such doped graphene materials are often utilized in the fields of semiconducting, catalysis, and energy storage.

2. Structure

Graphene is a 2D crystal of sp^2-hybridized carbon atoms, which can be modified by physical or chemical means to endow it with various properties and functions. This category of functionalization includes the introduction of structural defects into the 2D carbon framework (Fig. 2.3(a), the introduced heteroatoms are usually considered to be defects of

(a)   (b)   zig-zag   (c)

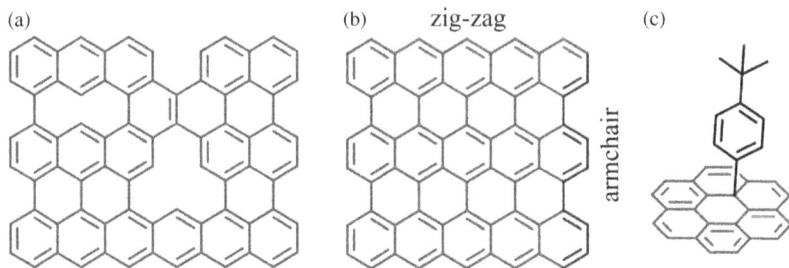

armchair

Figure 2.3.   Illustration of FG with a modified structure: (a) FG with defects, (b) zigzag and armchair edge of graphene, and (c) covalently functionalized graphene.

graphene, but here only structural defects are regarded, that is, carbon vacancies), modification of the edge structure, and covalent incorporation of other groups or molecules (Figs. 2.3(b) and 2.3(c)).

The structural defects are carbon vacancies, which can be divided into two categories: ordered and disordered. These structural defects can be obtained by controllable chemical synthesis or physical/chemical etching. These structural defects destroy the delocalization of graphene π-electrons and also affect the band structure of graphene. In addition, structural defects, which are similar to heteroatoms, will change the electronic structure of the surrounding carbon atoms and give rise to catalytically active sites. Therefore, defective FG is often used as a metal-free catalyst for various reactions. Furthermore, in single-atom catalysts, defective FG is also adopted as a support material, on which the defect sites are used to anchor metal atoms to achieve atomic-level dispersion of metals.

Previous studies have found that the structure of the edge also affects the properties of graphene. There are two kinds of typical edge structures in graphene — armchair and zigzag — which exhibit different electronic structures and chemical reactivities. The zigzag edge of a graphene nanoribbon possesses a unique electronic state that is near the Fermi level and localized at the edge carbon atoms, which is completely absent from the armchair edge, making the zigzag edge more chemically reactive. Therefore, manipulation of the edge structure of graphene can also be adopted to tailor the properties and achieve the functionalization of graphene.

In addition to destroying its inherent 2D structure by introducing carbon vacancies into graphene, incorporating other groups or molecules into graphene is also one of the dominant ways of achieving structural functionalization. The nature of graphene can be described as being similar to a polycyclic aromatic hydrocarbon (PAH) of quasi-infinite size; therefore, some organic reactions of PAHs are also applicable to graphene. As shown in Fig. 2.3(b), graphene can be functionalized by covalent incorporation with other groups through organic reactions as needed.

3. Dimension

In addition to the elemental composition and structure, the size of graphene is also closely related to its properties. Graphene synthesized by different approaches will be of different sizes. Graphene's magnetic, electronic, and optical properties can be fine-tuned by adjusting its dimension due to quantum confinement effect and the edge effect, which can be observed when the dimension of graphene is of the same magnitude as the Bloch electron wavelength (Fig. 2.4). As Fig. 2.4(a) shows, graphene quantum dots (GQDs) are graphene fragments with diameters smaller than 100 nm, and their bandgaps depend on the size and surface chemistry of GQDs. In addition, if one dimension of graphene is reduced to nanoscale, graphene nanoribbons (GNRs) are produced, which exhibit discrete energy levels and thus non-zero bandgaps (Fig. 2.4(b)).

4. Morphology

Since graphene is a 2D crystal, its functionalization can also be achieved by adjusting its micromorphology and macromorphology. As a consequence, the physical properties of graphene can be tailored by modification of its micromorphology. Previous studies have shown that the physical properties of graphene are determined by the number of layers, such as bilayer, trilayer, or multilayer.[5] In addition, the number of graphene layers also determines the thickness, specific surface area, and electronic properties. It has been reported that the stacking mode of bilayer or multilayer graphene (AA- and AB-stacking) and the twisted

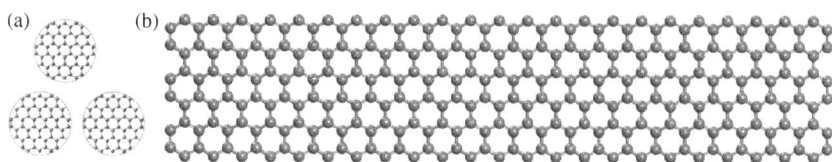

Figure 2.4.   Illustration of FG with modified dimensions: (a) graphene quantum dots and (b) graphene nanoribbons.[4]

angle of two layers would greatly affect the electronic properties. When sheets of bilayer or multilayer graphene in AA- or AB-stacking mode are rotated with respect to each other, a superlattice is created which drastically changes the electronic properties such as the tunable bandgap.[5] The most representative example is bilayer graphene. A study conducted in 2018 demonstrated that twisted bilayer graphene with an angle of about 1.1° displays intrinsic unconventional superconductivity at 1.7 K.[6] Therefore, micromorphology modification of graphene can be utilized to fine-tune its physical properties and achieve its functionalization. Furthermore, individual 2D graphene sheets can be assembled into various macroscopic structures to explore new functions and practical uses of graphene, such as graphene fibers, conductive transparent films, sponges, and foams. For example, graphene features the advantages of extremely high tensile strength (about 130 GPa), high Young's modulus (1 TPa), and outstanding electrical and thermal conductivity. Therefore, 1D graphene fibers, assembled from 2D graphene sheets, are expected to bring the ideal attributes of monolayer graphene into 1D fibrous materials on the macroscopic scale (Fig. 2.5(a)).[7] In addition, 3D structures can be fabricated by 2D graphene sheets, such as graphene sponges (Fig 2.5(b)).[8] In these 1D and 3D materials, the structure of graphene is usually not damaged, but the assembly of graphene is changed; the dimension of graphene is switched from 2D to 1D or 3D.

2.2.2 *Functionalized graphene composites*

As described in Chapter 1, graphene or FGs can be hybridized with other materials to create FGCs. Graphene can be used not only as a matrix

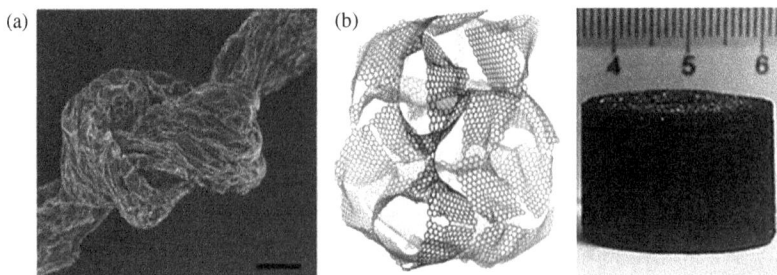

Figure 2.5. Functionalization of graphene via morphology modification: (a) graphene fiber[7] and (b) graphene sponge (left — micromorphology; right — macromorphology).[8]

material but also as a functional material in composite materials. Through hybridization in a composite, graphene-based materials with specific structures and functions can be obtained, and the physical and chemical properties of graphene or FG can be fine-tuned. Therefore, incorporation of graphene to create composites is also an effective means to functionalize graphene.

At present, the structures of (functionalized) graphene and functional materials in FGCs are classified into six different models, which are adopted from the structure models of graphene composites reported in electrochemical energy-storage devices: encapsulated, mixed, anchored, sandwich-like, layered, and wrapped.

(i)   *Encapsulated model*: Functional materials are encapsulated by (functionalized) graphene. Such a structure usually allows nanoscale dispersion of functional materials and enhances the chemical activity of functional materials.

(ii) *Mixed model*: Functionalized graphene and functional materials are synthesized separately and mixed mechanically. Such simple mechanical mixing is often effective in improving (functionalized) graphene's mechanical, electrical, and thermal properties.

(iii) *Anchored model*: Electroactive nanoparticles are anchored to (functionalized) graphene surface. Such a structure can not only improve the conductivity of functional materials but can also further enhance the performance of individual components through the interaction between (functionalized) graphene and other components.

(iv) *Sandwich-like model*: Functionalized graphene is used as a template to generate functional material (or functionalized) graphene sandwich structures. This structure allows the protection of the wrapped functional materials and effectively enhances the electrochemical performance of the composites, which is widespread in the energy field.

(v)   *Layered model*: Functional–material nanoparticles are alternated with (functionalized) graphene sheets to form a composite layered structure.

(vi) *Wrapped model*: Functional–material particles are wrapped by multiple (functionalized) graphene sheets. This wrapped structure is often used to enhance the electrical conductivity and structural strength of functional materials. In addition, the interrelationship among the structure, composition, performance, and application indicates that the properties of materials can be tailored to meet various application requirements by modifying the structure of materials.

It should be noted that the above-mentioned definitions of functionalized graphene and functionalized graphene composites can be adopted for determining the classification of graphene-based materials with functionalization. The representative samples of functionalized graphene is heteroatom doped graphene, graphene nanoribbons, graphene quantum dots, graphene nanomeshs, graphene oxide, and covalent functionalized graphene. On the other hand, graphene or functionalized graphene or both can be hybridized with other materials, such as small molecules, polymers, metals, inorganic compounds, and carbon nanotubes (CNTs), giving the "functionalized graphene composite" where there are no covalent bonds between (functionalized) graphene and other materials (Fig. 2.6.). The essential difference between functionalized graphene and functionalized graphene composites lies in the following aspects: (i) Functionalized graphene is a covalently bonded integrity that graphene itself is modified or covalently functionalized with other chemical groups. (ii) Functionalized graphene composite is a composite obtained by hybridizing (functionalized) graphene with other functional materials in a non-covalent manner. If one graphene material with a complex structure cannot be simply

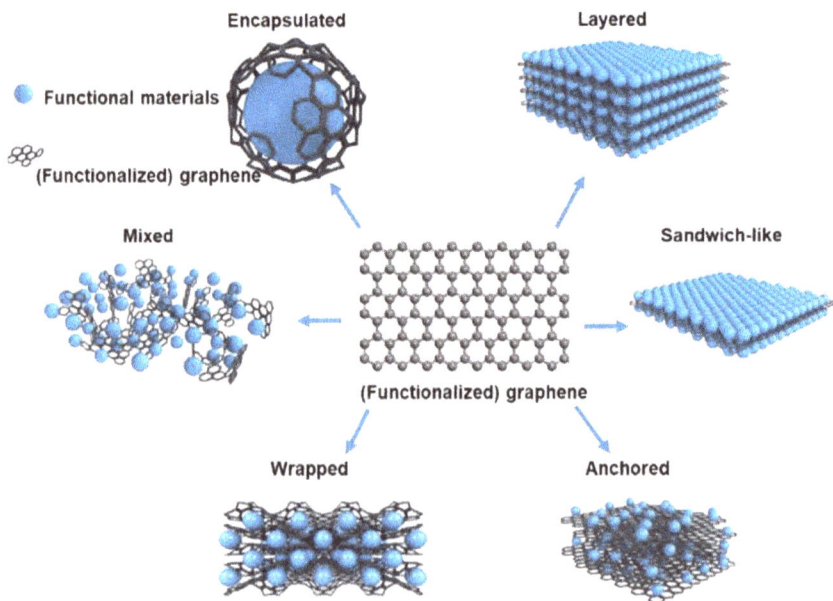

Figure 2.6.   Schematic diagram of the functionalized graphene composites.[9]

categorized as functionalized graphene or functionalized graphene composite, then it can be classified as a functionalized graphene material.

In practical applications, the functionalized products of graphene are very complicated. For instance, graphene is firstly oxidized to graphene oxide, and then other functional groups (such as polymers) are covalently functionalized on graphene oxide. Finally, the product is obtained after mixing with other materials such as carbon nanotubes. This material is classified as functionalized graphene material.

2.3 Preparation Strategies for Functionalized Graphene Materials

There are two strategies for preparing functionalized graphene materials: bottom-up and top-down approaches (Fig. 2.7). Both strategies have their own advantages and specific fields of application. Both strategies can also be employed to prepare the same functionalized graphene.

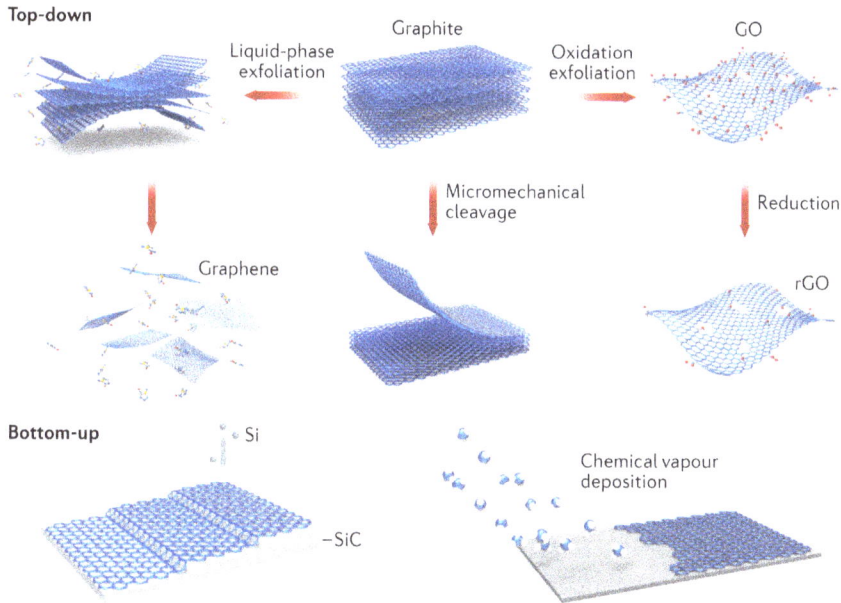

Figure 2.7. Schematic diagram of preparation of functionalized graphene materials by bottom-up and top-down strategies.[13]

2.3.1 *Bottom-up strategy*

The bottom-up strategy involves "polymerizing" starting compounds such as small molecules, polymers, and even graphene or functionalized graphene through physical or chemical methods to form an integrity, that is, functionalized graphene materials. The characteristic of this strategy is that the starting compounds and the basic building blocks of the products (functionalized graphene material) have similar or identical structures, indicating that the starting compounds can be regarded as the basic building blocks of the products (functionalized graphene material). Bottom-up approaches can be used to control the structure of functionalized graphene materials more precisely and are often used to prepare high-quality functionalized graphene or functionalized graphene composites with well-defined structures. For example, using nitrogen-substituted conjugated monomers as raw materials, nitrogen-doped graphene nanoribbons are synthesized on metal surfaces using a bottom-up strategy (Fig. 2.8(a))[10]; polycyclic aromatic hydrocarbons (PAHs) are used as raw materials to prepare wavy graphene nanoribbons by means of a bottom-up strategy (Fig. 2.8(b))[11]; and functionalized graphene, graphene oxide (GO), and copper hydroxide nanowires (CHNs) are used as raw materials to construct graphene oxide membranes containing nanopores using a bottom-up strategy (Fig. 2.8(c)).[12]

2.3.2 *Top-down strategy*

The top-down strategy of functionalized graphene materials involves the process of breaking down or reducing carbon materials, such as graphite, graphene, carbon nanotubes (CNTs), or fullerenes, by chemical or/and physical techniques into small fragments, during which functionalization is achieved simultaneously, or else the prepared small fragments themselves are functionalized graphene materials, such as graphene nanoribbons obtained by unzipping CNTs. The characteristic of this strategy is that the products (functionalized graphene material) have a similar or identical structure to the basic building blocks of the starting compounds, i.e., the products can be regarded as the basic building blocks of the starting compounds. The benefits of top-down methods include substrate transfer, price effectiveness, and high dependability compared to bottom-up approaches. However, top-down approaches usually suffer from drawbacks such as uncontrollable sizes and irregular edge structures. The representative example involves fabricating graphene oxide from graphite. Graphite has layers of graphene interconnected through weak Van der Waals forces. By means of oxidation

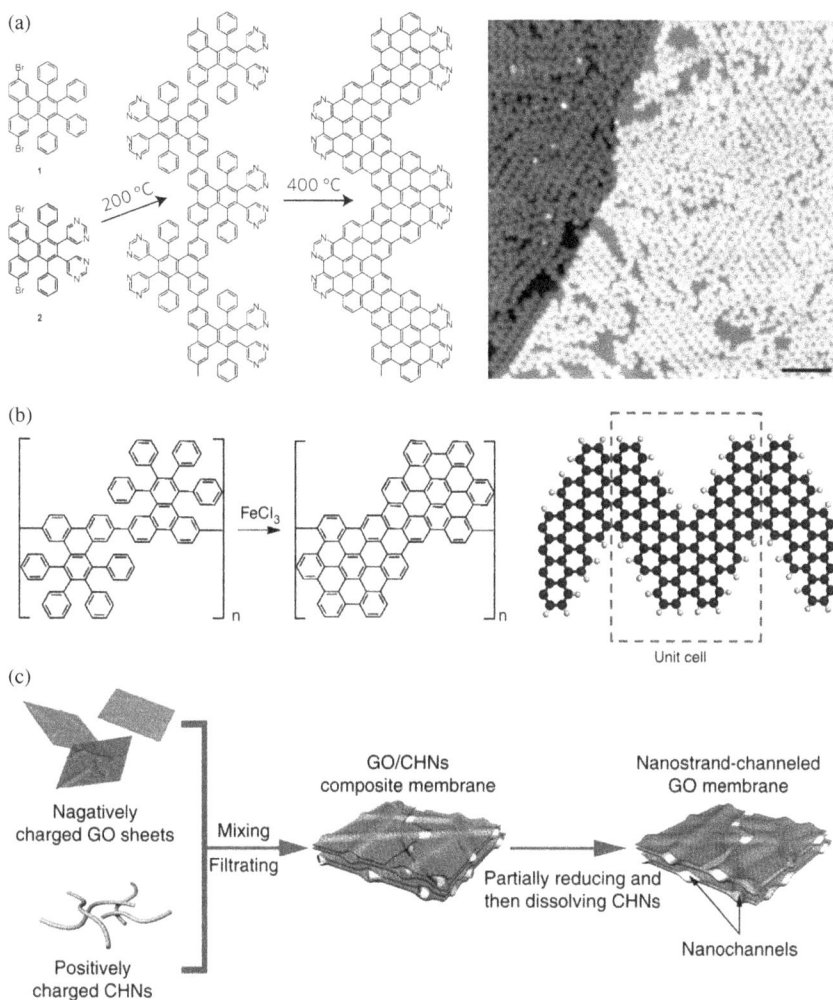

Figure 2.8. Schematic of functionalized graphene materials prepared by bottom-up strategies: (a) nitrogen-doped graphene nanoribbons synthesized on metal surfaces using nitrogen-substituted conjugated monomers as raw materials[10]; (b) wavy graphene nanoribbons fabricated in solution using polycyclic aromatic hydrocarbons as raw materials[11]; and (c) using functionalized graphene, graphene oxide, and copper hydroxide nanowires as raw materials to synthesize graphene oxide membranes containing nanopores.[12]

in aqueous solution, the stacked graphene sheets can be oxidized and peeled off to form dispersed graphene oxide sheets. As shown in Fig. 2.9, this process involves slicing and functionalizing (oxidizing) the graphene aggregates (graphite), producing single sheets of functionalized graphene.

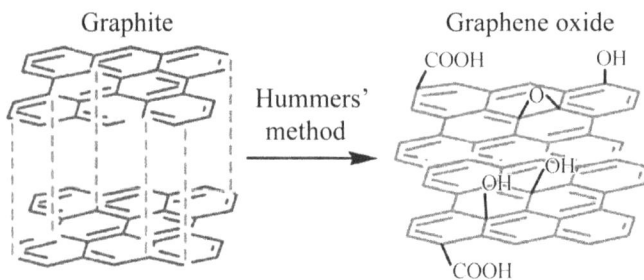

Figure 2.9.   Schematic of functionalized graphene materials synthesized by top-down strategies: graphene oxide prepared from graphite.[14]

It is certain that both the strategies can be used to prepare the same functionalized graphene, such as graphene nanoribbons with a defined size, which can be prepared by bottom-up polymerization of monomers or by top-down etching of graphene sheets. Another example is the preparation of graphene polymers (polymers containing building blocks of graphene) by bottom-up cross-coupling and graphitization of organic polymers or by top-down exfoliation and etching of graphite as shown in Fig. 2.10.[15] Graphene polymers (functionalized graphene as shown in Fig. 2.10) can be regarded as many graphene fragments connected to each other through polymeric chains. These graphene fragments are used as the basic building blocks to construct materials. The introduced functional groups and defects favors the functionalization, while the polymeric chains behave as bridges to create a variety of the structures, the products thus can be regarded as functionalized graphene. Functionalized graphene materials can be fabricated by top-down approaches as well, in which the graphite undergoes exfoliation and etching. Furthermore, porous polymeric networks synthesized by monomers through bottom-up methods are subject to specific chemical cross-linking (e.g., high-temperature heat treatment). Reactions such as decomposition and rearrangement of chemical bonds will occur, so many structures similar to those of graphene fragments are formed inside these polymers, after which graphene polymers are obtained.

In brief, functionalization not only facilitates the applications of graphene but also endows graphene with novel properties leading to its widespread applications. In practical applications, the preparation strategy of functionalized graphene materials is often chosen based on many factors such as application requirements, technical conditions, and cost-effectiveness.

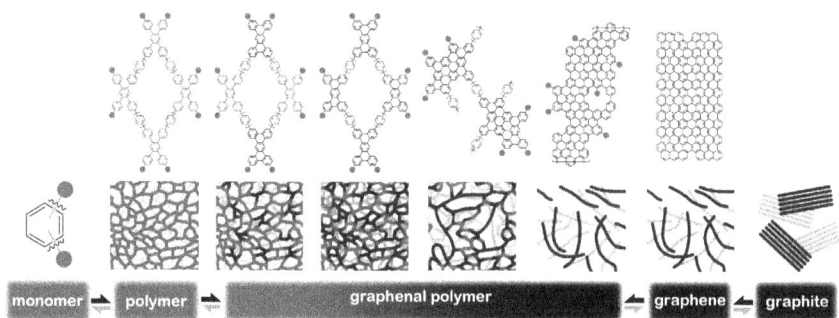

Figure 2.10. Schematic diagram of bottom-up and top-down strategies for preparing functionalized graphene (graphene polymer).[15]

2.4 Summary

In this chapter, the definition and classification of functionalized graphene materials are proposed, and their functionalization strategies are summarized. Chapters 3–5 will review and discuss in detail the preparation of various functionalized graphene materials on the basis of the functionalization strategy of graphene. Since the preparation of functionalized graphene composites is usually closely related to their applications, the preparation processes will be discussed along with the applications (see Chapter 6). At present, the various functionalized graphene materials are considered novel materials with unique development prospects, and they show great application potential in energy storage, biomedical engineering, environmental remediation, catalysis, water desalination, electromagnetic interference shielding, and optoelectronic devices. Chapter 6 will review the applications of functionalized graphene materials in these fields.

References

1. Georgakilas, V., Otyepka, M., Bourlinos, A. B., Chandra, V., Kim, N., Kemp, K. C., Hobza, P., Zboril, R., and Kim, K. S. (2012). Functionalization of graphene: Covalent and non-covalent approaches, derivatives and applications, *Chem. Rev.*, 112, 59.
2. Kuila, T., Bose, S., Mishra, A. K., Khanra, P., Kim, N. H., and Lee, J. H. (2012). Chemical functionalization of graphene and its applications, *Prog. Mater. Sci.*, 57(7), 1061–1105.

3. Wang, Y., Shao, Y., Matson, D. W., Li, J., and Lin, Y. (2010). Nitrogen-doped graphene and its application in electrochemical biosensing, *ACS Nano*, 4(4), 1790–1798.

4. Ma, Y. and Zhi, L. (2022). Functionalized graphene materials: Definition, classification, and preparation strategies, *Acta Phys.-Chim. Sin.*, 38(1), 2101004 (in Chinese).

5. Castro Neto, A. H., Guinea, F., Peres, N. M. R., Novoselov, K. S., and Geim, A. K. (2009). The electronic properties of graphene, *Rev. Mod. Phys.*, 81(1), 109–162.

6. Cao, Y., Fatemi, V., Fang, S., Watanabe, K., Taniguchi, T., Kaxiras, E., and Jarillo-Herrero, P. (2018). Unconventional superconductivity in magic-angle graphene superlattices, *Nature*, 556(7699), 43–50.

7. Xu, Z. and Gao, C. (2011). Graphene chiral liquid crystals and macroscopic assembled fibres, *Nat. Commun.*, 2(1), 571.

8. Wu, Y., Yi, N., Huang, L., Zhang, T., Fang, S., Chang, H., Li, N., Oh, J., Lee, J. A., Kozlov, M., Chipara, A. C., Terrones, H., Xiao, P., Long, G., Huang, Y., Zhang, F., Zhang, L., Lepró, X., Haines, C., Lima, M. D., Lopez, N. P., Rajukumar, L. P., Elias, A. L., Feng, S., Kim, S. J., Narayanan, N. T., Ajayan, P. M., Terrones, M., Aliev, A., Chu, P., Zhang, Z., Baughman, R. H., and Chen, Y. (2015). Three-dimensionally bonded spongy graphene material with super compressive elasticity and near-zero Poisson's ratio, *Nat. Commun.*, 6(1), 6141.

9. Raccichini, R., Varzi, A., Passerini, S., and Scrosati, B. (2015). The role of graphene for electrochemical energy storage, *Nat. Mater.*, 14(3), 271–279.

10. Cai, J., Pignedoli, C. A., Talirz, L., Ruffieux, P., Söde, H., Liang, L., Meunier, V., Berger, R., Li, R., Feng, X., Müllen, K., and Fasel, R. (2014). Graphene nanoribbon heterojunctions, *Nat. Nanotechnol.*, 9(11), 896–900.

11. Vo, T. H., Shekhirev, M., Kunkel, D. A., Morton, M. D., Berglund, E., Kong, L., Wilson, P. M., Dowben, P. A., Enders, A., and Sinitskii, A. (2014). Large-scale solution synthesis of narrow graphene nanoribbons, *Nat. Commun.*, 5(1), 3189.

12. Huang, H., Song, Z., Wei, N., Shi, L., Mao, Y., Ying, Y., Sun, L., Xu, Z., and Peng, X. (2013). Ultrafast viscous water flow through nanostrand-channelled graphene oxide membranes, *Nat. Commun.*, 4(1), 2979.

13. Wang, X.-Y., Narita, A., and Müllen, K. (2017). Precision synthesis versus bulk-scale fabrication of graphenes, *Nat. Rev. Chem.*, 2(1), 0100.

14. Zhou, C., Jiang, W., and Via, B. K. (2014). Facile synthesis of soluble graphene quantum dots and its improved property in detecting heavy metal ions, *Colloids Surf. B: Biointerfaces*, 118, 72–76.

15. Li, X., Song, Q., Hao, L., and Zhi, L. (2014). Graphenal polymers for energy storage, *Small*, 10(11), 2122–2135.

Chapter 3

Bottom-up Synthesis of Functionalized Graphene

The bottom-up approach is often used to construct functionalized graphene materials. The concept of bottom-up methodology is similar to the interlocking Legos that large quantities of smaller entities are constructed according to a predetermined route, giving rise to the formation of functionalized graphene materials with specific structures and properties. In these approaches, the starting materials used generally carry the same structure as that of the elementary building blocks (i.e., "smaller entities") of the functionalized graphene materials.

By means of this approach, the starting materials containing "substructures" can be designed and prepared, followed by the construction of functionalized graphene materials with well-defined structures according to the predetermined route. In this regard, this method is often used for the controllable synthesis of functionalized graphene materials.

In addition to the classical bottom-up synthesis described here, the synthesis of functional graphene materials by graphitization of other precursors (such as covalent organic frameworks (COFs), metal–organic frameworks (MOFs), biomass, and small organic molecules) can also be categorized as a bottom-up strategy. Similarly, the synthesis of functionalized graphene composites from (functionalized) graphene and other materials can also be regarded as a bottom-up strategy. This is because the methodology of these two approaches is the same as the classical bottom-up approach, which is done by starting with smaller entities and building them up to larger functional constructs.

The bottom-up synthesis of functionalized graphene is reviewed and discussed in detail in this chapter. According to the type of starting materials, the functionalized graphene prepared by the bottom-up strategy can be classified into the following categories: polycyclic aromatic hydrocarbons (PAHs), COFs, MOFs, biomass, small molecules, and other materials.

3.1 Functionalized Graphene Constructed by PAHs

PAHs are the most commonly used starting materials for bottom-up synthesis of functionalized graphene. PAHs are a large group of diverse organic compounds that are composed of two or more aromatic rings. According to the arrangements of benzene rings, PAHs can be divided into two categories: fused and non-fused PAHs. Fused PAHs are PAHs in which two conjugated benzene rings share two carbon atoms, such as naphthalene, anthracene, and pyrene. Non-fused PAHs are PAHs in which two conjugated benzene rings share one carbon atom, such as biphenyl.

Due to its unique advantages, bottom-up organic synthesis has become an indispensable means of preparing functionalized graphene with well-defined structures. As shown in Fig. 3.1, functionalized graphene with different structures can be constructed by adopting PAHs with specific structures as starting materials, including graphene nanosheets, graphene nanoribbons, graphene nanomesh, and doped graphene. Figure 3.1(a) illustrates a graphene nanosheet in which the aromatic rings are extended laterally; Fig. 3.1(b) shows a graphene nanoribbon, which is a linear extension of the aromatic rings; Fig. 3.1(c) presents a graphene nanomesh, in which the 2D extension of aromatic rings forms an ordered porous structure; and Fig. 3.1(d) exhibits a nitrogen-doped graphene nanosheet, which is a type of heteroatom-containing graphene (heteroatom like N, S, or O) synthesized by polymerization of PAHs and heterocyclic compounds such as pyridine, pyrrole, or thiophene.

Most importantly, bottom-up synthesis allows controllable preparation of functionalized graphene, and this also makes it possible to precisely control the edge structure of graphene. Different edge topologies allow graphene to exhibit different physical and chemical properties. As shown in Fig. 3.2, graphene typically features two prominent types of edge configurations, namely, armchair and zigzag edges. In addition, according to the chemical composition of the group, the graphene edge topologies can be classified into the following types: K-regions (convex

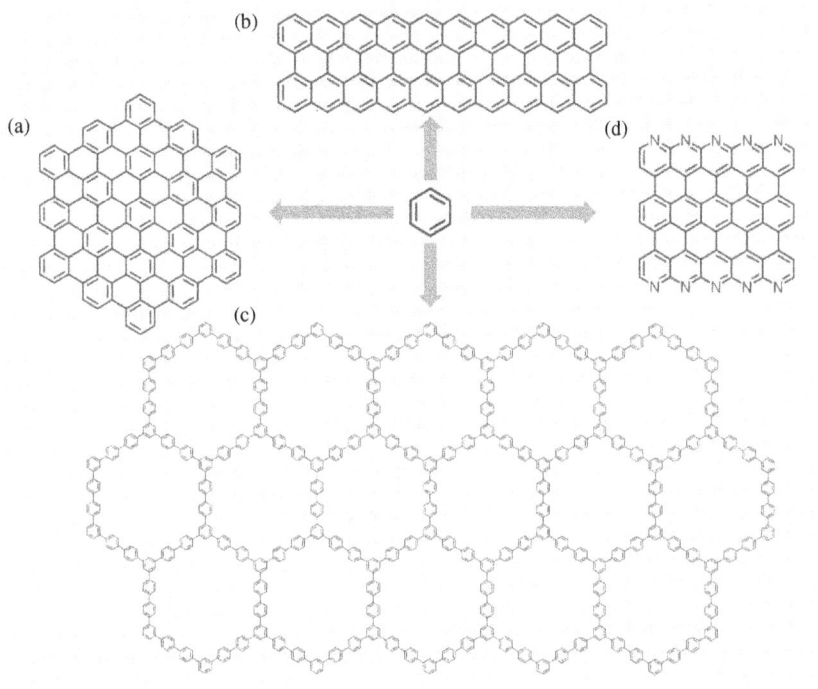

Figure 3.1. Functionalized graphene with different structures. (a) Graphene nanosheet (2D extension of benzene rings); (b) graphene nanoribbon (1D extension of benzene rings); (c) graphene nanomesh (conjugated framework modification); and (d) nitrogen-doped graphene nanosheet (elemental composition control).

armchair edges, that is, isolated C=C bonds that do not belong to the Clar sextet), bay-regions (concave armchair edges), L-regions (zigzag edges), cove-regions, and fjord-regions.

The bay-regions and K-regions at the edge of the graphene are highly chemically reactive and can be further derivatized and functionalized. In particular, they can be used to react with other molecules or moieties through C—H activation and the Diels–Alder reaction, producing conjugated groups to further extend the π-conjugated framework of the molecule (see Section 3.1.1 for details). In addition, the K-region, cove-region, and fjord-region cause non-planarity due to the prevailing steric hindrance.

The controllable synthesis of functionalized graphene with a well-defined structure offers the possibility to further reveal the properties of graphene materials, develop their application potential, and broaden the

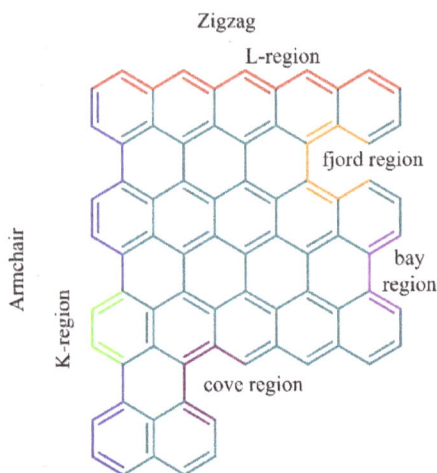

Figure 3.2.   Edge configurations of graphene.

scope of their application. For instance, although graphene has extremely high electron mobility and shows excellent prospects in the field of semi-conductors, its zero bandgap hinders it from being directly applied to semiconductors; the bandgap can be opened by a variety of methods. However, studies have found that graphene nanoribbons exhibit intrinsic bandgaps. Therefore, bottom-up synthesis of graphene nanoribbons with well-defined structures can be done, giving rise to intrinsic bandgaps. In addition, the structure of functionalized graphene can be precisely controlled by means of a bottom-up strategy, thereby enabling the precise control of its properties and performance.

3.1.1 *Strategies for synthesizing functionalized graphene from PAHs*

The synthesis of functionalized graphene from PAHs usually begins with the formation of conjugated small molecules with specific structures and functional groups. Using such small molecules as starting compounds, functionalized graphene can be prepared by the liquid-phase method or surface-assisted synthesis on the basis of specific strategies. By means of controlling the structure of the starting small molecules and applying the synthesis strategy, the structure of the product (functionalized graphene)

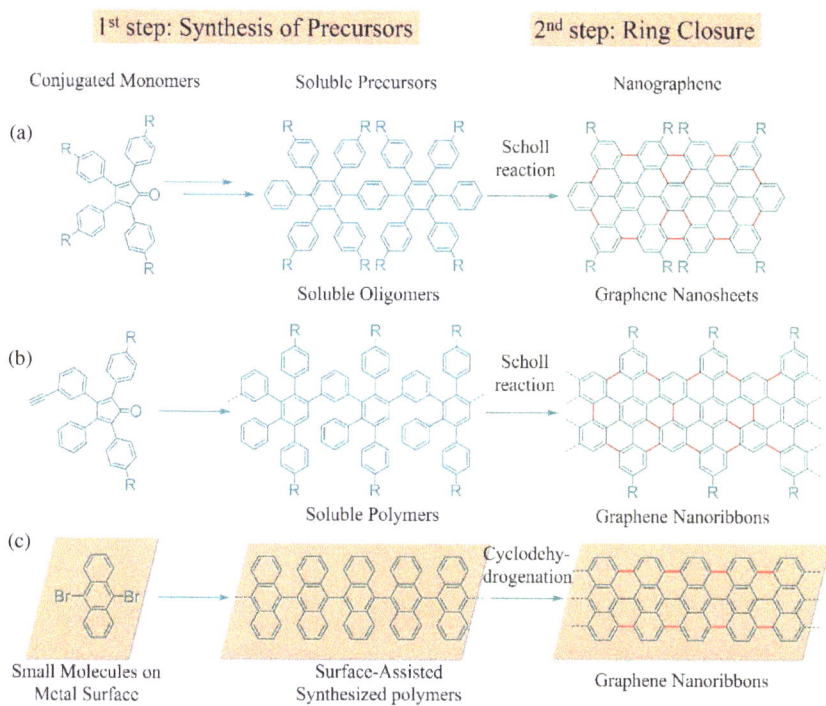

Figure 3.3. Schematic diagram of the two-stage strategy for synthesizing functionalized graphene: (a) graphene nanoribbons (b) liquid phase; and (c) two-stage strategy for surface-assisted synthesis of graphene nanoribbons.

can be manipulated. There are two types of synthesis strategies commonly used: two-stage synthesis and annulative π-extension.

3.1.1.1 *Two-stage synthesis of functionalized graphene*

The two-stage methodology is one of the most important strategies for synthesizing graphene nanosheets and graphene nanoribbons from PAHs. Just as the name implies, this strategy relies on two key stages: the formation of precursors and further ring closure ("stitching" or "graphitization") to yield the target functional graphene (Fig. 3.3).

The starting materials for preparing precursor molecules are conjugated monomers with functional groups and cross-linking groups. Precursor molecules are mainly divided into two categories, oligomers and polymers, which are determined by the target products of

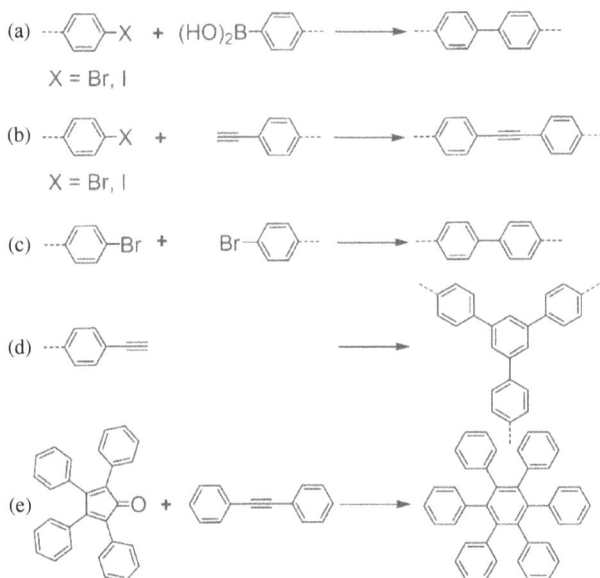

Figure 3.4. Commonly used reactions for the synthesis of precursors from conjugated monomers (mainly used in liquid-phase methods): (a) Suzuki–Miyaura cross-coupling reaction; (b) Sonogashira reaction; (c) Yamamoto reaction; (d) acetylene cyclotrimerization reaction; and (e) Diels–Alder reaction.

retrosynthesis. According to the synthesis method employed (liquid-phase synthesis and surface-assisted synthesis), the precursor molecules can also be classified as soluble ones and metal-surface-assisted ones. In liquid-phase synthesis, the reactants are required to be dissolved in the solvent, which means precursor molecules should be soluble.

There are many types of reactions through which the precursor molecules are synthesized from conjugated monomers. As shown in Fig. 3.4, in liquid-phase synthesis, the precursors are usually synthesized by cross-coupling reactions, acetylene cyclotrimerization reactions, and the Diels–Alder reaction. Cross-coupling reactions include the Suzuki–Miyaura cross-coupling reaction, Sonogashira reaction, and Yamamoto reaction. In surface-assisted synthesis, precursors are usually fabricated by surface-assisted diradical addition.

The functionalized graphene can be synthesized from the precursor through intramolecular ring closure (Fig. 3.3). In liquid-phase synthesis, the intramolecular Scholl reaction is often used for ring closure. This reaction, which is described as proceeding through either a radical cation pathway or an arenium cation pathway, is highly efficient in forming

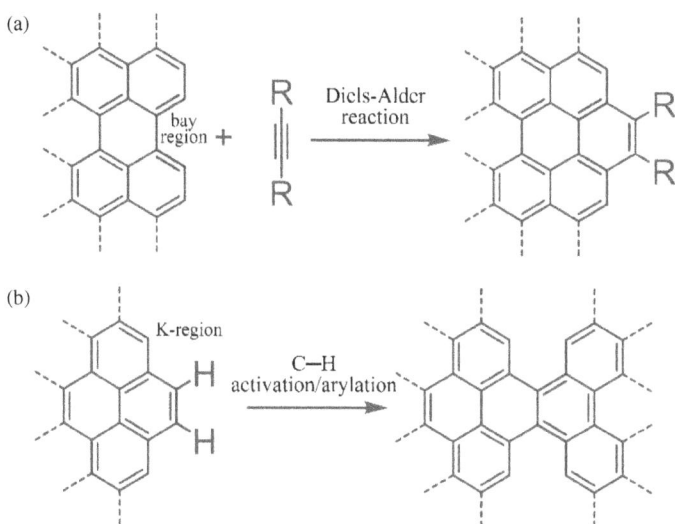

Figure 3.5. Reaction regions and types of annulative π-extension synthesis of functionalized graphene.

multiple carbon–carbon double bonds in one step and is powerful in synthesizing large PAHs. Although the yield of the intermolecular Scholl reaction is low, the intramolecular reaction works efficiently and the operation is simple, and thus it is widely used. In surface-assisted synthesis, the intramolecular dehydrocyclization reaction is often used; that is, the adjacent aromatic hydrocarbons in the precursor molecule are dehydrogenated and coupled to achieve ring closure at high temperature.

2. *Annulative π-extension synthesis of functionalized graphene*: Apart from the two-stage synthetic protocol, as described earlier, functionalized graphene can also be synthesized via annulative π-extension of existing PAHs. In this method, the highly active sites on the peripheries of graphene are utilized to react with other functional molecules to construct new aromatic rings and further extend the π-conjugated structure. As aforementioned, the graphene bay- and K-regions show high reactivity and react specifically with other molecules to generate new aromatic rings.

As shown in Fig. 3.5, four diene carbon atoms in the graphene bay-regions can undergo a "4 + 2" Diels–Alder reaction with the triple or double bonds of other molecules to construct a new aromatic ring. As shown in Fig. 3.6(a), the bay-regions of perylene and anthracene dimers (part of graphene nanosheets) can undergo Diels–Alder reactions with

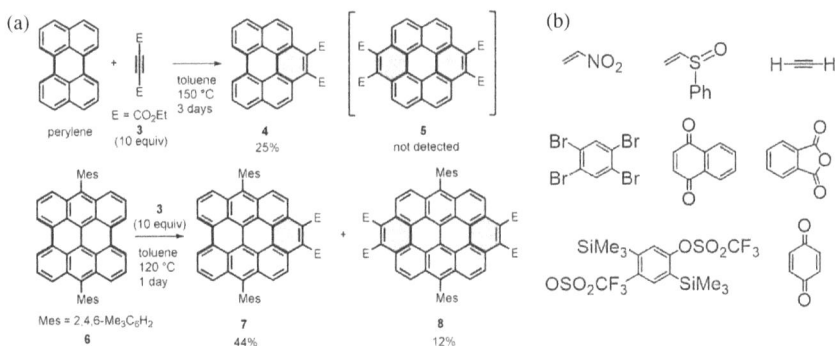

Figure 3.6. Functionalization of graphene bay-regions: (a) illustration of functionalization at graphene bay-regions by Diels–Alder reaction[1] and (b) illustration of functional molecules that can undergo Diels–Alder reaction with graphene bay-regions.

acetylene derivatives to produce new aromatic rings, further extending the π-conjugated structures of perylene and anthracene dimers.[1] There are many kinds of molecules that can undergo the Diels–Alder reaction with four diene carbon atoms in the graphene bay-regions, such as nitroethylene, vinylphenylsulfoxide, acetylene, 2,3,5,6-tetrabromobenzene, and p-benzoquinone (Fig. 3.6(b)).

As shown in Fig. 3.7, two C—H bonds in the K-regions of graphene can be activated or halogenated, and then fused with other conjugated molecules to extend the conjugated structure. The annulative π-extension reactions that occur at the K-regions mainly fall into the following categories: (i) direct C—H activation/arylation, as shown in Fig. 3.7(b), in which two carbon atoms in the K-region and tris(2-biphenylyl)boroxin (Boroesterified bay-region) are coupled under palladium catalysis, and then undergo intramolecular ring closure by the Scholl reaction to generate larger graphene nanosheets; (ii) one-step annulative π-extension, in which two carbon atoms in the K-region react directly with dibenzosilole to produce larger graphene nanosheets by the one-step reaction (Fig. 37(c)); (iii) monohalogenated K-region dimerization coupling, in which two hydrogen atoms on the C=C bonds of K-regions are replaced by a single halogen atom to generate monohalogenated functionalized graphene, and then two such molecules undergo a dimerization reaction to form larger conjugated structure [Fig. 3.7(a)]; and (iv) a reaction with dihalogenated aromatic molecules, in which the C=C bonds of K-regions and dihalogenated

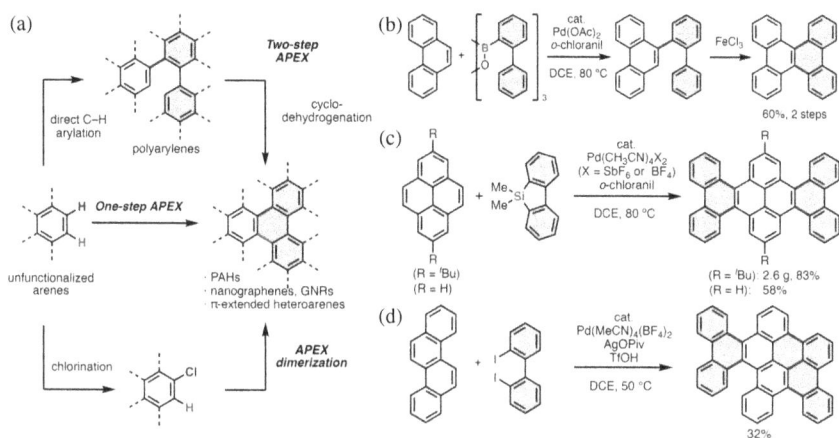

Figure 3.7.   Schematic of functionalization of graphene K-regions[2]: (a) C—H activation and one-step annulative π-extension; (b) direct aromatic hydrocarbon coupling reaction; (c) graphene nanosheets synthesized by one-step annulative π-extension; and (d) annulative π-extension reaction with diiodobiaryl π-extending reagents.

aromatic molecules undergo a coupling reaction under palladium catalysis to generate larger graphene nanosheets (Fig. 3.7(d)).

3.1.2 *Synthesis methods of functionalized graphene from PAHs*

The functionalized graphene constructed by the bottom-up strategy of PAHs can be divided into two categories, namely, liquid-phase synthesis and surface-assisted synthesis.

3.1.2.1 *Liquid-phase synthesis*

Liquid-phase synthesis, as the name implies, is the process of preparing functionalized graphene in solution. The preparation of functionalized graphene by this method can be done with ordinary equipment and techniques for organic synthesis, and the requirements for experimental conditions are not strict. In addition, the functionalized graphene prepared by this method usually shows good solubility in organic solvents and can be easily used for fabricating devices by spin coating and other methods. Therefore, liquid-phase synthesis is often used to prepare soluble graphene nanosheets and graphene nanoribbons. As mentioned in

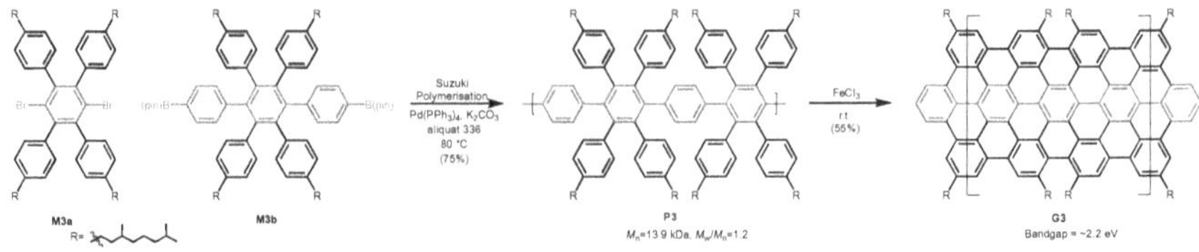

Figure 3.8.   Synthesis of graphene nanoribbons by liquid-phase method.[4]

Section 3.1.1, liquid-phase synthesis of functionalized graphene often adopts a two-stage strategy. As shown in Fig. 3.8, a linear armchair graphene nanoribbon is synthesized via a Suzuki-Miyaura polymerization of diiodobenzene **M3a** and bis(boronic ester) compound **M3b**, and followed by a Scholl reaction catalyzed by FeCl$_3$.[3] Due to the large number of alkyl side chains at adjacent positions of the aromatic periphery, the final product, graphene nanoribbons **G3**, presents excellent dispersity in organic solvents (such as dichloromethane, chloroform, and tetrahydrofuran (THF)). However, it is worth noting that this kind of functionalized graphene exhibits a larger conjugated structure, which is easy to accumulate and aggregate in the solution; the aggregation is more serious when the conjugated structure is large, the soluble groups are fewer, or the concentration is high, and it will hinder the future processing and applications of the products.

3.1.2.2 *Surface-assisted synthesis*

Surface-assisted synthesis is another efficient method for preparing functionalized graphene as illustrated in Fig. 3.9.[5] The method is as follows: (i) Dehalogenated (brominated or chloro) colligated monomers are deposited onto the surfaces of a metal (gold or copper) by vacuum sublimation using an evaporator (Knudsen cell type). (ii) At high temperatures (usually not lower than 180°C), the carbon–halogen bonds of colligated monomers are broken to generate diradical species, and the resulting surface-stable biradical intermediates (may form bonds, such as C–Ag bonds) are able to diffuse across the surface without quenching. (iii) Biradicals undergo radical addition reactions on the metal surfaces to form polymer intermediates. (iv) Under thermal treatment, surface-assisted cyclodehydrogenation of polymer intermediates establishes an extended fully aromatic system, that is, graphene nanosheets or nanoribbons.

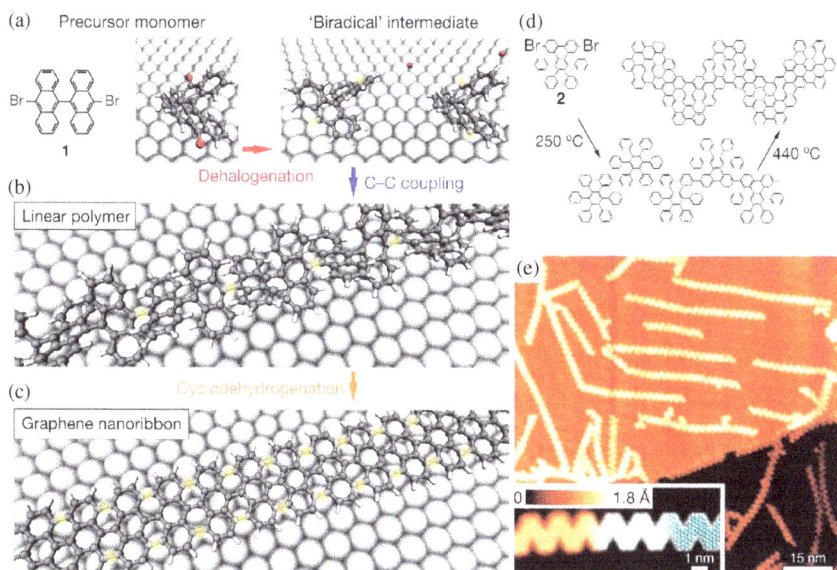

Figure 3.9.   Surface-assisted synthesis of graphene nanoribbons[5]: (a) thermal sublimation of the dibrominated colligated monomers removes their bromine substituents, yielding biradical intermediates; (b) the biradical intermediates undergo C—C coupling on the metal surfaces to form polymers; (c) at high temperature, the polymer undergoes intramolecular cyclodehydrogenation to generate graphene nanoribbons; (d) reaction scheme of surface-assisted synthesis of graphene nanoribbons; and (e) overview STM image of chevron-type GNRs fabricated on Au(111) surface.

It should be noted that the dehydrocyclization liberates hydrogen atoms, which will quench the resulting radicals, thereby terminating further reactions. Furthermore, only one of the two carbon-halogen bonds of the di-halogen functionalized monomers is decomposed, and the resulting halogenated radicals will also terminate the subsequent radical reactions. Under the influence of these two factors, the synthesized graphene nanoribbons via surface-assisted approach are usually short (<50 nm).

The yield of functionalized graphene prepared by surface-assisted synthesis is low, and usually, only monolayer nanographene with an area of less than 1 cm^2 can be obtained. The apparatus for surface-assisted synthesis is relatively expensive, and the synthesis requires an ultrahigh vacuum system to remove oxygen, water, or other contaminants from the environment and to quench the side reactions of radicals, polymer intermediates, and nanographene. Meanwhile, an ultrahigh vacuum

(a)

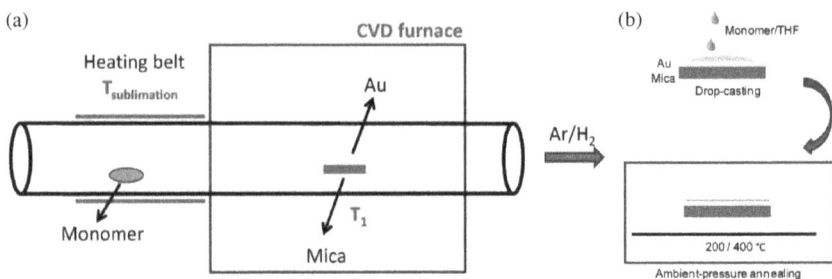

(b)

Figure 3.10. Schematic diagram of CVD system for synthesizing functionalized graphene: (a) synthesis of functionalized graphene on a metal surface by CVD method using colligated monomers as starting material and (b) deposition of monomers on the surface of a metal substrate by drop-casting method.

environment is also necessary for a scanning tunneling microscope (STM) and an atomic force microscope (AFM) to observe and characterize nanographene in high resolution. However, the reaction of colligated monomers on the surface does not require an ultrahigh vacuum environment. Therefore, in order to save costs, the classical chemical vapor deposition (CVD) method can be used to prepare functionalized graphene on metal surfaces using PAHs as starting materials. Compared with surface-assisted synthesis, this method does not require expensive apparatus and an ultrahigh vacuum system. As shown in Fig. 3.10(a), the preparation process of this method is similar to that of the surface-assisted synthesis: (i) The di-halogen colligated monomers are sublimated and deposited on the surfaces of the metal substrate in a tube furnace. (ii) At high temperatures, the monomers undergo coupling polymerization to form polymer intermediates. (iii) At high temperatures, the polymers undergo intramolecular cyclohydrogenation to form graphene nanoribbons.

However, the CVD synthesis of functionalized graphene again requires heat treatment and vacuum sublimation, which consumes a lot of energy, and this method cannot be used for colligated monomers with poor thermal stability or high molecular weight. In order to address this problem, a drop-casting method has been developed by Klaus Müllen's group to replace the vacuum sublimation step for depositing colligated monomers on the surface of metal substrate.[6] As shown in Fig. 3.10(b), a THF solution of monomers is drop-cast on a gold substrate placed on mica, forming a thin film of monomers. Subsequent annealing of the monomer film under ambient pressure induces homolytic carbon-halogen cleavage and polymerization of the resulting diradical intermediates to

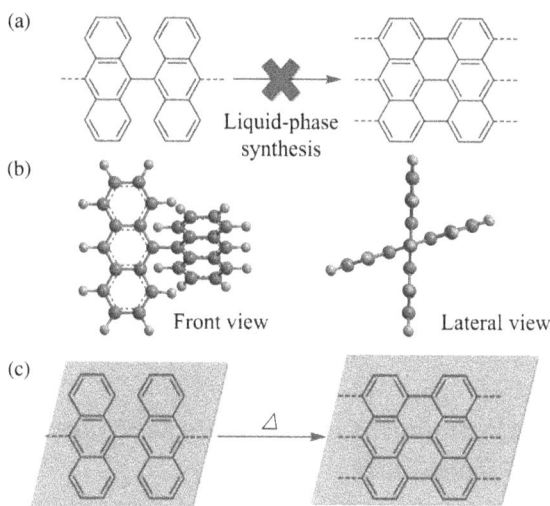

Figure 3.11. Advantages of surface-assisted synthesis over liquid-phase synthesis: (a) due to steric hindrance, the anthracene polymers in the liquid phase cannot achieve intramolecular ring closure; (b) the spatial structure between neighboring anthracene groups; and (c) intramolecular ring closure of the anthracene polymers on the metal surface results under thermal treatment.

form linear polymer precursors. Further annealing of the samples at 400°C provides GNRs over the whole surface area through surface-assisted intramolecular cyclodehydrogenation.

Although surface-assisted synthesis has a few obstacles, it shows certain advantages over liquid-phase synthesis. Due to the strong $\pi-\pi$ interactions between the functionalized graphene prepared by the liquid-phase method, the functionalized graphene are easy to aggregate in dispersions, especially for the graphene nanoribbons or nanosheets of a large size. Although the introduction of a large number of alkyl chains at the edge of graphene can effectively improve its solubility in organic solvents, these non-electroconductive chains will deteriorate the electronic properties of graphene, leading to poor performance in devices. On the contrary, graphene nanoribbons or nanosheets without alkyl chains can be directly prepared by means of surface-assisted synthesis; in this manner, the best performance of functionalized graphene in devices can be achieved.

In addition, polymers prepared from colligated monomers cannot undergo intramolecular ring closure in the liquid phase due to steric hindrance. As shown in Figs. 3.11(a) and 3.11(b), because of significant steric hindrance, the neighboring two anthracene groups are not on the same plane, and thus the intramolecular ring closure cannot be carried out in the liquid phase to obtain graphene nanoribbons. On the contrary, when surface-assisted synthesis is used, anthracene polymers are formed on the surface of the metal substrate (Fig. 3.11(c)). In this polymeric product, each anthracene group lies flat on the surface of the metal substrate, and the angle between them is 0°, indicating that they are in the same plane. Therefore, these anthracene polymer intermediates formed on the surface of metal substrates can effectively undergo intramolecular dehydrocyclization to generate graphene nanoribbons.

3.1.3 *Control strategies for constructing functionalized graphene from PAHs*

3.1.3.1 *Size control of functionalized graphene*

Size control of functionalized graphene can be achieved by controlling the structures of the conjugated precursor monomers, followed by appropriate polymerization. As mentioned earlier in this chapter, graphene nanoribbons can be prepared by extending aromatic structures along the 1D direction; graphene nanosheets can be produced by simultaneously extending aromatic structures along 2D directions. Compared with graphene nanosheets, graphene nanoribbons have a more complex structure and have gained more interest among researchers. In this regard, this section mainly focuses on the size control of graphene nanoribbons.

Due to advances in experimental methods, researchers have been able to control the width and boundary configuration of graphene nanoribbons with atomic precision. However, synthesis of graphene nanoribbons with controllable length is somehow difficult. It is well known that different widths provide graphene nanoribbons with different electrical properties, and therefore the size of graphene nanoribbons is mainly measured by the width. As shown in Fig. 3.12, the width of the graphene nanoribbon is defined with the integer N, which indicates the number of dimers (two carbon sites) for the armchair nanoribbons and the number of zigzag lines for the zigzag nanoribbons. Graphene nanoribbons with different widths can be prepared by controlling the molecular structure of monomers and using specific polymerization methods (Fig. 3.13). For nanoribbons with identical edge geometries, their width should be well defined. If the edge geometries of nanoribbons are inconsistent and the widths are different, it is recommended to consider the largest width as the width.

N=7 armchair GNR

N=4 zigzag GNR

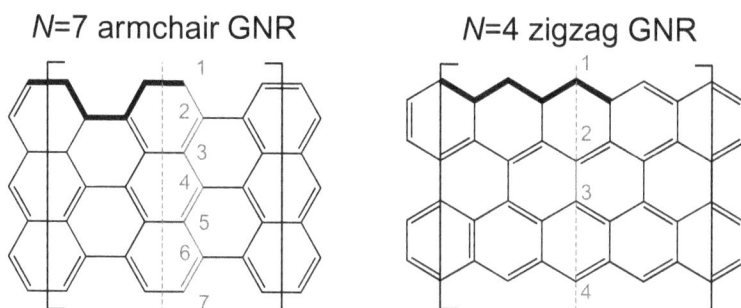

Figure 3.12. Illustration of the width defination of armchair and zigzag graphene nanoribbons.[4]

3.1.3.2 *Edge control of functionalized graphene*

The electronic structures of graphene nanoribbons are not only influenced by their width but also closely related to their edge topologies. Therefore, in order to tailor the electronic characteristics of graphene nanoribbons, a large number of graphene nanoribbons with a variety of edge topologies are fabricated, as shown in Fig. 3.14. The graphene nanoribbons that have been prepared so far mainly present the following edge structures: straight GNRs (armchair or zigzag edge), chevron-type GNRs, cove-type GNRs, "kinked" GNRs, necklace-like GNRs, and chiral GNRs. To highlight the edge topologies of graphene nanoribbons, alkyl chains on the edges are ignored. Moreover, in order to adjust the solubility, hydrophilicity, and hydrophobicity of graphene nanosheets or nanoribbons, as well as the interactions with other materials, the peripherical atoms can be modified by different functional groups, such as alkyl chains, glycol chains, chlorine atoms, trifluoromethyl, and luminescent groups.

3.1.3.3 *Defect control of functionalized graphene*

Defect-free graphene is a 2D material composed of a hexagonal lattice. Defects can be intentionally introduced into graphene through a bottom-up strategy; the typical defects are non-six-membered ring structures, such as five-membered rings, seven-membered rings, or eight-membered rings, that is, nanoporous structures of different sizes and shapes (Fig. 3.15). One representative example is nanoporous graphene (NPG). Due to its unique structure, NPG has great application potential in many fields such as field-effect transistors, atomically thin molecular sieves, ion transport, gas separation, and water purification. Its performance in

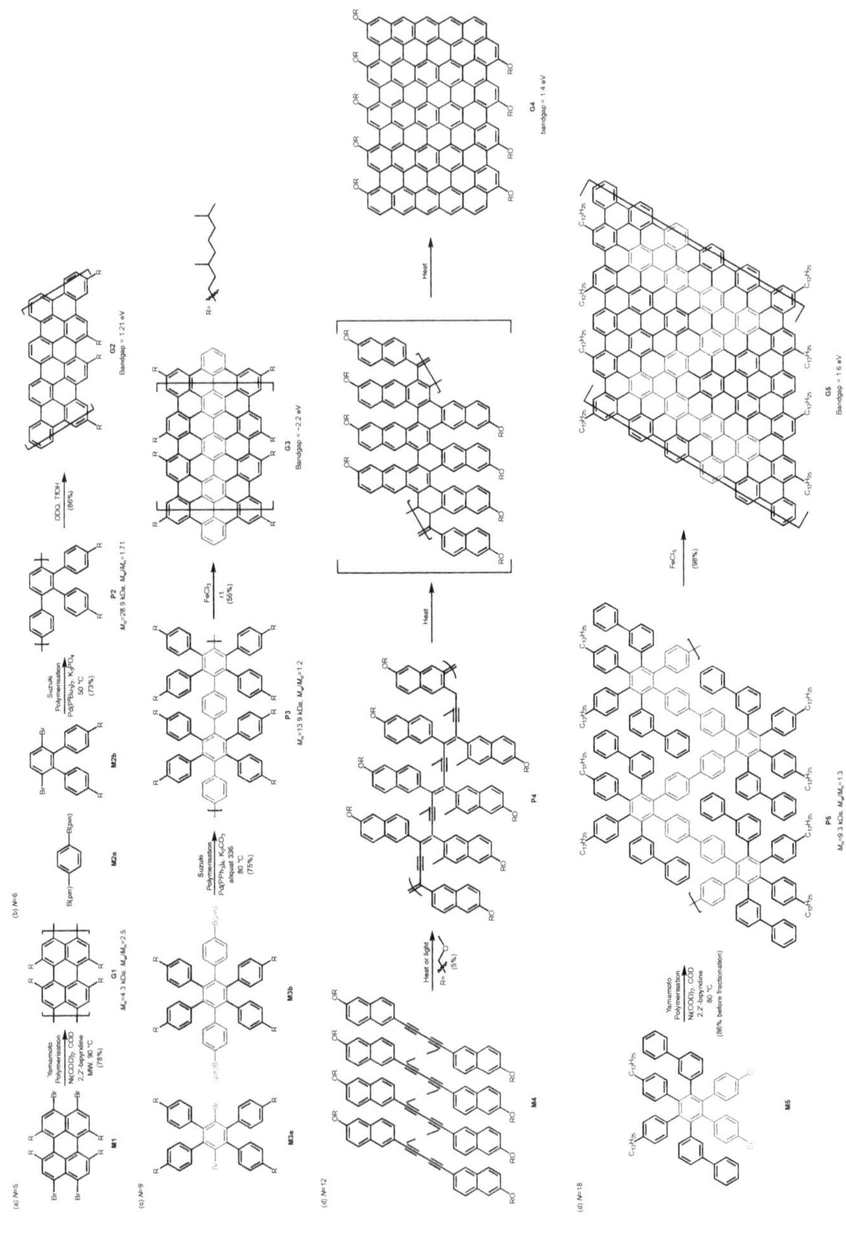

Figure 3.13. Synthesis of GNRs with different widths.[4]

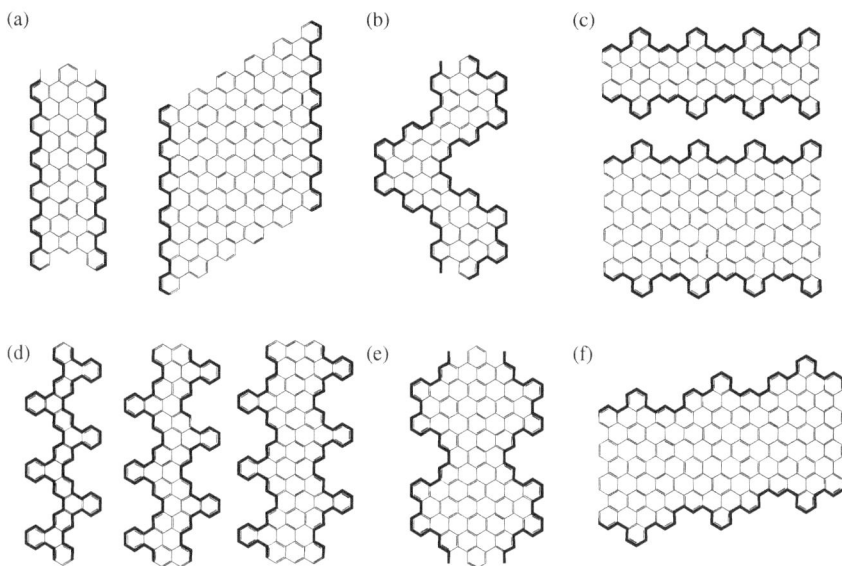

Figure 3.14. Edge topologies of GNRs[7]: (a) straight GNRs; (b) chevron-type GNR; (c) cove-type GNRs; (d) "kinked" GNRs; (e) necklace-like GNR; and (f) chiral GNR.

these applications is closely related to factors such as the size, shape, and periodic distribution of pores, and the precise control of these porous structures can be achieved by bottom-up approaches. At present, a variety of NPGs with different structures have been synthesized through the bottom-up strategy. The manipulation of these diversified porous structures is also achieved by the design of different monomers, the selection of appropriate polymerization, and the control of molecular ring closure.

3.1.3.4 *Heteroatom doping of functionalized graphene*

Heteroatom doping is a decent option to adjust the chemical composition and structural properties of graphene. The introduction of foreign heteroatoms into graphene allows the electronic, magnetic, and catalytic properties of graphene to be modulated. Hence, heteroatom doping is expected to be a useful way to boost applications in the fields of semiconductors, electronic devices, energy storage, and catalysis.

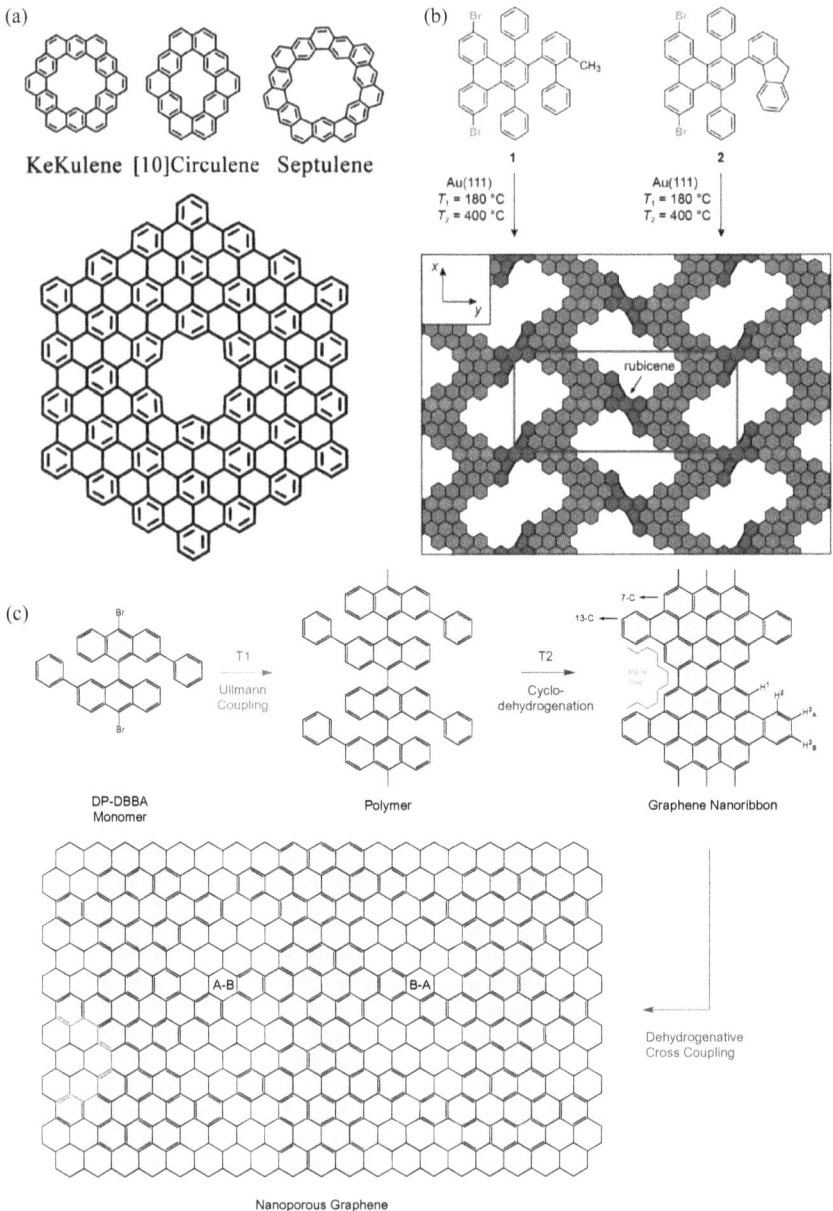

(a)

KeKulene [10]Circulene Septulene

(b)

Au(111)
T_1 = 180 °C
T_2 = 400 °C

Au(111)
T_1 = 180 °C
T_2 = 400 °C

rubicene

(c)

DP-DBBA
Monomer

T1

Ullmann
Coupling

Polymer

T2

Cyclo-
dehydrogenation

Graphene Nanoribbon

Nanoporous Graphene

Dehydrogenative
Cross Coupling

Figure 3.15. Examples of graphene with well-defined micropores synthesized via bottom up approaches: (a) graphene nanorings and graphene nanosheets with micropores, (b) and (c) nanoporous graphene prepared by surface-assisted synthesis.[8,9]

Heteroatoms commonly used in graphene doping include nitrogen (N), boron (B), phosphorus (P), and sulfur (S). The performance of doped graphene in these fields is closely related to the type, doping density, and distribution of heteroatoms, as well as the chemical structure after doping. For example, when applied in the field of supercapacitors, nitrogen doping can not only raise the surface wettability but also increase the redox reactivity of the electrolyte surface, therefore providing dramatic pseudocapacitance.[10] In addition, pyridinic-N graphene is reported to present high catalytic activity for redox reactions.[11] The heteroatom types, mass transfer sites, and chemical environment of heteroatoms can all be precisely controlled via a bottom-up strategy. Similarly, the doped graphene can be precisely fabricated by controlling the structure and elemental composition of the colligated monomers as the starting compounds, followed by controlling the polymerization and ring closure steps (Fig. 3.16).[12,13]

3.2 CVD Synthesis of Functionalized Graphene

CVD is a commonly used method to fabricate large-area, high-quality graphene, and it's also extensively employed to prepare different types of graphene-based transparent conductive films as well as functionalized graphene with various macroscopic structures.

3.2.1 *Overview of the CVD method*

The CVD process of fabricating graphene is as follows: (i) The carbon source (e.g., methane) is transported by carrier gas to the high-temperature zone, and the decomposed fragments or carbon radicals are transported onto the metal substrate through surface adsorption. (ii) The decomposed fragments or carbon radicals undergo dehydrogenation to form carbon atoms, which migrate on the surface of the metal substrate to form graphene. The mechanism of graphene growth by CVD varies depending on the metal substrate. At present, a variety of metals have been demonstrated to be good substrates for graphene growth, such as Ni, Cu, Ru, Ir, Pt, Co, Pb, and Re, and conventionally, substrates commonly used in the growth of graphene through CVD include Ni, Cu, and other insulating substrates.

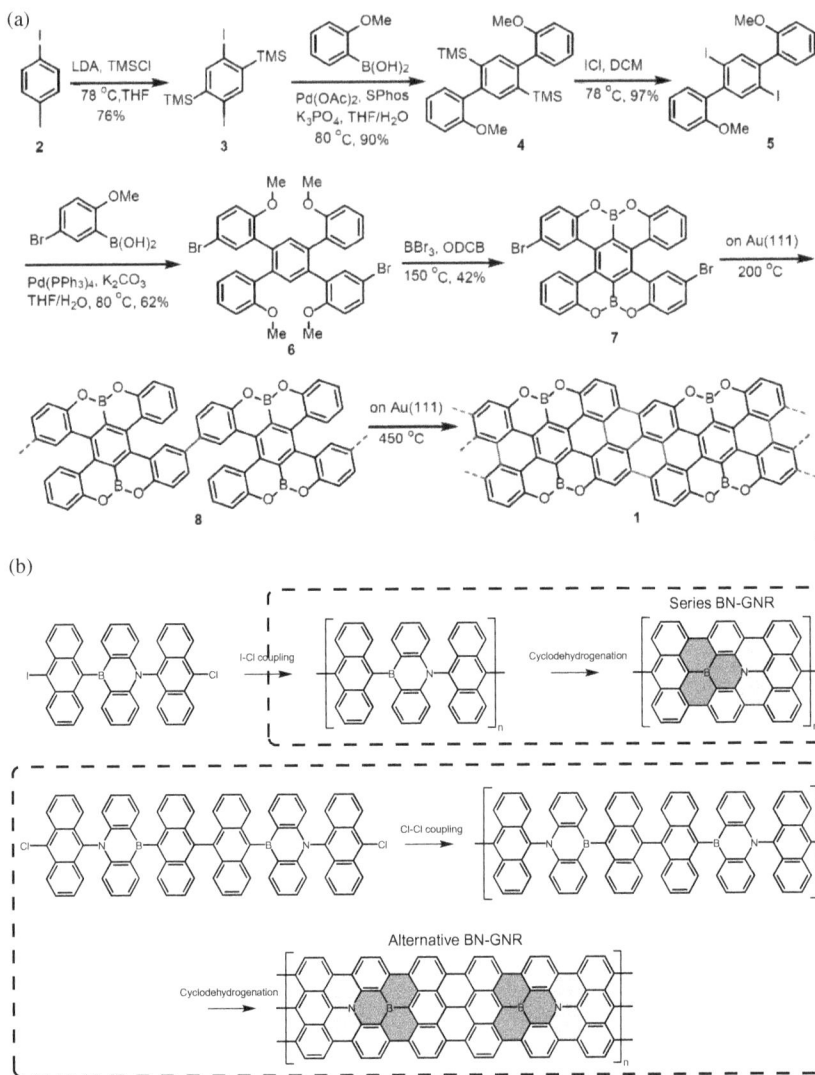

Figure 3.16. Fabrications of heteroatom-doped GNRs: (a) boron-oxygen-doped[12] and (b) boron-nitrogen-doped GNRs.[13]

Note: LDA: lithium diisopropylamide; TMSCl: trimethylsilyl chloride; SPhos: 2-dicyclohexylphos-phino-2´,6´-dimethoxybiphenyl; DCM: dichloromethane; BN-GNR: boron-nitrogen-doped GNR.

3.2.1.1 *Ni substrates*

Ni is a highly successful metal in the CVD synthesis of carbon nanotubes and, thus, it is no surprise that early research on the CVD synthesis of graphene exploited Ni. The growth of graphene on the surface of Ni substrate is considered to be a carbon segregation–precipitation process. Ni has a relatively high C solubility; as the carbon precursor decomposes, it produces C species (catalytically aided by the Ni), which then dissolve in the Ni film at elevated temperatures. This is followed by a cooling down step during which the solubility of C in the Ni decreases with rising temperature; C atoms diffuse out from the Ni–C solid solution and precipitate on the Ni surface and form graphene films. Polycrystalline Ni films would result in large single-crystal domains during high-temperature annealing. Single-crystal Ni(111) has a good lattice match with graphene, and it has been demonstrated to be an ideal substrate for epitaxial growth of graphene. Since the graphene films growth on Ni is governed by the carbon segregation process, the hydrocarbon concentration and cooling rate have a great influence on the layer numbers and quality of graphene. Therefore, the graphene layer numbers tend not to be homogeneous due to excess C dissolving out at grain boundaries, leading to multilayer nucleation.

3.2.1.2 *Cu substrates*

The growth process of graphene on the surface of Cu substrates is different from that of Ni substrates. The C solubility in Cu is extremely low, even if the hydrocarbon concentration is high and the growth time is long; only a trace amount of C dissolves into Cu. Therefore, the C species for the formation of graphene mainly comes from the catalytic decomposition of hydrocarbons on the copper surface. The underlying Cu foil beneath the formed graphene island would be preserved from hydrocarbons, leading to the end of the growth. Therefore, graphene growth on Cu substrates is a self-limiting surface reaction, which is often used to prepare high-quality single-layer graphene. Ruoff's group verified this mechanism by taking isotopically labeled hydrocarbons as carbon sources and utilizing Raman spectroscopy.[14] However, Kong's group carefully compared the growth kinetics of CVD-synthesized graphene on Cu surfaces at atmospheric pressure, low pressure, or under ultrahigh vacuum conditions; interestingly, the results revealed that graphene growth using a Cu catalyst under atmospheric pressure conditions with higher carbon source concentrations is not self-limiting as observed in the case of low-pressure CVD, and a large amount of multilayer graphene is obtained.[15] Therefore, the growth mechanism of graphene by CVD on Cu substrates still needs to be further explored.

3.2.1.3 *Insulating substrate*

In conventional CVD synthesis, a polymer-assisted transfer procedure of graphene sometimes leaves unwanted organic and metallic impurities, which worsens the condition of the electrical properties of the transferred graphene, due to which it cannot be comparable with graphene obtained by mechanical exfoliation. In this regard, researchers have devoted much effort toward developing the technology of direct graphene growth (without transfer) on the surface of insulating substrates.

The insulating substrates used in CVD synthesis of graphene can be categorized into metals and non-metals. Metals such as Ni and Cu can catalyze the growth of high-quality graphene, so it is a good choice to involve these metal films or vapors in graphene growth on the surfaces of insulating substrates. In 2011, Tour's group from Rice University report a transfer-free method of synthesizing bilayer graphene directly on SiO_2 substrates by carbon diffusion through a layer of nickel.[16] The 400 nm nickel layer was deposited on the top of SiO_2 substrates and used as the catalyst. During the annealing process at 1000°C, the carbon sources on the top of the nickel decomposed and diffused into the Ni layer. When cooled to room temperature, high-quality bilayer graphene was formed between the Ni layer and the insulating substrates. The Ni films were removed by etchants, and bilayer graphene was then directly obtained, eliminating any transfer process. It is believed that bilayer graphene is directly synthesized on the SiO_2 substrate by the diffusion and precipitation of carbon in the Ni layer. Furthermore, a Cu layer is also used to grow high-quality graphene between the metal layer and SiO_2 substrate.[17] It is proposed that the C can diffuse through the Cu layer at the grain boundaries.

In all the aforementioned CVD synthesis procedures, graphene is fabricated by means of pre-depositing the metal catalysts on the insulating substrates. However, CVD graphene can also be fabricated on insulating substrates that do not physically come into contact with metal catalysts such as Cu or Ni. One representative example is that Cu foils are placed in the upstream gas flow and the insulating substrates are placed downstream away from the Cu foil.[18] The upstream Cu vapor plays the role of floating catalyst, resulting in the more complete decomposition of the hydrocarbons, thereby forming multilayer graphene with few defects and amorphous carbons. However, it is revealed that the crystalline quality of the resulting graphene is effectively governed by the distance between the insulating substrates and the Cu foils. It is believed that the CH_x species are generated from the catalytic decomposition of CH_4 on the Cu foil and transported downstream to the SiO_2 surface; in addition, the subliming Cu particles generated from Cu foil immigrate on top of insulating substrates, leading to the remote catalyzation of graphene. Recently, a Korean research group successfully synthesized large-size monolayer graphene on the surface of non-metal substrates by suspending Cu foil on top of SiO_2 substrates (the gap is about 50 μm), and the overall quality of the resulting graphene was comparable to that of graphene grown by regular metal-catalyzed CVD on Cu foil.[19] It is believed that Cu vapor in this research plays a similar role to catalyze the growth of graphene, and the resulting monolayer graphene does not contain Cu metal as confirmed by X-ray Photoelectron Spectroscopy (XPS). However, the carrier mobility of graphene synthesized by this method is still only 800 cm²/(V·s), which indicates that the resulting graphene still has a large number of structural defects.

Another category of approaches for growing graphene on insulating substrates is by means of metal-free catalysis. Under appropriate CVD growth conditions, graphene can be directly grown on the surface of an insulating substrate that matches the graphene lattice structure. For example, graphene with fewer defects can be grown on the 300-nm SiO_2/Si substrates.

3.2.1.4 *Liquid substrate*

In addition to the aforementioned solid metal and non-metal substrates, liquid substrates can also be used to synthesize graphene by CVD methods. An early demonstration was the use of liquid Ga surface for the synthesis of uniform high-quality monolayer graphene using CVD.[20] The low vapor pressure of growth substrates provides a smooth surface without grain boundaries and favors the growth of high-quality graphene with a low nucleation density. As a comparison, the typical nucleation densities on liquid Ga and on solid polycrystalline Cu are about 1/1000 μm² and 1/10 μm², respectively, indicating even better uniformity for graphene grown over Ga. The graphene shows a high crystalline quality with electron mobility reaching levels as high as 7400 cm²/(V·s) under ambient conditions. A systematic study with liquid metals (such as liquid Cu or liquid In) showed that they are highly suited for strictly single-layer graphene.[21] This is because, during cooling from the CVD process, the surface metal solidifies, quickly blocking the precipitation of the absorbed carbon. As a result, growth is a self-limited catalytic process and is robust to variations in growth parameters.

3.2.2 *Macromorphology control of graphene by CVD synthesis*

Based on the above-mentioned discussion, it can be concluded that CVD methods are often used to prepare high-quality graphene. In the typical CVD process, graphene is deposited on the surface of various substrates, and finally the substrates are removed to obtain graphene. In this regard, it is anticipated that the substrates can also be used as templates to control the macromorphology of graphene, that is, using substrates with different shapes to grow graphene with different structures on the surface. The grown graphene almost completely replicates the structure and size of the substrate. Therefore, CVD methods are often used to synthesize continuous high-quality graphene of different shapes, that is, functionalized graphene with controllable macromorphology. The most typical example is graphene foam (GF), and the general procedure consists of the following: (i) synthesis of a 3D substrate, (ii) growth of graphene on the surface of the 3D substrate by CVD, and (iii) removal of the substrate to obtain a GF. Other materials can be incorporated into these basic steps to fabricate the GF composites. The GF prepared by this method is a bulk material with a 3D network structure and a large quantity of pores. This structural feature gives it the characteristics of high electrical conductivity, large specific surface area, and interconnected internal pores, and it has great application advantages in the fields of energy storage, catalysis, and pollutant adsorption.

Due to its low price and availability in various sizes/pore densities and thicknesses, commercial Ni foam with a pore size of ~200 to 500 µm is the most frequently employed template. This is because, compared with other substrates, the process of CVD graphene growth on Ni substrate is easy to control, the resulting graphene is of a high quality, and the Ni substrate is easy to remove. As shown in Fig. 3.17(a), the Ni nanowires are assembled into 3D Ni nanowire foams by vacuum filtration or mild pressing; the 3D Ni nanowire foam is then heated in a tube furnace with CH_4 gas acting as a carbon source, allowing for sufficient dissolution of the carbon source and for the graphene to fully develop on the Ni catalyst.[22] After CVD, 3D GF is obtained by removing the Ni catalyst by sequential chemical etching using 1 M $FeCl_3$ and 3 M HCl.

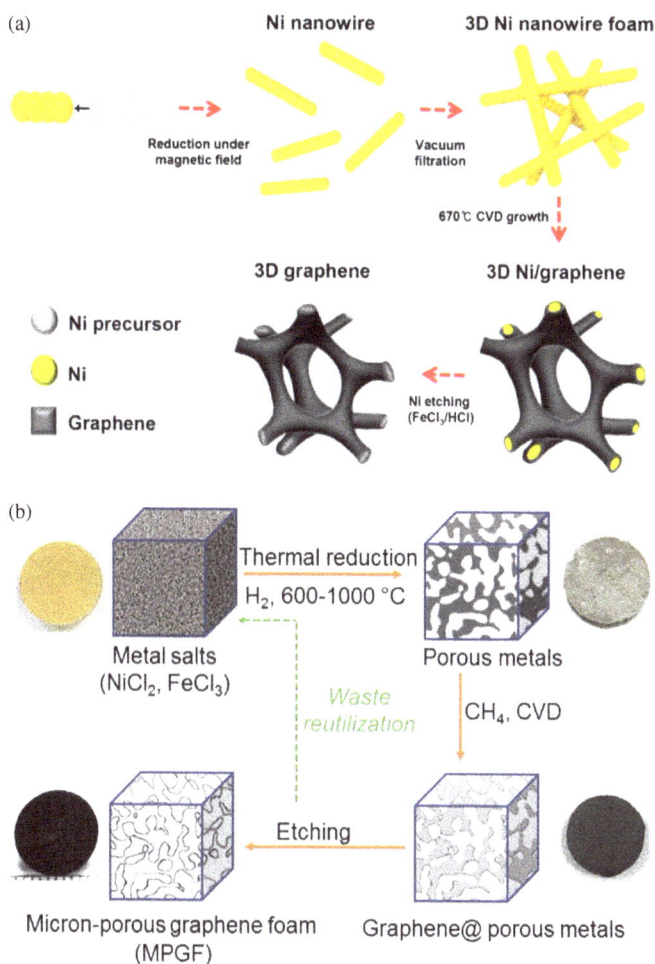

Figure 3.17. Schematic illustration of the procedure used to produce 3D GF: (a) synthesis of GF with 3D Ni nanowire foam as a template[22] and (b) synthesis of microporous 3D GF from a metallic salt precursor.[23]

3D Ni substrates can also be prepared using nickel salt precursors. As shown in Fig. 3.17(b), NiCl$_2$ powders are pressed into chips, which undergo hydrogen reduction typically at 600–1000°C in a tube furnace, resulting in microporous Ni foam chips. During the hydrogen reduction, the chips of metallic salt are reduced to microporous metal chips with a slight shrinkage but no change in shape. After reduction, the temperature

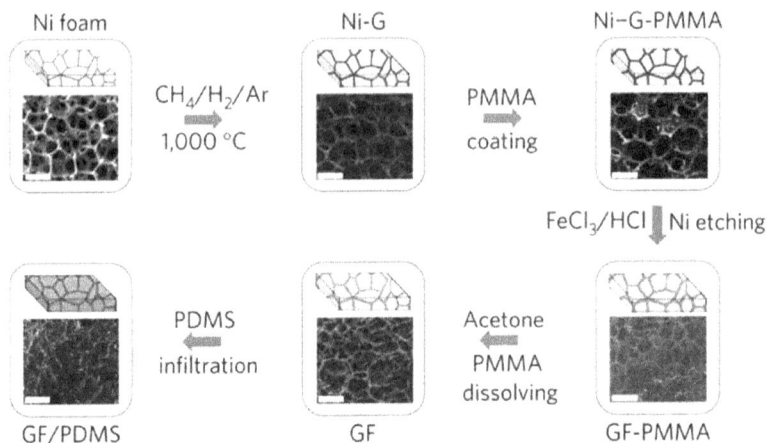

Figure 3.18. Schematic diagram of synthesis of a GF and integration with PDMS by the CVD method.[24]

is continuously raised to 1000°C for direct CVD growth of graphene by using methane as the carbon source. Finally, microporous GF chips are obtained by etching away the Ni templates in 1 M $FeCl_3$ solution or 2 M HCl solution. It should be pointed out that the nickel chloride waste from the etching of Ni can be easily dried and reutilized for the next batch of synthesis without complicated post-treatment.

As shown in Fig. 3.18, Ni foam, a porous structure with an interconnected 3D scaffold of nickel, is used as a template for CVD growth of GFs at 1000°C with CH_4 as the carbon source.[24] Because of the difference between the thermal expansion coefficients of nickel and graphene, ripples and wrinkles are formed on the graphene films. These ripples and wrinkles probably result in an improved mechanical interlocking with polymer chains and consequently better adhesion, when a GF is integrated with a polymer to form composite materials. To obtain GFs, before etching away the nickel skeleton with a hot HCl (or $FeCl_3$) solution, a thin layer of poly(methyl methacrylate) (PMMA) is deposited on the surface of the graphene films as a support to prevent the graphene network from collapsing during nickel etching. After the PMMA layer is carefully removed by hot acetone, a GF, a monolith of a continuous and interconnected graphene 3D network, is obtained. In order to improve the conductivity of the GF, the as-prepared GF is infiltrated into poly(dimethyl siloxane) (PDMS), followed by drying to fabricate a GF/PDMS composite. The electrical

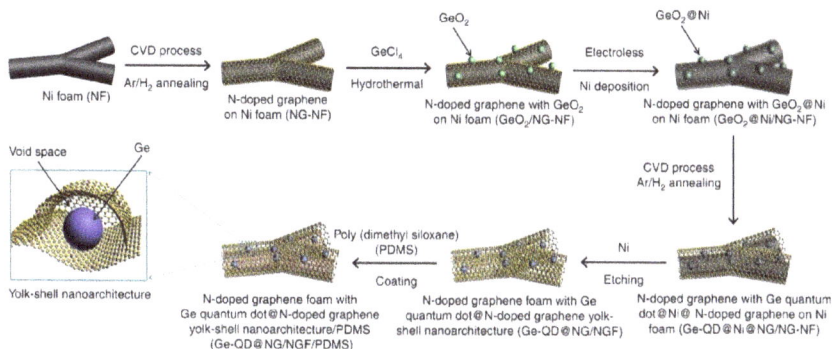

Figure 3.19. Schematic illustration of the preparation of Ge-QD@NG/NGF/PDMS yolk-shell nanoarchitecture by the CVD method.[25]

conductivity of this composite is ~10 S/cm, which is ~6 orders of magnitude higher than that of the chemically derived graphene-based composite.

The CVD method can also be adapted to synthesize multilayer GF composites and to achieve the doping of graphene at the same time. As shown in Fig. 3.19, a 3D interconnected porous nitrogen-doped graphene foam (NGF) with encapsulated Ge quantum dot@nitrogen-doped graphene yolk-shell nanoarchitecture (Ge-QD@ NG/NGF) is synthesized using nickel foam as the template.[25] In this research, 3D interconnected porous Ni foam is selected as the template for N-doped graphene growth, carbon and nitrogen are deposited into the Ni foam by decomposing pyridine at 900°C. N-doped graphene is well distributed and deposited on the 3D interconnected porous Ni foam by CVD. The homogeneous non-aggregated GeO_2 nanoparticles are then deposited using $GeCl_4$ and loaded in a 3D interconnected porous N-doped graphene network with a Ni foam (GeO_2/NG-NF) matrix. The GeO_2/NG-NF is then coated with a Ni thin layer using electroplating deposition. Thereafter, the GeO_2@Ni/NG-NF nanoarchitecture is catalyzed for conformal N-doped graphene growth by CVD at 650°C. After that, Ge quantum dots are generated through the reduction of GeO_2 and thermally annealed at 650°C, and acid etching is used to remove the Ni foam and sacrificial layer to obtain the Ge-QD@ NG/NGF yolk-shell nanoarchitecture. As a final step, in order to test the flexibility and electrochemical properties of the nanoarchitecture, a thin layer of PDMS is uniformly coated on the surface of the Ge-QD@NG/ NGF yolk-shell nanoarchitecture (Ge-QD@NG/NGF/PDMS).

3.3 Synthesis of Functionalized Graphene from Other Materials

In addition to those aforementioned materials, other materials can also be utilized to synthesize functionalized graphene based on a bottom-up strategy. These materials mainly include COFs, MOFs, biomass, and organic molecules. The fundamental idea of synthesizing functionalized graphene based on these materials is that these materials as starting materials undergo carbonization by heat treatment under different temperatures and conditions, and various forms of functionalized graphene are finally obtained.

3.3.1 *Graphene derived from COFs*

COFs are a class of materials that form periodic ordered 2D or 3D structures through reversible polymerization of specific small organic molecules (Fig. 3.20).[26] The synthesis of COFs is also based on bottom-up approaches, and their elemental composition and molecular structure are determined by the small-molecule precursors used. Therefore, the structure and properties of COFs, as well as their performance in applications, can also be tuned by precursor control. COFs are excellent materials for the synthesis of functionalized graphene due to their abundant elemental composition, diverse geometric configurations, and ordered porous structure. Functionalized graphene is usually synthesized by the carbonization of COFs at high temperatures. Since COFs are mainly prepared via a bottom-up strategy, the composition and structure of the resulting functionalized graphene can also be tuned by controlling the small-molecule precursors.

As shown in Fig. 3.21, a 2D covalent triazine-based framework (CTF) can be transformed to a 3D N-doped graphene at high temperature.[27] At about 400°C, terephthalonitrile is trimerized into 2D CTF and then evolves into 3D N-doped graphene at a higher temperature (500–700°C). The cross-linked structure and carbonization degree of 2D CTF can be controlled by the carbonization temperature. The synthesized 3D N-doped graphene has abundant pore distributions, high specific surface area (up to 2237 m^2/g), excellent electrical conductivity, and high nitrogen content (up to 12.44%). The XPS analyses show that there are three types of nitrogen configurations: pyridinic nitrogen, pyrrolic nitrogen, and quaternary

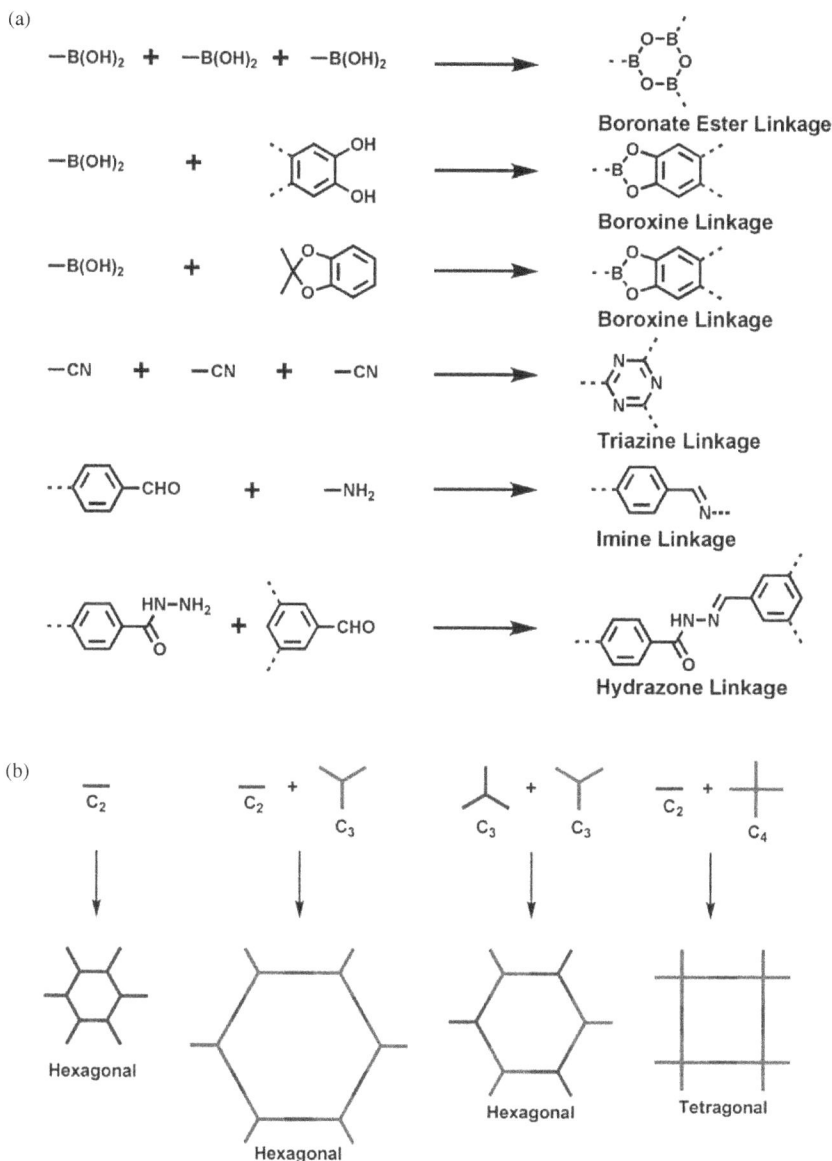

Figure 3.20. Schematic representation of synthesis and molecular model of COFs[26]: (a) dynamic chemical reactions for the synthesis of COFs and (b) the combination of building blocks with different geometries to design 2D COFs.

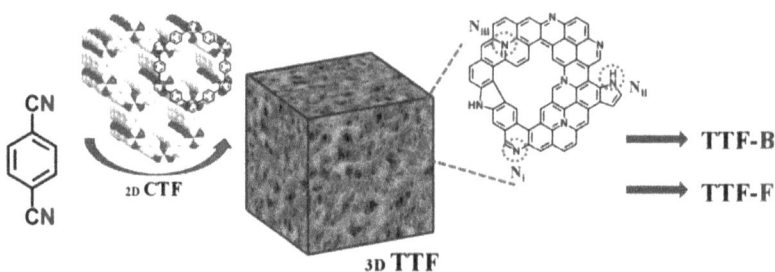

Figure 3.21. Schematic of the bottom-up construction of 3D N-doped graphene from COFs: the terephthalonitrile is trimerized into 2D CTF and then evolves into 3D N-doped graphene at high temperatures.[27]

nitrogen. Furthermore, nitrogen and boron co-doped 3D graphene (TTF-B) is synthesized by high-temperature treatment (600°C) using a mixture of 2D CTF and boron oxide as the raw materials. Similarly, if a mixture of 2D CTF and ammonium fluoride is used as the raw material, 3D graphene co-doped with nitrogen and fluorine (TTF-F) is fabricated after high-temperature treatment.

COF materials possess abundant pore channels, and these pores are able to accommodate metal ions to form compounds. The high-temperature treatment of these compounds could result in functionalized graphene containing metal ions, which can be used in the fields of catalysis and energy storage. The doping of graphene with metals offers catalytic sites with high activity or energy storage sites with redox properties. As shown in Fig. 3.22, COF materials based on 2D polyamine (Schiff base) incoporate catalytic metal ions (Fe^{III}, Co^{II}, Ni^{II}) into the cavities of their laminar structures, which favours corrugated graphene formation upon controlled thermal treatment. Small-molecule monomers containing amino and aldehyde moieties are catalyzed by acetic acid to form 2D COFs linked by Schiff base groups. This 2D COF consists of uniform micropores, which can adsorb metal ions Fe^{III}, Co^{II}, or Ni^{II} into its cavities via the imine groups. Therefore, the 2D COF is stirred in a methanol solution of iron acetylacetonate, cobalt acetylacetonate, or nickel acetylacetonate, respectively, followed by washing and drying, and then a 2D COF loaded with metal ions is obtained. The composite of 2D COF and metal is pyrolyzed in a N_2 atmosphere at 900°C for 4 h, leading to the formation of a hierarchical N-doped porous graphene incorporating metal ions.

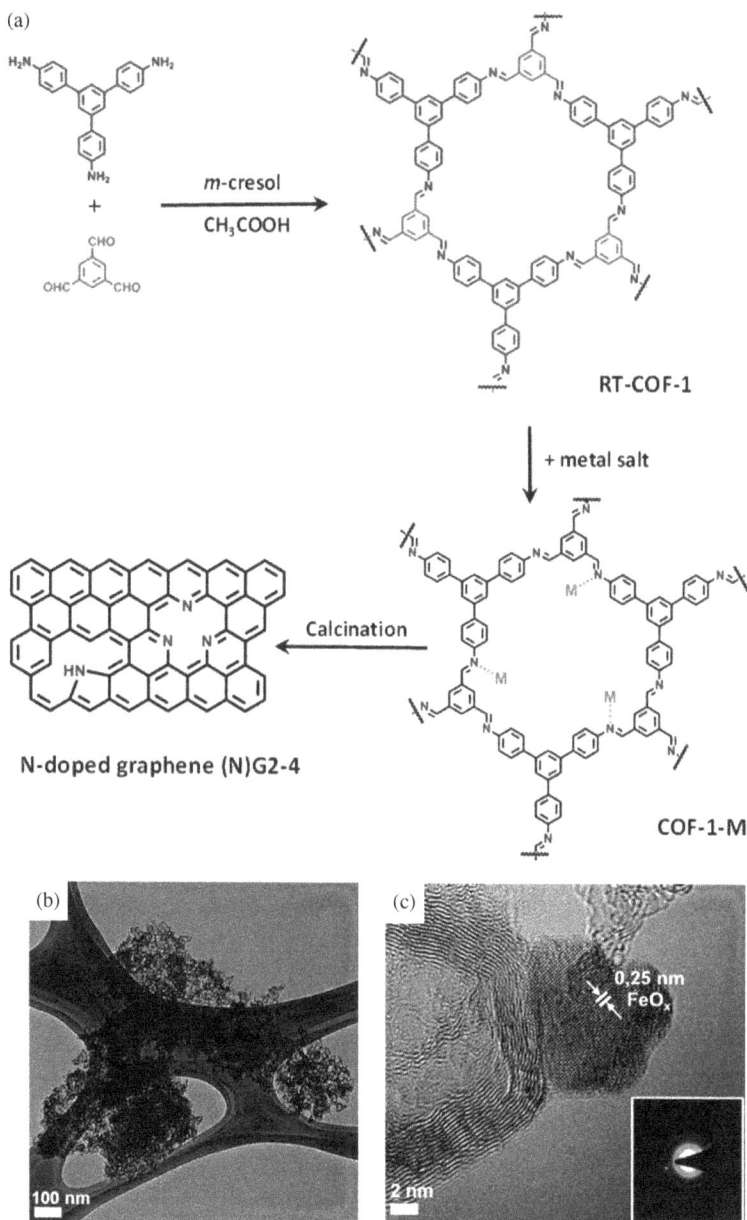

Figure 3.22. Schematic of synthesis of N-doped porous graphene from COFs[28]: (a) 2D COF incorporating with metal ions is transformed into 2D N-doped graphene at high temperature and (b–c) high-resolution transmission electron microscopy (HRTEM) of 2D N-doped graphene.

3.3.2 *Functionalized graphene derived from MOFs*

Similar to COFs, MOFs are 1D, 2D or 3D periodic frameworks, which are usually synthesized via bottom-up approaches (Fig. 3.23).[29] However, MOFs are formed by the self-assembly of metallic centers and bridging organic linkers. Similarly, the elemental composition, structure, properties, and application performance of MOFs can also be effectively tuned by controlling small-molecule ligands and metal ions. MOFs also have a good elemental composition and diverse morphological structures, so they are also ideal materials for synthesizing functionalized graphene. By adjusting the composition and structure of MOFs, functionalized graphene can be controllably formed by high-temperature carbonization.

Functionalized graphene derived from MOFs has the following advantages: (i) It has a tunable structure, composition, and function because in the process of preparing MOFs, metal ions/clusters and organic ligands can be selectively adjusted according to performance

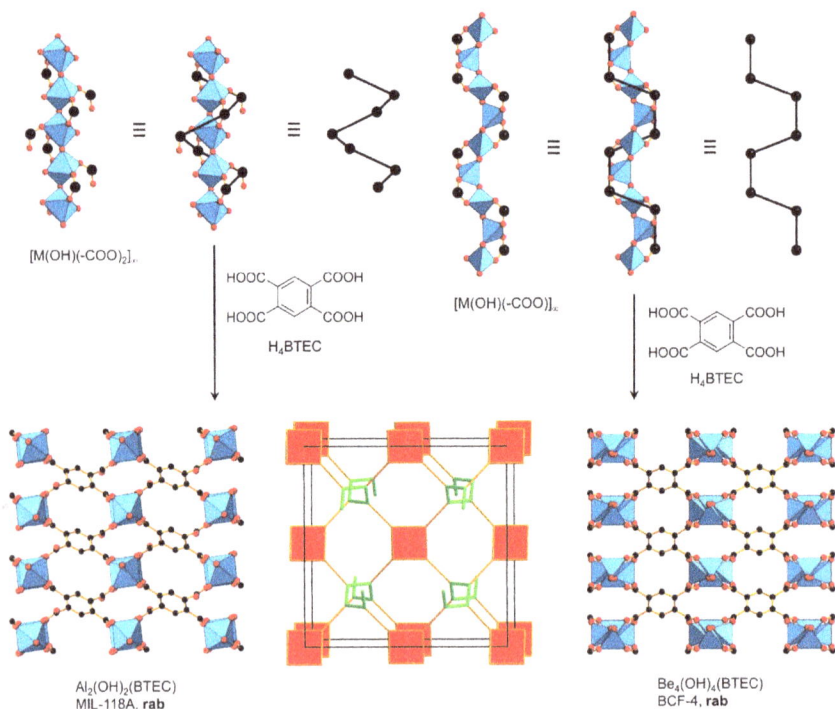

Figure 3.23.   Illustration of synthetic routes and structure of MOFs.[29]

requirements. (ii) It is easy to fill the carbon framework with metal or non-metal elements, which is mainly due to MOFs' large specific surface area, diverse pore distribution, and ordered pore channels. (iii) The highly ordered structure is retained after carbonization, and it is easy to directly form a graphene structure, which is attributed to the ordered structure of MOFs. Therefore, the synthesis strategy of functionalized graphene derived from MOFs is able to control the material structure and composition and satisfy the various demands in applications.

By carefully designing and synthesizing MOFs and controlling the high-temperature carbonization protocol, functionalized graphene can be prepared by direct carbonization of MOFs. As shown in Fig. 3.24, two- to six-layered GNRs can be directly prepared by rod-shaped MOFs

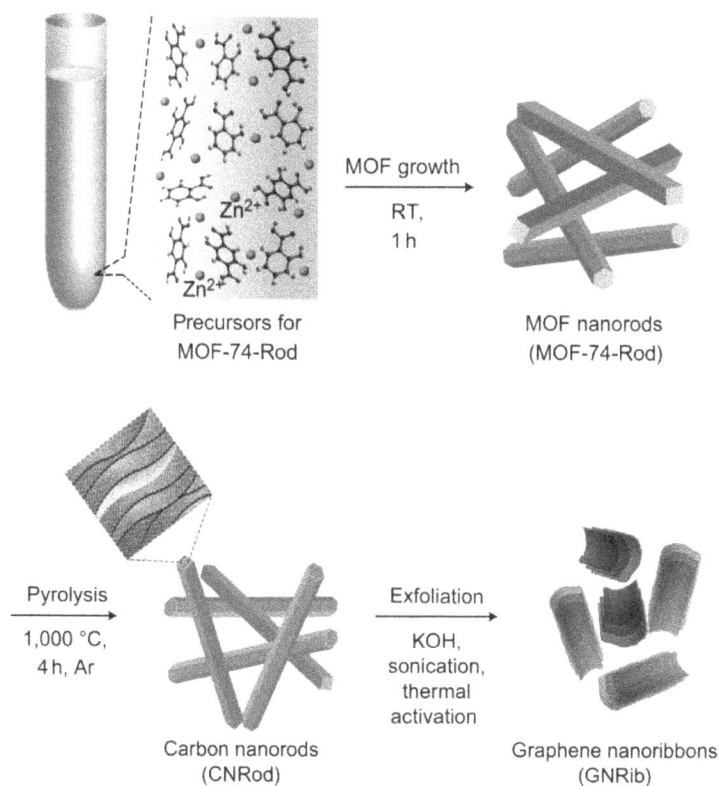

Figure 3.24. Scheme of the synthesis of rod-shaped MOF-74 following the preparation of GNRs by high-temperature heat treatment.[30]

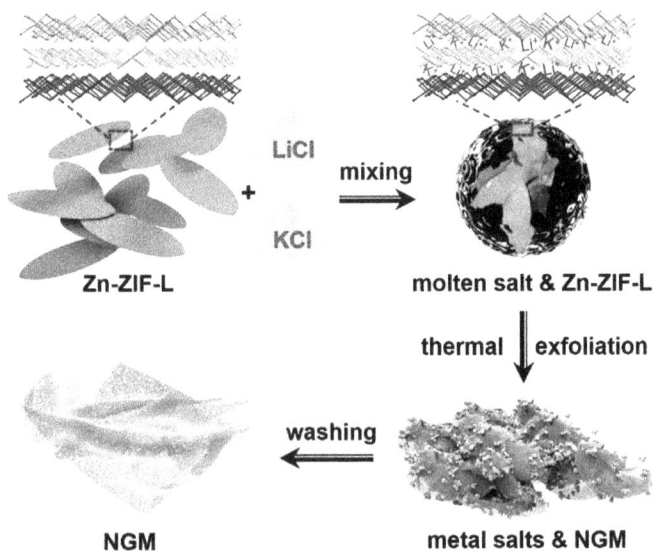

Figure 3.25.   Illustration of thermal exfoliation of Zn-ZIF-L to produce N-doped graphene nanomesh.[31]

(MOF-74) after carbonization and exfoliation.[30] The room-temperature reaction of zinc acetate and 2,5-dihydroxyterephthalic acid in the presence of salicylic acid as a modulator leads to the formation of rod-shaped MOF-74. Carbonization of rod-shaped MOF-74-Rod at 1000°C in an argon flow yields 1D carbon nanorods via a morphology-preserved thermal transformation process. KOH treatment of carbon nanorods under ultrasonication treatment followed by thermal activation at 800°C results in the formation of GNRs (50–70 nm wide and 100–150 nm long).

In addition, graphene nanosheets can also be synthesized through carbonization of MOFs with 2D sheet-like morphology. As shown in Fig. 3.25, a zeolite imidazolate framework (ZIF, an MOF material) with nanoscale leaf-like 2D structure can be carbonized and exfoliated to produce a N-doped graphene nanomesh.[31] The Zn-containing ZIF nanoleaves (Zn-ZIF-L) are prepared by using zinc nitrate and 2-methylimidazole precursors in an aqueous solution, which exhibits a leaf-like shape and a lateral dimensional size of about 5 nm. Direct carbonization of Zn-ZIF-L at a typical temperature of 800°C produces a leaf-like N-doped porous carbon. The exfoliator and etching agent of LiCl/KCl is used and mixed with Zn-ZIF-L precursors. Then, the mixture is heat treated under the protection of a nitrogen atmosphere to 900°C to produce a nitrogen-doped graphene nanomesh. The possible exfoliation process is proposed as follows: When the temperature is above 369°C, LiCl/KCl is molten and the resulting small metal ions are inserted into the porous carbon layers. Then, as the temperature is raised over 700°C, the carbon ring of Zn-ZIF-L forms leaf-like porous carbon; meanwhile, LiCl/KCl evaporates the exfoliated layer structure of carbonized Zn-ZIF-L. The graphene nanomesh presents a 2D morphology, a large specific surface area (up to 1329.5 m²/g), and abundant N-doping sites, and these structural features contribute to an unprecedented catalytic activity.

3.3.3 *Biomass-derived functionalized graphene*

Low-cost biomass not only features a variety of microstructures but also contains a variety of elements, which makes it an ideal raw material for the synthesis of functionalized carbon materials. The fabricated functionalized carbon materials formed by carbonization of biomass with specific structures and compositions not only possess tunable active sites but also exhibit superior electrical conductivity, high specific surface area, and/or abundant porosity. Biomass that can be used to prepare functionalized carbon materials includes eggshell membranes,[32] seaweed,[33] hemoglobin,[34] amaranth waste,[35] and silk.[36] Biomass with various inherent porous structures can be utilized to produce diverse carbonaceous materials with desirable hierarchical porous superstructures via different methods. By selecting biomass materials with appropriate microstructures, functionalized graphene can be obtained via carbonization.

Wheat straw is an abundantly available biomass due to its enormous worldwide output (about 350 million tons per year over the world). However, it is not effectively utilized. It is reported that wheat straw can be used as a raw material to synthesize high-quality graphene nanosheets via a combined hydrothermal and graphitization approach.[37] As shown in Fig. 3.26, the wheat straw is first cut into small pieces (approximately 3 cm in length), washed with distilled water, and then dried before use. The washed wheat straw is immersed in a Teflon-lined stainless autoclave wherein a homogeneous 3 M KOH solution is contained, and then the autoclave is heated at a temperature of 150°C for 6 hours. The following high-temperature pyrolysis and chemical activation are carried out at

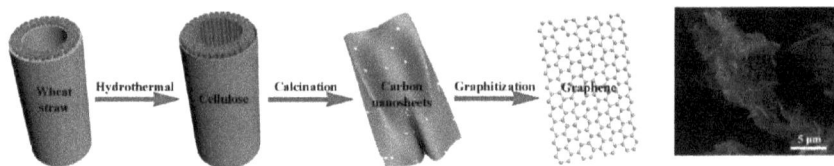

Figure 3.26.   Schematic diagram of functionalized graphene prepared by graphitization of wheat straw.[37]

800°C for 3 h under the protection of a N_2 atmosphere. After that, the carbonaceous product is thoroughly washed with HCl solution to remove the residual KOH and dried in a vacuum oven at 80°C for 12 hours. Finally, the as-prepared product is thermally treated at 2600°C for 5 min using a graphite furnace, giving rise to graphene nanosheets. The resulting graphene nanosheets show favorable features such as ultrathin nanosheet frameworks (2−10 atomic layers), high graphitization (up to 90.7%), graphite-like interlayer spacing (0.3362 nm), and a mesoporous structure.

Eggplant has an inherent sheet-like microstructure, so a one-step carbonization method can be utilized to fabricate sheet-like porous carbon materials from it with a high specific surface area (Fig. 3.27).[38] In this work, the carbonization temperature was found to be crucial for determining the microstructure and pore structure of carbon materials. When the carbonization temperature is higher than 700°C, thick corrugated sheets of eggplant transform into loose and thinner sheets after carbonization, and the sheet thickness is in the range of 100−200 nm. When the carbonization temperature is 900°C, the specific surface area of sheet-like porous carbon material reaches 950 m²/g. However, when the carbonization temperature is further increased to 1000°C, the specific surface area of the resulting material is decreased to 133 m²/g, which may be caused by the expansion and even collapse of the microporous structure inside the material due to the high temperature. This sheet-like porous carbon material exhibits excellent electrochemical performance when applied as an electrode material for supercapacitors.

3.3.4   *Functionalized graphene derived from carbonization of organic molecules*

Using small organic molecules and polymers as raw materials, functionalized graphene can also be obtained by heat treatment at different

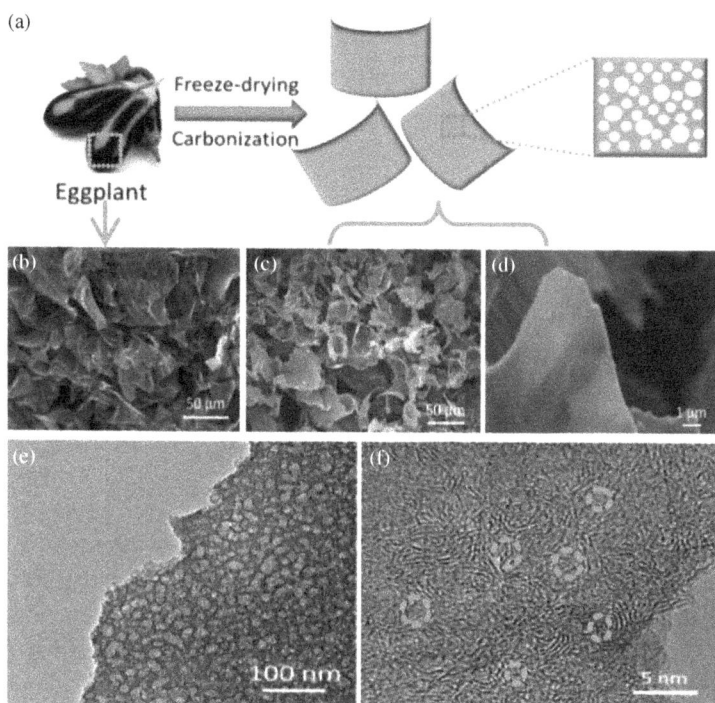

Figure 3.27. Schematic of synthesis of functionalized graphene derived from eggplant.[38]

temperatures. By adjusting the elemental composition and structure of small organic molecules, the corresponding structure and composition of functionalized graphene can be tuned. As shown in Fig. 3.28, monolayer graphene is synthesized via direct solid-state pyrolytic conversion of a sodium carboxylate material, such as sodium gluconate or sodium citrate, in the presence of Na_2CO_3.[39]

Sodium gluconate and Na_2CO_3 powder are ground thoroughly in a 1:20 molar ratio using an agate mortar and pestle. The resulting mixture is heated at 950°C under the protection of Ar. After cooling naturally to room temperature under a flow of Ar, the remaining traces of Na_2CO_3 salt are removed with dilute hydrochloric acid. Finally, the pyrolysis products are purified by washing with distilled water and absolute ethanol during vacuum filtration until the filtrate gives a neutral pH value. Subsequent drying at 80°C for 12 h yields a pure monolayer graphene powder. The primary reaction taking place during the preparation of monolayer

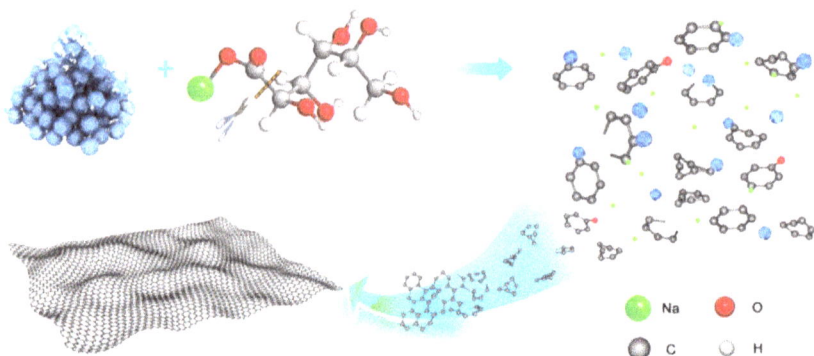

Figure 3.28. Schematic outline of the general solid-state pyrolytic conversion strategy for preparation of the monolayer graphene. The blue spheres represent the Na_2CO_3 salt.[39]

graphene via the general solid-state pyrolytic conversion method involves thermal cyclodehydration and in-plane carbon reconstruction. Upon heating, the sodium carboxylate monomers become covalently cross-linked in the Na_2CO_3 medium and eliminate water molecules and through a cyclodehydration reaction. Initially, the highly polarized covalent bond between the carboxyl carbon and its nearby carbon atom is broken, resulting in the formation of disordered carbonaceous radicals, which may consist of cyclized carbon atoms. At elevated temperatures, the cyclized carbon atoms are rearranged in the 2D direction, thereby producing a single graphene layer. Generally, the pyrolytic conversion from the sp^3-bonded carbon atoms into the sp^2 network always leads to the formation of amorphous carbon fragments. However, in this heat treatment system, the Na_2CO_3 phase exhibits strong interactions with the carbon π-electron species and thus drives the cyclized carbonaceous radicals toward in-plane rearrangement. In addition, the *in situ* generated Na_2CO_3 clusters are evenly distributed within the carbon matter, thereby allowing for full contact, which is responsible for the complete conversion from carbonaceous radicals to monolayer graphene. The average thickness of monolayer graphene obtained by this method is about 0.5 nm, which is indicative of the extremely high yield of monolayer graphene.

In addition to small molecules, polymers can also be used to fabricate functionalized graphene by pyrolysis. As shown in Fig. 3.29, spatially non-uniform and disconnected graphene was fabricated when PAN film coated with a Ni layer was pyrolyzed, resulting in flake-like graphene.

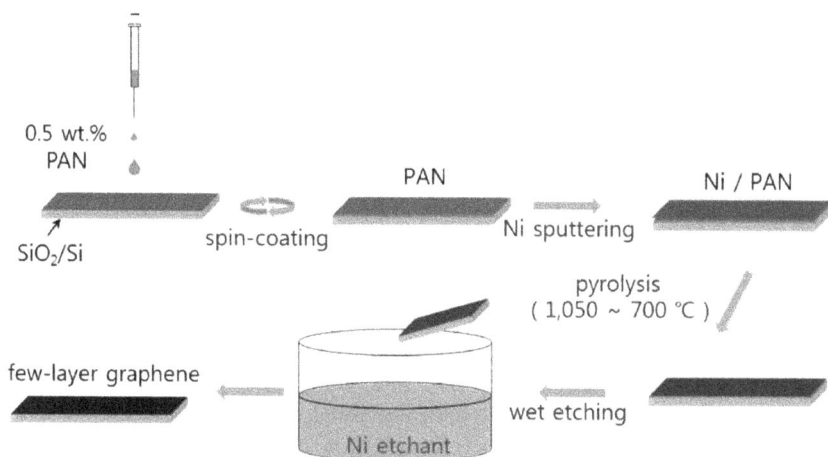

Figure 3.29. Schematic diagram of pyrolysis of polyacrylonitrile to prepare flake-like graphene.[40]

PAN (M_w = 150,000) dissolved in N,N-dimethylformamide (DMF) is spin coated on SiO_2/Si wafers.[40] Subsequently, a Ni layer is then coated on the spin-coated PAN/SiO_2/Si substrate using a magnetron sputtering system. The Ni-coated PAN/SiO_2/Si samples are pyrolyzed at high temperature (700–1500°C) under vacuum. After the pyrolysis, the coated Ni layer is removed with a Ni etchant to obtain flake-like multilayer graphene. Therefore, very non-uniform, flake-like graphene is fabricated by the pyrolysis of the Ni-coated PAN films due to the dewetting of the catalytic Ni films during the pyrolysis process. An agglomeration of micrometer- or nanometer-sized Ni particles is observed after pyrolysis at 950°C, and the formation of the Ni particles from the Ni film is attributed to the dewetting phenomenon. By reducing the temperature and optimizing the conditions, the dewetting phenomenon can be alleviated or suppressed, and the spatial uniformity of graphene can be improved.

3.4 Summary

In summary, by designing and selecting the chemical structures of the building blocks, controllable construction of functionalized graphene can be achieved via a bottom-up strategy. The key factor of this strategy is to design and select the building blocks purposefully according to the needs, together with appropriate synthesis methods, to realize the synthesis of functionalized graphene. The characteristics of well-defined structures and compositions of functionalized graphene makes it possible to establish the structure-property relationship in its applications, which is conductive to further optimizing its structure and fully exploiting its application potential. However, most methods for synthesizing functionalized grpahene menthioned here are not suitable for mass production. Obstacles such as long preparation steps, scale-up effects, and high costs hinder the large-scale preparation of functionalized graphene by bottom-up approaches. In this regard, these issues are yet to be addressed.

In addition to the bottom-up strategy discussed in this chapter, from the perspective of preparation principles, most of the approaches for synthesizing functionalized graphene composites, from graphene or functionalized graphene and other materials, can be classified as bottom-up strategies. However, the preparation of functionalized graphene composites is usually closely related to their applications; thus, their preparation and applications will be discussed in detail in Chapter 6.

References

1. Ito, H., Ozaki, K., and Itami, K. (2017). Annulative π-extension (APEX): Rapid access to fused arenes, heteroarenes, and nanographenes. *Angew. Chem. Int. Ed.*, 56(37), 11144–11164.
2. Ito, H., Segawa, Y., Murakami, K., and Itami, K. (2019). Polycyclic arene synthesis by annulative π-extension. *J. Am. Chem. Soc.*, 141(1), 3–10.
3. Yang, X., Dou, X., Rouhanipour, A., Zhi, L., Räder, H. J., and Müllen, K. (2008). Two-dimensional graphene nanoribbons. *J. Am. Chem. Soc.*, 130(13), 4216–4217.
4. Yoon, K.-Y. and Dong, G. (2020). Liquid-phase bottom-up synthesis of graphene nanoribbons. *Mater. Chem. Front.*, 4(1), 29–45.
5. Cai, J., Ruffieux, P., Jaafar, R., Bieri, M., Braun, T., Blankenburg, S., Muoth, M., Seitsonen, A. P., Saleh, M., Feng, X., Müllen, K., and Fasel, R. (2010). Atomically precise bottom-up fabrication of graphene nanoribbons. *Nature*, 466(7305), 470–473.
6. Chen, Z., Berger, R., Müllen, K., and Narita, A. (2017). On-surface synthesis of graphene nanoribbons through solution-processing of monomers. *Chem. Lett.*, 46(10), 1476–1478.
7. Shekhirev, M. and Sinitskii, A. (2017). Solution synthesis of atomically precise graphene nanoribbons. *De Gruyter*, 2(5).

8. Moreno, C., Vilas-Varela, M., Kretz, B., Garcia-Lekue, A., Costache, M. V., Paradinas, M., Panighel, M., Ceballos, G., Valenzuela, S. O., Peña, D., and Mugarza, A. (2018). Bottom-up synthesis of multifunctional nanoporous graphene. *Sci.*, 360(6385), 199–203.

9. Jacobse, P. H., McCurdy, R. D., Jiang, J., Rizzo, D. J., Veber, G., Butler, P., Zuzak, R., Louie, S. G., Fischer, F. R., and Crommie, M. F. (2020). Bottom-up assembly of nanoporous graphene with emergent electronic states. *J. Am. Chem. Soc.*, 142(31), 13507–13514.

10. Chang, J., Gao, Z., Liu, X., Wu, D., Xu, F., Guo, Y., Guo, Y., and Jiang, K. (2016). Hierarchically porous carbons with graphene incorporation for efficient supercapacitors. *Electrochim. Acta*, 213, 382–392.

11. Jin, J., Fu, X., Liu, Q., Liu, Y., Wei, Z., Niu, K., and Zhang, J. (2013). Identifying the active site in nitrogen-doped graphene for the VO^{2+}/VO_2^+ redox reaction. *ACS Nano*, 7(6), 4764–4773.

12. Wang, X.-Y., Urgel, J. I., Barin, G. B., Eimre, K., Di Giovannantonio, M., Milani, A., Tommasini, M., Pignedoli, C. A., Ruffieux, P., Feng, X., Fasel, R., Müllen, K., and Narita, A. (2018). Bottom-up synthesis of heteroatom-doped chiral graphene nanoribbons. *J. Am. Chem. Soc.*, 140(29), 9104–9107.

13. Kawai, S., Nakatsuka, S., Hatakeyama, T., Pawlak, R., Meier, T., Tracey, J., Meyer, E., and Foster, A. S. (2018). Multiple heteroatom substitution to graphene nanoribbon. *Sci. Adv.*, 4(4), eaar7181.

14. Li, X., Cai, W., An, J., Kim, S., Nah, J., Yang, D., Piner, R., Velamakanni, A., Jung, I., Tutuc, E., Banerjee, S. K., Colombo, L., and Ruoff, R. S. (2009). Large-area synthesis of high-quality and uniform graphene films on copper foils. *Sci.*, 324(5932), 1312.

15. Bhaviripudi, S., Jia, X., Dresselhaus, M. S., and Kong, J. (2010). Role of kinetic factors in chemical vapor deposition synthesis of uniform large area graphene using copper catalyst. *Nano Lett.*, 10(10), 4128–4133.

16. Peng, Z., Yan, Z., Sun, Z., and Tour, J. M. (2011). Direct growth of bilayer graphene on SiO_2 substrates by carbon diffusion through Nickel. *ACS Nano*, 5(10), 8241–8247.

17. Lee, C. S., Baraton, L., He, Z., Maurice, J.-L., Chaigneau, M., Pribat, D., and Cojocaru, C. S. (2010). Dual graphene films growth process based on plasma-assisted chemical vapor deposition. In *Proceedings of SPIE 7761, Carbon Nanotubes, Graphene, and Associated Devices III*, p. 77610P.

18. Teng, P.-Y., Lu, C.-C., Akiyama-Hasegawa, K., Lin, Y.-C., Yeh, C.-H., Suenaga, K., and Chiu, P.-W. (2012). Remote catalyzation for direct formation of graphene layers on oxides. *Nano Lett.*, 12(3), 1379–1384.

19. Kim, H., Song, I., Park, C., Son, M., Hong, M., Kim, Y., Kim, J. S., Shin, H.-J., Baik, J., and Choi, H. C. (2013). Copper-vapor-assisted chemical

vapor deposition for high-quality and metal-free single-layer graphene on amorphous SiO_2 substrate. *ACS Nano*, 7(8), 6575–6582.

20. Wang, J., Zeng, M., Tan, L., Dai, B., Deng, Y., Rümmeli, M., Xu, H., Li, Z., Wang, S., Peng, L., Eckert, J., and Fu, L. (2013). High-mobility graphene on liquid p-block elements by ultra-low-loss CVD growth. *Sci. Rep.*, 3(1), 2670.

21. Zeng, M., Tan, L., Wang, J., Chen, L., Rümmeli, M. H., and Fu, L. (2014). Liquid metal: An innovative solution to uniform graphene films. *Chem. Mater.*, 26(12), 3637–3643.

22. Min, B. H., Kim, D. W., Kim, K. H., Choi, H. O., Jang, S. W., and Jung, H.-T. (2014). Bulk scale growth of CVD graphene on Ni nanowire foams for a highly dense and elastic 3D conducting electrode. *Carbon*, 80, 446–452.

23. Lu, L., De Hosson, J. T. M., and Pei, Y. (2019). Three-dimensional micron-porous graphene foams for lightweight current collectors of lithium-sulfur batteries. *Carbon*, 144, 713–723.

24. Chen, Z., Ren, W., Gao, L., Liu, B., Pei, S., and Cheng, H.-M. (2011). Three-dimensional flexible and conductive interconnected graphene networks grown by chemical vapour deposition. *Nat. Mater.*, 10(6), 424–428.

25. Mo, R., Rooney, D., Sun, K., and Yang, H. Y. (2017). 3D nitrogen-doped graphene foam with encapsulated germanium/nitrogen-doped graphene yolk-shell nanoarchitecture for high-performance flexible Li-ion battery. *Nat. Commun.*, 8(1), 13949.

26. Feng, X., Ding, X., and Jiang, D. (2012). Covalent organic frameworks. *Chem. Soc. Rev.*, 41(18), 6010–6022.

27. Hao, L., Zhang, S., Liu, R., Ning, J., Zhang, G., and Zhi, L. (2015). Bottom-up construction of triazine-based frameworks as metal-free electro-catalysts for oxygen reduction reaction. *Adv. Mater.*, 27(20), 3190–3195.

28. Romero, J., Rodriguez-San-Miguel, D., Ribera, A., Mas-Ballesté, R., Otero, T. F., Manet, I., Licio, F., Abellán, G., Zamora, F., and Coronado, E. (2017). Metal-functionalized covalent organic frameworks as precursors of supercapacitive porous N-doped graphene. *J. Mater. Chem. A*, 5(9), 4343–4351.

29. Schoedel, A., Li, M., Li, D., O'Keeffe, M., and Yaghi, O. M. (2016). Structures of metal–organic frameworks with rod secondary building units. *Chem. Rev.*, 116(19), 12466–12535.

30. Pachfule, P., Shinde, D., Majumder, M., and Xu, Q. (2016). Fabrication of carbon nanorods and graphene nanoribbons from a metal–organic frame-work. *Nature Chem.*, 8(7), 718–724.

31. Xia, W., Tang, J., Li, J., Zhang, S., Wu, K. C.-W., He, J., and Yamauchi, Y. (2019). Defect-rich graphene nanomesh produced by thermal exfoliation of metal–organic frameworks for the oxygen reduction reaction. *Angew. Chem. Int. Ed.*, 58(38), 13354–13359.

32. Mohammad-Rezaei, R., Razmi, H., and Dehgan-Reyhan, S. (2014). Preparation of graphene oxide doped eggshell membrane bioplatform modified Prussian blue nanoparticles as a sensitive hydrogen peroxide sensor. *Colloids Surf. B Biointerfaces*, 118, 188–193.

33. Liu, L., Yang, X., Lv, C., Zhu, A., Zhu, X., Guo, S., Chen, C., and Yang, D. (2016). Seaweed-derived route to Fe_2O_3 hollow nanoparticles/N-doped graphene aerogels with high lithium ion storage performance. *ACS Appl. Mater. Interfaces*, 8(11), 7047–7053.

34. Huang, C., Bai, H., Li, C., and Shi, G. (2011). A graphene oxide/hemoglobin composite hydrogel for enzymatic catalysis in organic solvents. *Chem. Commun.*, 47(17), 4962–4964.

35. Gao, S., Geng, K., Liu, H., Wei, X., Zhang, M., Wang, P., and Wang, J. (2015). Transforming organic-rich amaranthus waste into nitrogen-doped carbon with superior performance of the oxygen reduction reaction. *Energy Environ. Sci.*, 8(1), 221–229.

36. Hu, K., Gupta, M. K., Kulkarni, D. D., and Tsukruk, V. V. (2013). Ultra-robust graphene oxide-silk fibroin nanocomposite membranes. *Adv. Mater.*, 25(16), 2301–2307.

37. Chen, F., Yang, J., Bai, T., Long, B., and Zhou, X. (2016). Facile synthesis of few-layer graphene from biomass waste and its application in lithium ion batteries. *J. Electroanal. Chem.*, 768, 18–26.

38. Li, Z., Lv, W., Zhang, C., Li, B., Kang, F., and Yang, Q.-H. (2015). A sheet-like porous carbon for high-rate supercapacitors produced by the carbonization of an eggplant. *Carbon*, 92, 11–14.

39. Zhu, Y., Cao, T., Cao, C., Ma, X., Xu, X., and Li, Y. (2018). A general synthetic strategy to monolayer graphene. *Nano Res.*, 11(6), 3088–3095.

40. Kwon, H.-j., Ha, J. M., Yoo, S. H., Ali, G., and Cho, S. O. (2014). Synthesis of flake-like graphene from nickel-coated polyacrylonitrile polymer. *Nanoscale Res. Lett.*, 9(1), 618.

Chapter 4

Top-down Synthesis of
Functionalized Graphene

4.1 Overview

A top-down strategy most often involves the splitting of graphene from a variety of carbon sources (graphite, carbon nanotube, carbon black, and others), and the most prominent top-down procedures used are chemical oxidation, electrochemical exfoliation, and physical exfoliation. In the top-down synthesis of graphene, graphene can be functionalized spontaneously, such as incorporation of functional groups or control of shape and size. In contrast to bottom-up synthesis, the top-down strategy has the following advantages: (i) abundant raw materials, such as graphite, carbon nanotubes (CNTs), and other carbon sources; (ii) a variety of preparation processes; and (iii) simpler implementation at large scales. It is not surprising that top-down techniques are used to produce most of the commercially available graphene. Therefore, the development of new methods and improvements in existing top-down processes are of utmost importance to achieve successful graphene commercialization and industrial acceptance.

As discussed in Chapters 2 and 3, the preparation of functionalized graphene composites is closely related to their practical applications, and their preparation and applications will be discussed in Chapter 6. This chapter mainly focuses on top-down approaches to preparing functionalized graphene and representative non-covalently functionalized graphene composites. According to the carbon source used in the current top-down strategy, the resulting functionalized graphene can be categorized into graphitic functionalized graphene, graphitic non-covalently functionalized graphene composites, and non-graphitic functionalized graphene. In short, graphitic functionalized graphene refers to functionalized graphene prepared by various physical or chemical methods using graphitic materials as raw materials; graphitic non-covalently functionalized graphene composites are composites made of graphene and other materials by non-covalent bonding while using graphite as raw material; non-graphitic functionalized graphene refers to functionalized graphene prepared via various physical or chemical methods while using non-graphitic materials (such as CNTs, carbon black, and organic carbon sources) as raw materials.

4.2 Graphitic Functionalized Graphene

4.2.1 *Introduction to graphite*

Graphite is a layered material formed by graphene through van der Waals forces and π-π interaction. It shows excellent physical and chemical properties such as high-temperature resistance, oxidation resistance, corrosion resistance, thermal shock resistance, high strength, good toughness, self-lubrication, high thermal conductivity, and electrical conductivity. In addition, it is also well known that the reserves of graphite are abundant. China is by far the leading global producer and exporter of graphite and is the country with the most abundant natural graphite reserves in the world. Abundant graphite makes the large-scale production of functionalized graphene by a top-down strategy possible and encourages applications of functionalized graphene in many fields. Based on the morphologies and preparation methods, graphitic materials are mainly divided into the following three types: pure graphite, expanded graphite, and intercalated graphite.

1. *Pure graphite mainly refers to natural and artificial graphite.* Based on the different crystallization states, natural graphite is mainly divided into three types: dense crystalline graphite, flake graphite, and aphanitic graphite. Artificial graphite is a graphite material obtained by graphitizing carbonaceous raw materials at high temperatures. The pure graphitic materials usually used for synthesizing functionalized graphene are required to have an excellent graphitic crystal structure, which is beneficial for exfoliation and functionalization.

2. *Expanded graphite is a modified graphite in which the interlayer space of flake graphite is significantly expanded by chemical or physical methods.* It not only retains the inherent properties of graphite, such as high-temperature resistance, oxidation resistance, corrosion resistance, self-lubrication, and radiation resistance, but also has excellent adsorption and catalytic properties. Therefore, expanded graphite is currently widely used as an electrode, sealing material, oil-absorbing material, fire retardant, and antistatic material. Flexible graphite paper prepared from expanded graphite particles has been widely used in many sealing fields and is called the contemporary "seal of the King."

3. *Intercalated graphite, also known as graphite intercalation compound (GIC), is a material prepared by the intercalation of diverse non-carbonaceous reactants via physical or chemical routes. These materials are crystalline compounds that retain a lyered graphitic structure.* In the carbon layer of graphite sheets, the bonding energy of carbon atoms is 345 kJ/mol and the distance between atoms is 0.142 nm. The graphite layers are held together by the weak Van der Waals attraction with a bonding energy of 16.7 kJ/mol and an interlayer distance of 0.3354 nm between the adjacent layers. The layers of graphite are weakly bound and the space is large, so that various chemical substances (such as atoms, molecules, ions, or ion clusters) can be inserted into the space between graphite layers, after which GICs are finally obtained.

4.2.2 *Graphitic functionalized graphene*

By means of a top-down strategy and taking graphite as the raw material, covalently bonded functional groups and defects can be introduced into graphene sheets, thus controlling the element composition, structure, size, and shape of graphene. Therefore, the properties of the synthesized functionalized graphene depend on the methods of functionalization.

4.2.2.1 *Chemical oxidation*

The most typical graphitic functionalized graphene synthesized by the top-down strategy is graphene oxide (GO), which is derived from graphite oxide. Graphite oxide was first prepared in 1859 and is a material

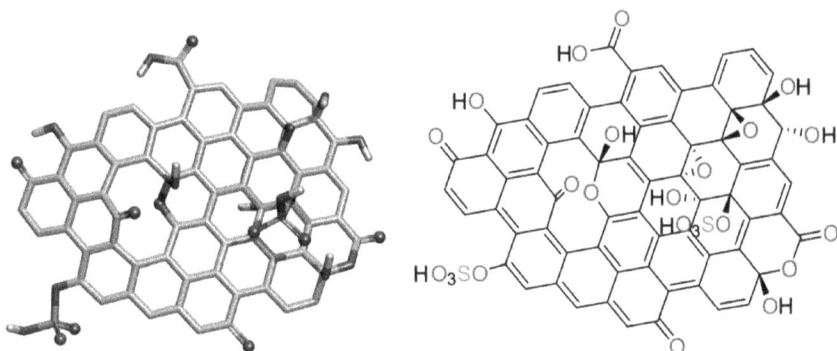

Figure 4.1.   Molecular structure and characterization of GO: Structural model of GO.[1]

synthesized by the chemical oxidation of graphitic materials with strong acids and strong oxidants.

Graphite oxide has a layered structure compared to graphite. However, despite the degree of oxidation being different, various oxygen-containing functionalities, such as hydroxyl, carboxyl, and epoxide groups, are covalently incorporated into graphite oxide sheets. Owing to hydrophilic oxygen functionalities, graphite oxide can simply disperse in water or other polar solvents. With the assistance of water or other polar solvents, graphite oxide is exfoliated to multilayer, few-layer, and monolayer GO (Fig. 4.1).

To date, the chemical oxidation methods to prepare GO mainly include the Brodie method, the Staudenmaier method, and the Hummers' method. Thanks to the ease and short time of execution, Hummers' method is widely adopted to prepare GO. As shown in Fig. 4.2, in general, Hummers' method uses H_2SO_4, $KMnO_4$, and $NaNO_3$ to prepare graphite oxide, which is then exfoliated to obtain GO. The oxidized graphite is a viscous slurry containing concentrated H_2SO_4 and other oxides, which needs to be washed before exfoliation. The washed graphite oxide is exfoliated into GO by ultrasonication or stirring in a solvent. The general solvent adopted here is water; in addition, solvents such as N-methylpyrrolidone (NMP), *N,N*-Dimethylformamide (DMF), and tetrahydrofuran (THF) can also be used.

GO inherits a large quantity of oxygen-containing functional groups from graphite oxide layers; therefore, GO can be well dispersed in many

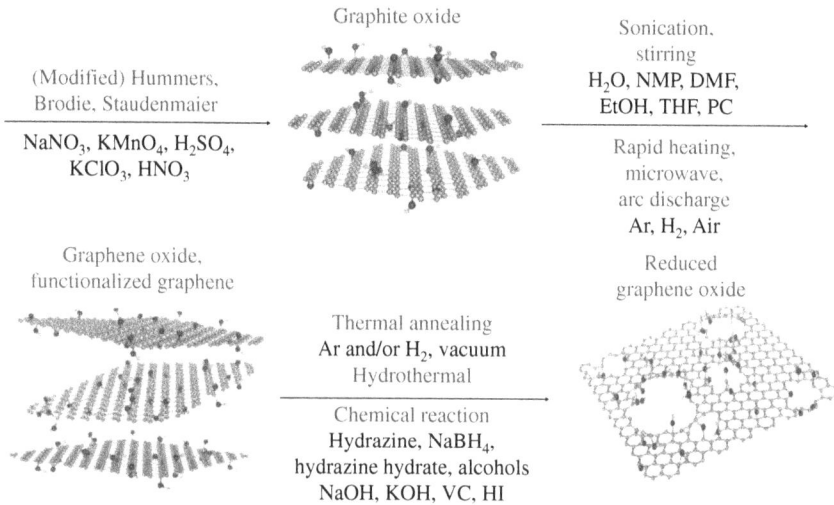

Figure 4.2.   Schematic diagram of the preparation of GO and rGO by a top-down strategy.[3]

different solvents to form a stable dispersion. It is worth noting that in the oxidation process, oxygen-containing functional groups are incorporated and the sp^2-conjugated structure of graphene is damaged; thus, defects and sp^3-hybridized carbons are introduced into the conjugated carbon framework of graphene. This will lead to a decrease in the conductivity of the functionalized graphene. However, reduced graphene oxide (rGO) can be obtained by reduction of GO via heat treatment, chemical methods, or hydrothermal methods, and the inherent conjugated structure and electrical conductivity of graphene are partially restored. By controlling the reduction conditions, several defects and oxygen-containing functional groups can be intentionally retained in rGO to obtain functionalized rGO as shown in Fig. 4.2. These residual defects and oxygen-containing functional groups are proven to play an important role in applications such as catalysis, supercapacitor preparations, and battery electrode preparations.

In addition, the oxygen-containing functionalities of GO, such as hydroxyl, carboxyl, and epoxide groups, are able to react with other groups to further functionalize GO. Therefore, GO synthesized directly by oxidation of graphite is defined as first-generation functionalized graphene. Graphene materials further derived from GO are classified as second-generation functionalized graphene (materials). Third-generation

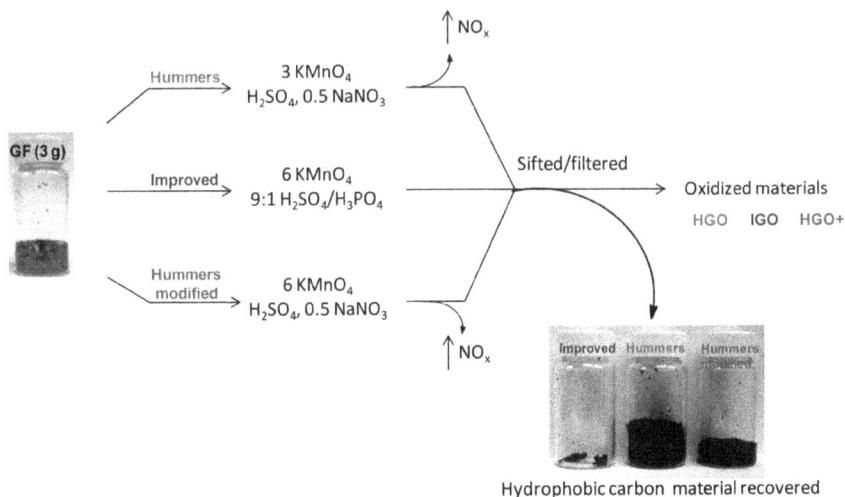

Figure 4.3.   Schematic diagram of GO prepared by the improved Hummers method.[4]

functionalized graphene (materials) refers to those materials fabricated from second-generation functionalized graphene.

In recent years, in order to improve the preparation efficiency of GO, reduce the difficulty of preparation, and reduce environmental pollution, a variety of new approaches have emerged. In 2010, Marcano *et al.* modified the solvent system to a 9:1 mixture of concentrated H_2SO_4/H_3PO_4 based on Hummers' method (Fig. 4.3), excluding $NaCO_3$ and increasing the amount of $KMnO_4$ required for the reaction.[4] In contrast to Hummers' method, this improved method yields a higher fraction of GO and does not generate toxic gases such as NO_2 and N_2O_4. In addition, the temperature in the reaction process is easily controlled, which is important for large-scale production of GO.

The structure and properties of GO depend significantly on the type of graphite used or oxidation protocol, and they are also closely related to the quenching and purification procedures in the preparation process. In 2012, Dimiev *et al.* used non-aqueous solvents to quench and purify GO (Fig. 4.4).[5] The prepared GO contained a significant amount of vacancy defects terminated by ketone groups, and the size of the aromatic domains did not exceed 5–6 benzene rings. The oxidized sp^3 area was dominated by epoxides, and covalent sulfates and alcohols were present in smaller amounts. The freshly synthesized GO powder was either colorless or pale

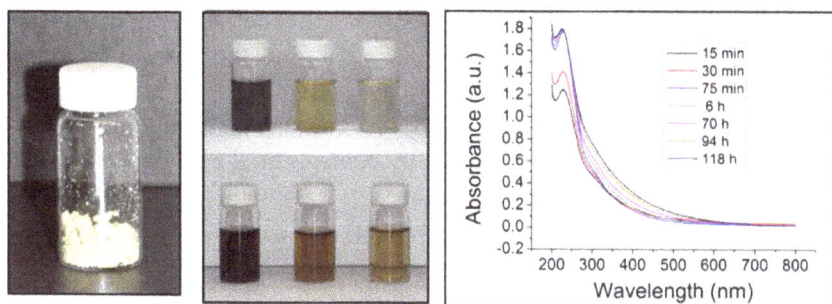

Figure 4.4.   Photographs and UV-Vis spectra of aqueous solutions of light-colored GO samples.[5]

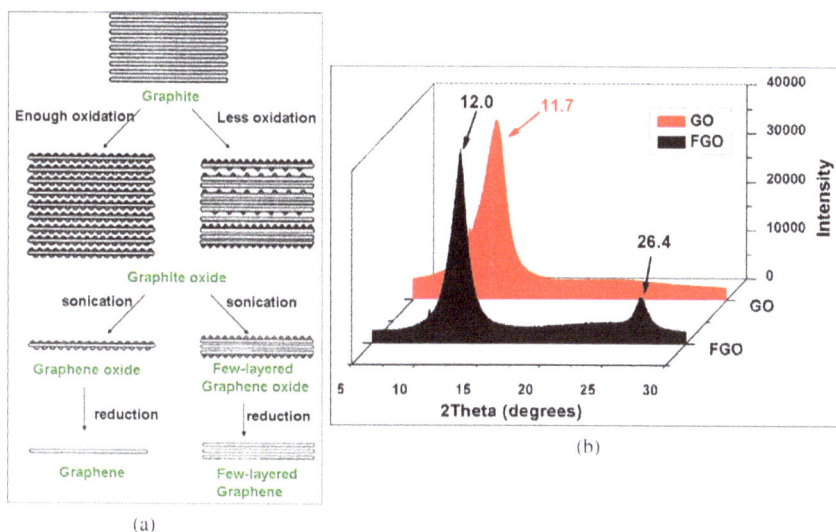

Figure 4.5.   Synthesis of single-layered/few-layered GO by chemical oxidation[6]: (a) outlined oxidization/intercalation process for the preparation of few-layered GO and GO and (b) XRD patterns of few-layered GO and GO.

yellow. After washing copiously, the color gradually darkened and changed from yellow to dark brown, suggesting significant chemical transformations in the course of purification that resulted in increased conjugation of the π-system.

In order to further increase the degree of oxidation of GO, the synthesis of single-layered and few-layered GO was investigated (Fig. 4.5), and the differences between them were compared. The study indicated that with the increment of oxidizers or the reaction time, the average layer area of the obtained single-layered GO became smaller (from 59,000 nm^2 to 550 nm^2) and the particle size presented a Gaussian distribution. The few-layered GO with 3−4 layers obtained by controlling the oxidation exfoliation process showed similar dispersion behavior in water to the single-layered GO. However, the few-layer GO showed much better electrical conductivity than its single-layered counterpart after the same reduction treatment. These results suggest that, compared with few-layered GO, the π-conjugation in single-layered GO is more seriously disrupted, resulting in more defects. The higher electrical conductivity of few-layered GO may be attributed to its less disrupted π-conjugation due to less functionalization, a larger sheet size, and probably its multilayered structure.

4.2.2.2 *Electrochemical method*

Although the chemical oxidation method has long been the predominant means for the synthesis of GO, the preparation time is long and the strong oxidizers used are dangerous, causing environmental pollution. For a green and efficient preparation of GO, a variety of synthesis approaches have been developed, e.g., the electrochemical method, which utilizes electrochemical reactions to achieve intercalation, exfoliation, and oxidation or some other functionalization of graphite (Fig. 4.6).[7]

One of the key steps in the electrochemical synthesis of graphene is the electrochemical exfoliation of graphite. The crucial point of this method is the electrochemical intercalation of graphite electrodes. Similar in principle to graphite exfoliation, the electrochemical synthesis of graphene generally involves the usage of graphite electrodes with an electrolyte bath using an aqueous or organic solution of sulfate or lithium salt. Under the effect of the electric field force, graphite is completely peeled off, after which few-layered graphene can be directly synthesized and single-layered graphene can also be prepared. Electrochemical synthesis of graphene exhibits the following advantages and characteristics: (i) The preparation is mild but not harsh (ambient temperature and pressure). (ii) The investment required for the equipment is low, the electrolyte is recyclable, and the general cost is low. (iii) This method is free of toxic chemicals, contamination, and concentrated H_2SO_4, and is green and environmental-friendly as well. (iv) Various types of graphite can be

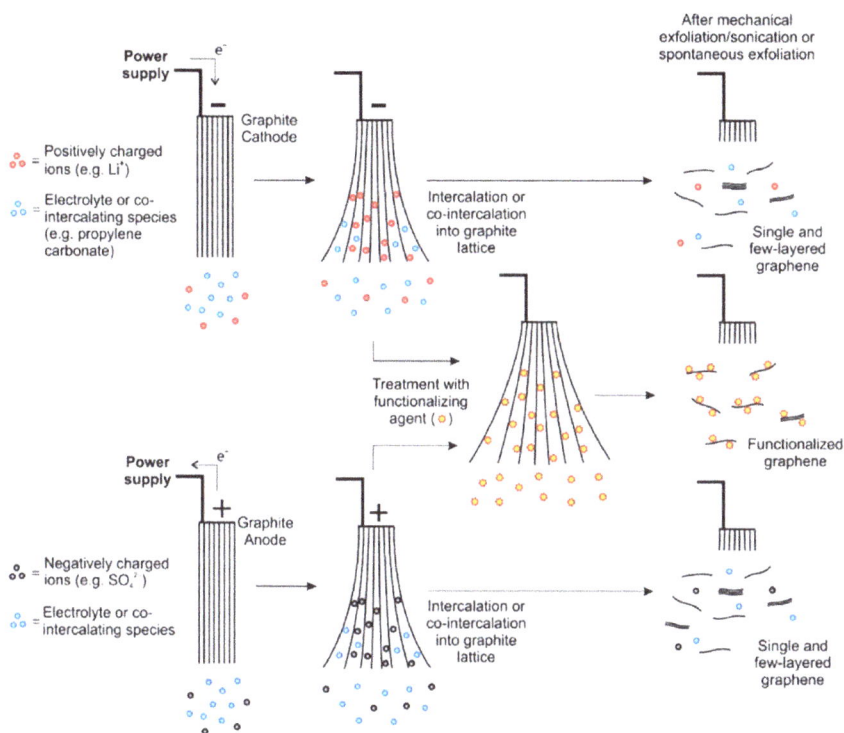

Figure 4.6. Schematic diagram of intercalation and exfoliation mechanisms of (functionalized) graphene by the electrochemical method.[7]

adopted as raw materials, including earthy graphite and high-grade flake graphite deposits. (v) The synthesized graphene is of a high quality with fewer defects due to which reduction and heat treatment are not necessary. (vi) The number of layers and the size of 2D graphene sheets are controllable. (vii) The synthesized graphene shows excellent solubility, good processability, and can be easily used to fabricate functional coatings with properties of thermal conduction, electrical conduction, anti-corrosion, electromagnetic shielding, etc. (viii) Functionalized graphene materials can be synthesized by *in situ* functionalization.

Graphene prepared by the electrochemical method has the advantages of abundant raw materials and low cost, and it is widely used in many fields such as electronics, illumination, energy, environment, water treatment, biomaterial, chemical industry, and weaponry.

The electrochemical method can also be used for the synthesis of functionalized graphene. As shown in Fig. 4.6, the procedure of preparing functionalized graphene by this method includes two main steps: (i) electrochemical intercalation and exfoliation of graphite and (ii) functionalization of graphene, including oxidation.

The electrochemical intercalation-assisted exfoliation method involves intercalating various ions, such as Li^+, SO_4^{2-}, OH^-, and HSO_4^-, between graphite layers under an electric field. The diameters of these ions are slightly larger than those of the graphite interlayer space (0.3354 nm). The intercalated ions result in the expansion of graphite layers, and subsequent sonication produces electrochemically exfoliated graphene.

Depending on the power supply applied on graphite electrodes, there are cathodic (applying a negative bias) and anodic (using a positive bias) exfoliation methods. In the case of anodic intercalation, the negatively charged ions, such as SO_4^{2-}, are intercalated into the positively charged graphite anode under an applied electric field to form surfactant-intercalated graphite (electrode) through ionic interaction, which expands the van der Waals gap between the graphite layers. At the same time, solvents or other compounds can also intercalate into the graphite interlayer together with negative ions. When the graphite anode is treated as the cathode, electrochemical exfoliation of the surfactant-intercalated graphite electrode occurs due to electrostatic repulsion of the anionic surfactant (e.g., Li^+ and Et_4N^+) and the negatively charged electrode. Similarly, intercalation of cointercalating species such as solvents into the graphite interlayer can also sometimes occur.

After graphite is intercalated, exfoliation occurs under some reactions. This mainly includes two forms: the cointercalating species (e.g., water) will react and transform to other substances (e.g., water is converted into hydrogen or oxygen) at a suitable electrical potential (when the cointercalating species is water, the cathode potential is less than 0 V and the anode potential is greater than 1.23 V), making the intercalated graphite further expand and then exfoliate; in other cases, the electrochemically expanded graphite needs to be further mechanically exfoliated (e.g., sonication).

In the process of electrochemical intercalation-assisted exfoliation of graphite, the spontaneous functionalization of graphene can be also achieved through which functionalized graphene can be obtained directly. As shown in Fig. 4.7, in a strong acid electrolyte, OH^-, O^-, and SO_4^{2-} can be quickly intercalated into the layers of the graphite anode and oxidized when the graphene is exfoliated, resulting in GO with a small amount of

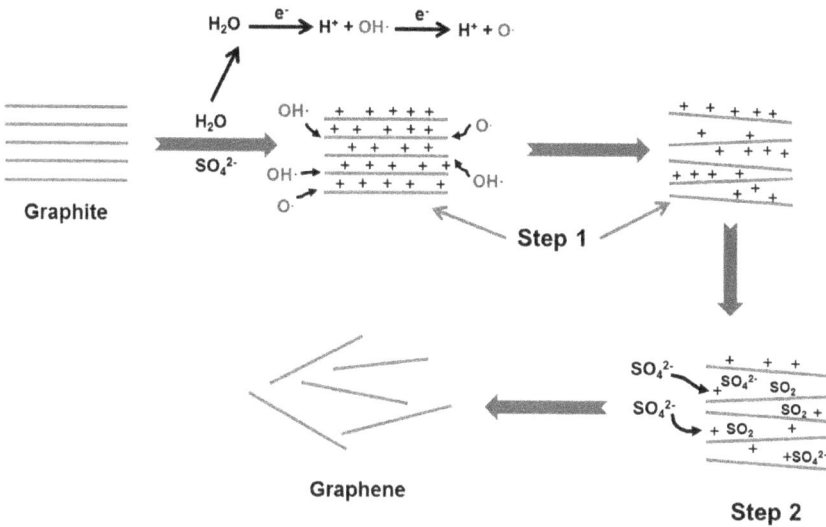

Figure 4.7. Schematic illustration of the synthesis of GO with low oxygen content by electrochemical exfoliation.[8]

oxygen-containing functional groups.[8] In addition to the strong acid environment, high voltage can also be adopted to accelerate the intercalation of ions and to oxidize the exfoliated graphene to produce GO. However, the aforementioned reaction conditions of electrochemical exfoliation are mainly used for the synthesis of expanded graphite. In these processes, GO is only a by-product with a low yield. To prepare GO in large quantities by electrochemical methods, the raw materials and reaction conditions need to be optimized.

In 2017, a two-step electrochemical intercalation and oxidation approach was reported to produce GO on a large laboratory scale.[9] As shown in Fig. 4.8, this two-step approach comprises (i) forming a stage-1 GIC in concentrated sulfuric acid and (ii) oxidizing and exfoliating the stage-1 GIC in an aqueous solution of 0.1 M ammonium sulfate. This two-step approach leads to GO with a high yield (70 wt.%), good quality (>90%, monolayer), and reasonable oxygen content (17.7 at.%). Moreover, the as-produced GO can be subsequently deeply reduced (3.2 at.% oxygen; C/O ratio 30.2) to yield highly conductive (54,600 S/m) reduced GO. This is indicative that GO prepared by this method presents a lower density of defects.

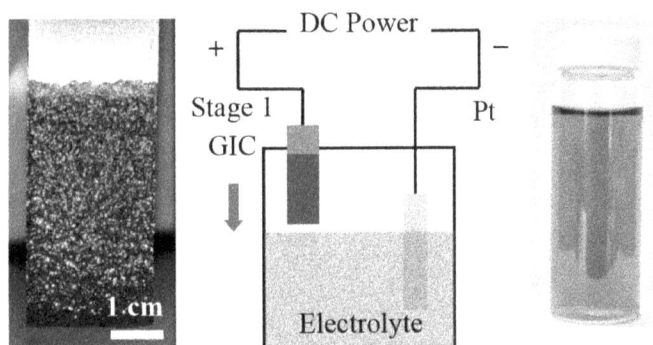

Figure 4.8. Schematic illustration of two-step electrochemical intercalation and oxidation process for the production of GO by using ammonium sulfate as the electrolyte.[9]

In 2018, the Institute of Metal Research, Chinese Academy of Sciences, proposed a similar electrochemical method, by which GO is synthesized in high efficiency Fig. 4.9(a).[10] This method is not only scalable, green, and safe but also suitable for large-scale production. In addition, the synthesized GO has a similar structure and properties to GO prepared by traditional methods. The synthesis of GO contains two sequential processes: (i) The commercial flexible graphite paper (FGP, Fig. 4.9(b)) is subjected to electrochemical intercalation in concentrated H_2SO_4 to form blue-colored stage-1 GIC paper (GICP, Fig. 4.9(c)). (ii) Then, the GICP is used as an anode for the electrochemical reaction in diluted H_2SO_4 (50 wt.%) electrolyte. Very surprisingly, the blue-colored GICP dipped in diluted H_2SO_4 is oxidized quickly to yellow-colored GO (Fig. 4.9(d)) within a few seconds along with exfoliation. After vacuum filtration and cleaning with water, the filter cake is exfoliated in water by sonication to form electrochemically synthesized GO dispersion (EGO, Fig. 4.9(e)).

The mechanism of this anodic electrocatalytic oxygen evolution reaction of water occurs under an applied voltage as shown in Fig. 4.9(f). The radical intermediates, *OH, *O, and *OOH, are generated in turn at an active site (*) on the electrode surface accompanied by water electrolysis, and these radicals are adsorbed on the active sites. In the case of graphitic anode, the adsorbed reactive *OH, *O, and *OOH can react with the carbon lattice that is highly positively charged to form covalently bonded oxygen-containing functional groups. The oxidation degree of the GO sheets can be easily tuned by changing the concentration of H_2SO_4 in the

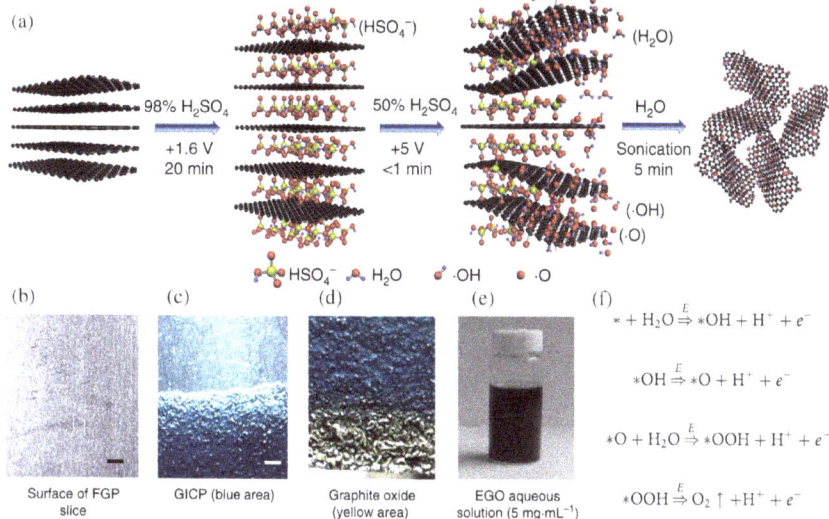

Figure 4.9. Synthesis of GO by the electrochemical method[10]: (a) schematic illustration of the synthesis process of GO by water electrolytic oxidation; (b–e) photos of the raw material and the products obtained at each step; (b) FGP; (c) GICP (blue area) obtained after electrochemical intercalation of FGP; (d) graphite oxide (yellow area) obtained by water electrolytic oxidation of the GICP; (e) well-dispersed GO aqueous solution obtained by sonication of the graphite oxide in water; and (f) anodic electrocatalytic oxygen evolution reaction of water active sites on the anode.

electrochemical oxidation process, and the number of layers and lateral size can be tuned by sonication time.

Compared with the traditional Hummers and K_2FeO_4 methods, the electrochemical oxidation rate of this method is over 100 times faster, and the cleaning and purification of GO in this method are also much easier, needing ten times lesser water. In addition, H_2SO_4 is not consumed in the reaction, and no other substances are generated, and thus it can be fully recycled for electrochemical reactions. Therefore, this electrochemical oxidation method combines the advantages of safety, ultrafast synthesis, easy control, environmental friendliness, no metal ion contaminations, and ease of scaling up, paving the way for industrial production and applications of GO sheets at a low cost.

In order to ensure sufficient oxidation of graphite, concentrated H_2SO_4 and other strong oxidants are commonly used. This is because although the ordinary water electrolysis process can also oxidize graphite (Fig. 4.7) to produce GO, the bubbles such as hydrogen gas generated in this process will result in the quick exfoliation of the graphite layer, and then the exfoliated graphite sheet cannot be oxidized adequately nor can it be further oxidized in the following process. The usage of strong oxidants such as concentrated H_2SO_4 can greatly promote the oxidation of graphite in this process. However, the use of concentrated H_2SO_4 or other strong oxidants makes the production of GO a potential risk and produces serious environmental pollution. Therefore, green and efficient approaches for preparing GO have been developed, and the main strategy is to enhance the oxidation of graphite by means other than using strong oxidants.

The preparation setup is presented in Fig. 4.10(a) with commercial rectangular graphite foil as the anode and platinum foil as the counter electrode.[11] Fig. 4.10(b) presents a visual schematic illustration of the preparation process and the mechanism. p-phthalic acid (PTA) and NaOH are dissolved in water as the electrolytes, and a constant voltage (10 V) is applied until the anode is exhausted within 6–8 h. The color of the electrolyte gradually changes from transparent to yellowish, then to brown, and finally to black as the anodization continues. The stripped products in the electrolyte undergo a treatment of bath sonication for 1 h to further increase the yield and reduce the thickness. In the end, the final product GO is obtained by repeatedly washing with high-rate centrifugation. In this manner, the GO is obtained with a remarkably high yield of 87.3% and an outstanding concentration of 8.2 mg/mL over 6 months without aggregation.

The mechanism is as follows: NaOH aqueous solution favors the dissolution of PTA, but as the reaction proceeds, the pH around the graphitic anode is reduced due to the electrolysis of water, which is responsible for the localized PTA precipitation and partial coverage on the graphite anode. The partial PTA coverage markedly hinders the electrode reaction, thereby leading to less bubble generation, which allows prolonged oxidation of the exposed anode surface by oxygen free radicals (HO· and O·), thus enhancing the oxidation degree of graphite. Sonication after oxidation further exfoliates GO sheets to obtain few-layered and single-layered GO. In addition, this strategy leads to sheet exfoliation and gradual anode thinning. Based on this unique mechanism, GO is realized using non-oxidant electrolytes, and the product's problems of multiple layers, low

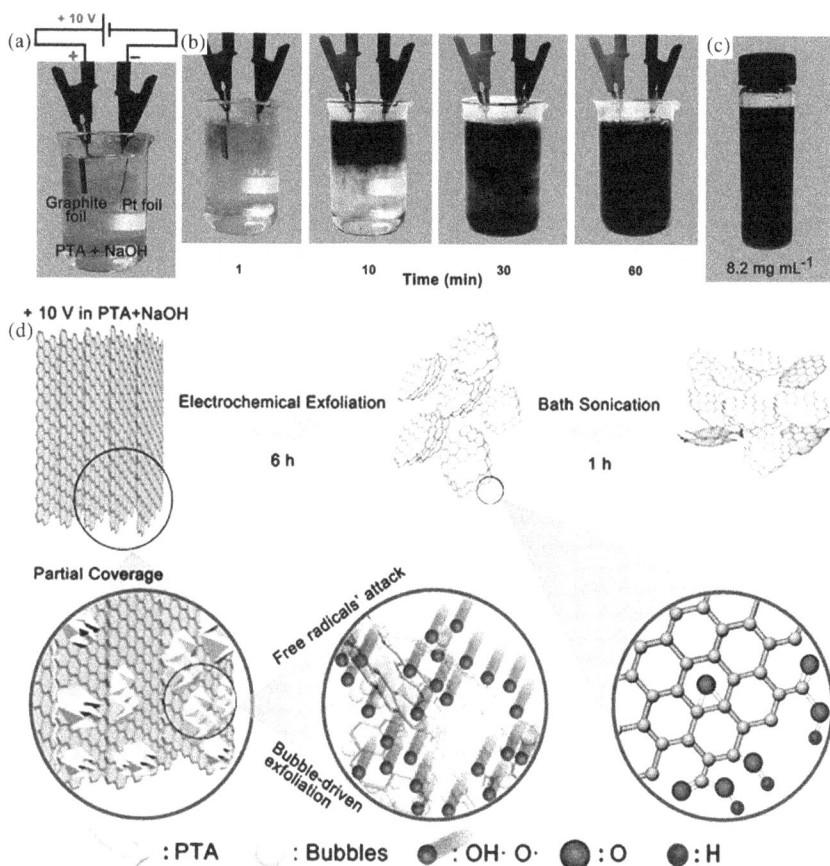

Figure 4.10. Schematic illustration of green and efficient preparation of GO[11]: (a) experimental setup; (b) the evolution of the electrochemical process; (c) the obtained GO solution with a concentration of 8.2 mg/mL; and (d) the graphite anode is selectively reacted with free radicals when the graphite surface is partially covered by PTA molecules, leading to adequate oxidation.

quality, and poor solubility caused by inadequate oxidation of conventional electrochemical methods can be addressed.

As described in Fig. 4.6, in addition to the synthesis of GO, exfoliation and functionalization of graphene can be achieved spontaneously while adding other reactants in the stage of electrolysis. As shown in Fig. 4.11, single-stage simultaneous electrochemical exfoliation and functionalization of graphene was reported by Dryfe's group from the University of Manchester.[12] In this research, when the graphene was

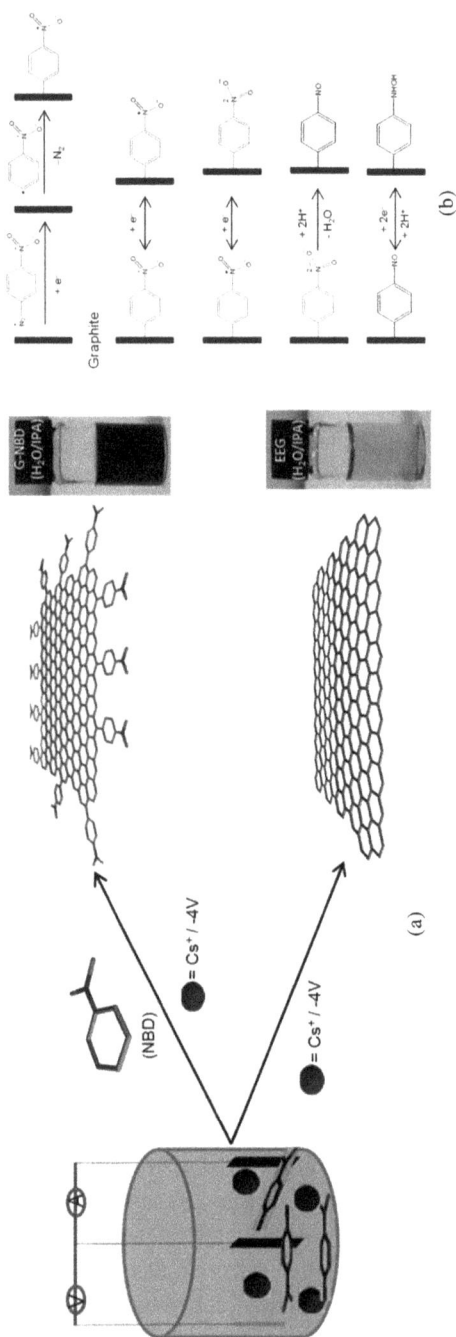

Figure 4.11.   Schematic illustration of functionalized graphene prepared by the electrochemical method[12]: (a) single-pot electrochemical exfoliation and functionalization of graphene and (b) reactions that represent the electrochemical reduction of NBD cations at the graphite electrode and the subsequent grafting and reduction steps.

electrochemically exfoliated, diazonium functionalization of graphene was performed. As shown in Fig. 4.11(a), electrochemical exfoliation of graphite (a graphite rod is used as the cathode and a Ag electrode is used as the anode) was performed in the presence of a diazonium compound in the exfoliation solution (0.1 M $CsClO_4$ and 1 mM 4-nitrobenzenediazoniumtetrafluoroborate (NBD) in an anhydrous dimethyl sulfoxide (DMSO)). Since the ionic diameter of Cs^+ (0.338 nm) is similar to the interlayer spacing of graphite, Cs^+ can be intercalated into the layers of the graphite cathode during electrolysis. At the same time, the diazonium salt NBD accepts electrons from the graphite cathode, and $-N_2$ is removed to generate nitrogen gas, which leads to the formation of a radical. The resulting active radicals then rapidly react with the sp^2-hybridized C-atoms at the graphite edge, and the nitro groups on NBD are reduced to NHOH moieties; finally, covalently functionalized graphene is obtained. The degree of functionalization can be tuned by the concentration of diazonium salt in the electrolyte. The major advantage of such an approach is that it allows the functionalization of the graphene sheets before they get the chance to re-aggregate. The N_2 generated during *in situ* diazonium reduction may also aid the separation of functionalized graphene sheets. Furthermore, such processes may present an opportunity to selectively functionalize the more reactive graphene edge. Edge functionalization prevents damage to the basal plane, which largely retains the physicochemical properties of pristine graphene. Edge functionalization also increases the dispersibility of graphene and is attractive for various potential applications including catalysis, polymer composite formation, and energy storage applications.

Studies have shown that the use of other electrolytes, such as ionic liquids (ILs), can also achieve simultaneous intercalation-assisted exfoliation and functionalization of graphene. As shown in Fig. 4.12, two graphite rods are immersed as electrodes into the solution of 1-octyl-3-methyl-imidazolium hexafluorophosphate and water (volume ratio 1:1); then a static potential of 15 V is applied for 6 h at room temperature. On the cathode, graphite is exfoliated to ionic liquid functionalized graphene nanosheets (GNSs[IL]).[13] On the basis of the analysis, the functionalization mechanism is speculated to be as follows. During the electrochemical reaction process, imidazolium ions, which are the cations of the ionic liquid, are reduced on the graphite cathode, which means that an electron is added to the ionic liquid molecule. The reduction of the cation, in principle, leads to the formation of the 1-octyl-3-methylimidazolium free radical. This highly active free radical reacts rapidly with sp^2-hybridized

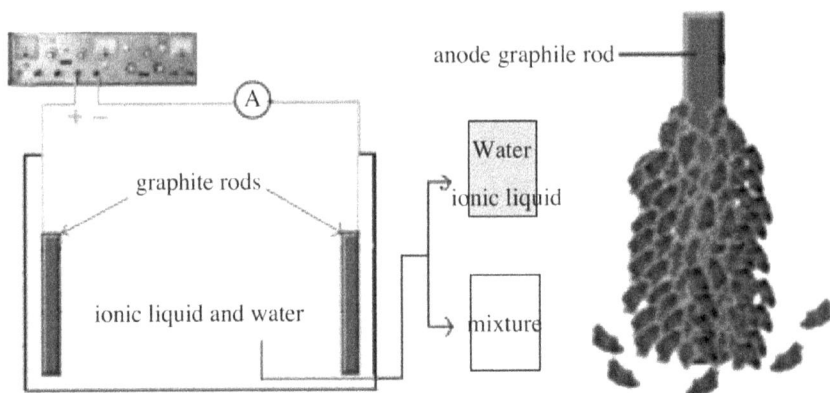

Figure 4.12. Synthesis of functionalized graphene in an ionic liquid system[13]: experimental setup diagram (left) and the exfoliation of the graphite anode (right).

carbon of graphene to produce functionalized graphene GNSs[IL]. The obtained functionalized graphene no longer disperses in water, but readily forms stable and homogeneous dispersion in a polar aprotic solvent, such as *N,N*-dimethylformamide (DMF), dimethyl sulfoxide (DMSO), and *N*-methylpyrrolidone (NMP) solution, after brief ultrasonic treatment. Therefore, this potential property makes GNSs[IL] the ideal candidate for advanced filler materials and for the synthesis of functional graphite/polymer composites.

4.2.2.3 *Arc discharge process*

The arc discharge process is a powerful technique for synthesizing carbon nanomaterials such as fullerenes, CNTs, and graphene, as well as functionalized graphene. This technique involves establishing a constant direct current (DC) or alternating current (AC) between a pair of graphite electrodes under a buffer gas (inert gas or hydrogen) at a certain pressure (positive or negative pressure). A very high electric field can be generated under extremely high voltage (V) conditions leading to a sudden spark with the generated arc plasma vaporizing the electrode material. During this process, the temperature of the anode is higher than that of the cathode, and the anode graphite is continuously consumed to generate carbon nanomaterials, such as graphene, which are deposited on the cathode and chamber walls. During the arc discharge process, the anode graphite is evaporated into carbon atoms or small carbon clusters, which can be expressed as $N_C = n_C + n_{C^+} + n_{C^{2+}} + n_{C^-} + 2n_{C_2} + 3n_{C_3} + 4n_{C_4} + 5n_{C_5}$, where N_C is the total number of carbon sources evaporated, C is a carbon atom, C^+ and C^{2+} are carbocations, C^- is a carbanion, and C_2, C_3, C_4, and C_5 are small carbon clusters.[14]

Since the temperature in the arc-discharging region between two electrodes is extremely high, the graphene synthesized by this method has the advantages of high crystallinity, few defects, excellent electrical conductivity, and high thermal stability. Either DC or AC is available during the arc discharge method, and the common choice is low voltage and high current. In the case of DC, the product would deposit on the graphite cathode, while in the case of AC, two electrodes behave as the cathode and the anode alternately, resulting in an extremely high temperature between two electrodes. This means that the diffusion rate of carbon atoms and clusters around the arc discharge region will increase, and thereby the probability of a collision between all the carbon species and gas molecules reduces. The schematic of the AC arc discharge apparatus and the influence of different buffer gases on the graphene products are shown in Fig. 4.13.[15] A buffer gas such as H_2 has the highest thermal conductivity and leads to a steep temperature gradient of plasma in the furnace. The cooling rate is fast for hydrogen gas, so carbon clusters are

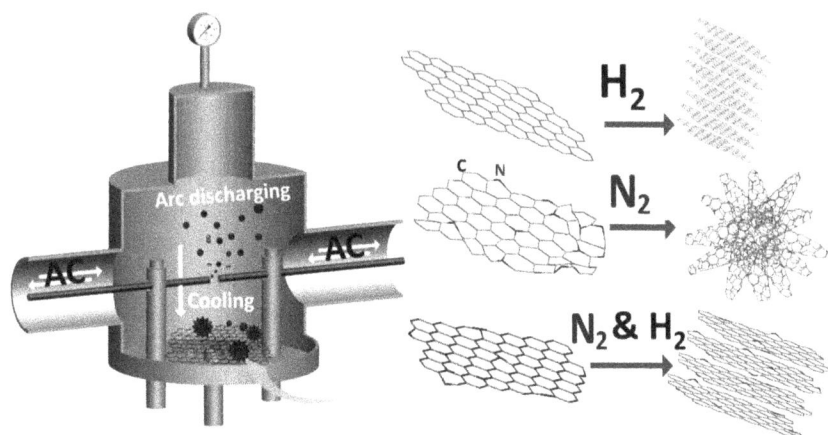

Figure 4.13. Schematic of AC arc discharge apparatus and the influence of different buffer gases on the graphene product.[15]

easily formed before atoms deposit to form a crystal structure. These clusters do not have enough energy and time to arrange themselves into a long-distance ordered crystalline and instead form a disordered structure like amorphous carbon or thick graphene sheets. To settle such a deficiency, inert gas like N_2 with low thermal conductivity will be brought in. N_2 is thought to be a key component in the synthesis of graphene for producing curved structures by doping N atoms into the lattice and altering the temperature gradient, whereas H_2 can effectively restrain the formation of dangling bonds and closed structures. Consequently, N-doped graphene prepared by a proper ratio of N_2 to H_2 exhibits an intact structure with a low heteroatom content.

Different nitrogen sources can not only tune the degree of nitrogen doping in graphene but also control the type of nitrogen doping. As shown in Fig. 4.14, when nitrogen is used as the nitrogen source (nitrogen is introduced with the buffer gases: $He/N_2/H_2$), the nitrogen-doped graphene synthesized by the arc discharge method mainly contains graphitic-N (N-Q). In contrast, when melamine is used as the nitrogen source (evenly mixed melamine with graphite powders as anode; He/N_2 mixtures as buffer gas), the nitrogen-doped graphene prepared by the arc discharge method contains both N-Q and pyridinic-N (N-6), which is attributed to the difference in pyrolysis temperature of different nitrogen sources. When N_2 is used as the nitrogen source, the high temperature at stage I (3000–5000 K) makes the nitrogen molecules decompose into nitrogen atoms. However, there is no sufficient energy to convert the nitrogen molecules into N atoms at stage II (1000–3000 K). Therefore, the nitrogen atoms only exist at stage I, and the rest of the arc chamber is filled with nitrogen molecules. At stage I, the existing N atoms combine with carbon atoms or small carbon clusters to form N—C configurations. At stage II, the carbon clusters doped with N atoms or without N atoms reunite to form the doped graphene. In these cases, most of the nitrogen atoms are doped at basal planes of graphene to form N-Q rather than at the edges to form N-6. However, N atoms can exist at stage I and stage II simultaneously when melamine is used as the nitrogen source because melamine can be easily decomposed into nitrogen atoms above 773 K. Therefore, nitrogen atoms of stage II will combine with carbon atoms, which are situated at the edges of the reunited carbon clusters, to form the N-6 graphene. When melamine acts as the nitrogen source, the content of nitrogen atoms in N-doped graphene can reach 2.88 at.%.

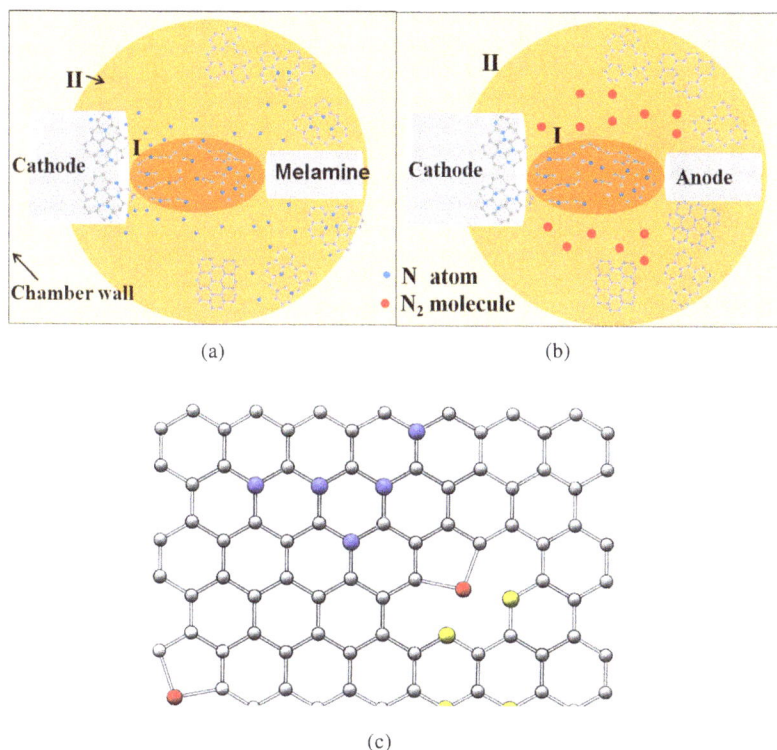

Figure 4.14. Schematic illustration of N-doped graphene sheets synthesized by the arc discharge method[17]: (a–b) the doping mechanism of N-doped graphene sheets by using melamine (a) and N_2 (b) as nitrogen sources, and (c) three bonding configurations of N doping, i.e., graphitic- (N-Q, blue), pyrrolic- (N-5, red), and pyridinic-N (N-6, yellow) bonding configurations.

By adjusting the anodic carbon source and nitrogen source, the nitrogen content of N-doped graphene can be further increased. As shown in Fig. 4.15, commercial graphite powder, GO, and polyaniline (PANi) are utilized as anodes and pure graphite is employed as the cathode; the resulting N-doped graphene flakes via the arc process show a remarkably increased doping level ($\approx 3.5\%$ N).[16] This is because the decomposed products of the anodic carbon precursors (graphite, GO, and PANi) are injected directly into the hot plasma zone, and studies have shown that the nucleation and growth of carbon nanostructures may take place in

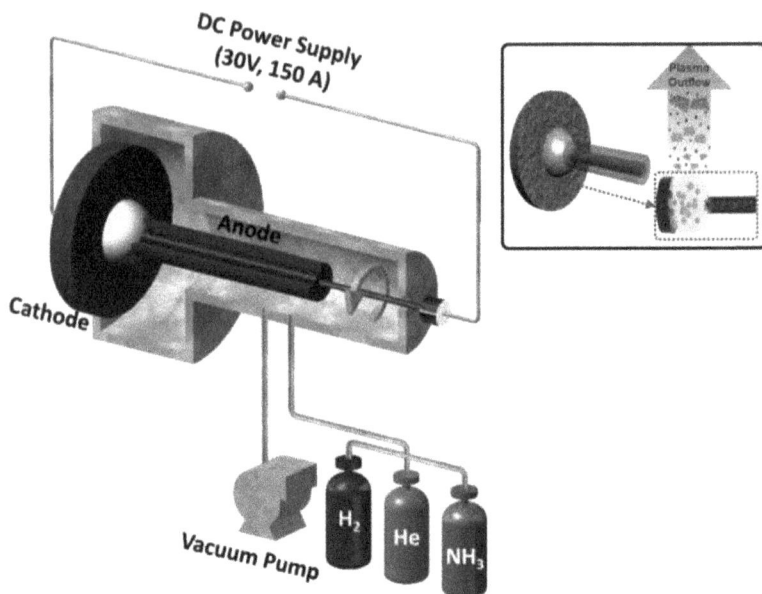

Figure 4.15.   Schematic of the arc discharge apparatus using GO and PANi as the anode (inset illustrates discharge process in which graphite anode is evaporated).[16]

proximity to the plasma discharge core when graphene is synthesized by the arc discharge method.

In addition, functionalized graphene materials can also be prepared by an aqueous single-reactor arc discharge process (Fig. 4.16).[18,19] The experimental setup used in this method is roughly the same as that adopted in conventional arc discharge, but the two graphite electrodes are submerged in deionized water for arc discharge, and the current and voltage used are much lower than the conventional arc discharge process. The difference in the experimental setup and working environment also leads to different synthesis mechanisms of (functionalized) graphene compared to conventional arc discharge. Graphitic carbon structures can be obtained by either exfoliation of graphite or evaporation and recombination of carbon molecules in an arc discharge process. The conventional arc discharge process, which uses high current (>15 A), can achieve a sublimation temperature of 4000 K, resulting in the evaporation of carbon molecules and recombination into specific carbon structures. In contrast, the graphitic carbon nanostructures can be obtained without high energy consumption,

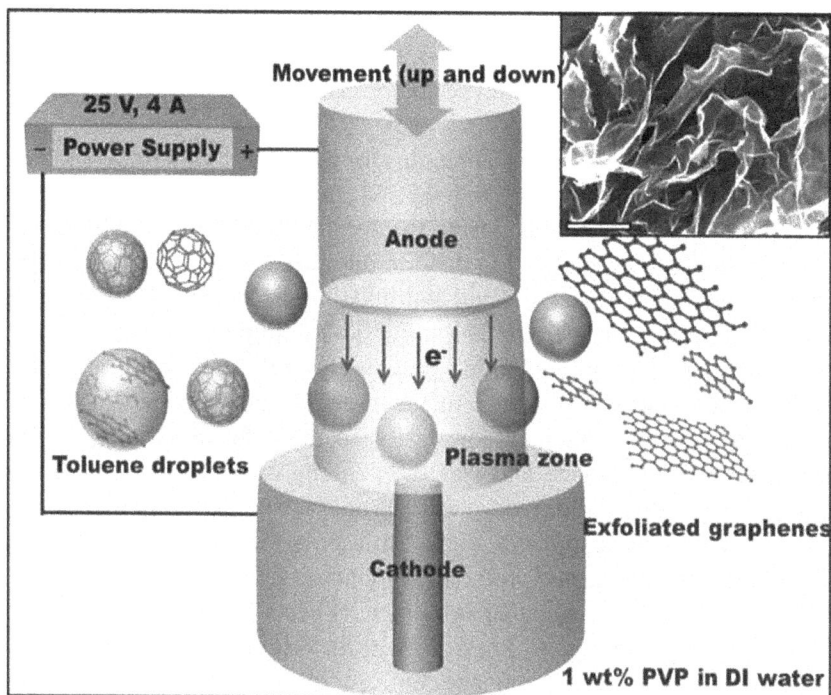

Figure 4.16. Schematic illustration of the synthesis of GO film and GO spheres by the aqueous arc discharge method.[18]

wherein the arc discharge power is controlled by simply adjusting the current levels below 10 A and the temperature generated by the arc discharge is lower than 4000 K. Therefore, it is not enough to evaporate graphite to carbon atoms and other small carbon structures. However, the temperature of the plasma zone induced by arc discharge is increased rapidly, thereby causing the nearby graphite electrodes to thermally expand; rapid heating can induce water cavitation in this zone (such a process is often assisted by ultrasound). The water cavitation is found to be 17 MPa to the surface of graphite electrodes at 80°C, and 2D graphene is produced by thermal expansion and mechanical tension by water cavitation. In addition, the arc discharge in water generates a plasma zone composed of electrons and highly reactive O^+, O^{++}, and H^+ ions. Those reactive species result in the oxidation of graphitic carbon nanoparticles or the formation of gas species such as CO/CO_2. Therefore, the aqueous arc discharge process is able to control the number of graphene layers as well as the degree of oxidation of GO by simply altering the arc discharge power. In addition, the morphological features of GO can be manipulated by the aqueous arc discharge conditions.

As shown in Fig. 4.16, the apparatus for the aqueous single reactor arc discharge process is able to produce 3D crumpled graphene spheres in oil-in-water (O/W) emulsion during the arc discharge. Two graphite electrodes are arc discharged (25 V, 4 A) in deionized water (18.2 MΩ), and the electrodes are exfoliated and partially oxidized to GO sheets. GO sheets float to the surface of water and self-assemble into large GO films. When a water solution containing polyvinylpyrrolidone (PVP) at a concentration of 1% w/v is added to the process, the O/W emulsion can encapsulate the 2D graphene sheets inside the toluene droplets or the interface between toluene and water depending on the difference in the solvation energy and the hydrophobic properties of the graphene layers. The O/W emulsions can exert significant force to induce the deformation of the entrapped graphene, producing higher curvatures. The deformation of the 2D MLGs is accelerated as toluene is evaporated by heating from the arc discharge process followed by capillary compression. The combination of the O/W emulsion and the arc discharge process produces unique morphologically distinct individual 3D crumpled graphene spheres.

4.2.2.4 *Ball milling method*

The ball milling approach is another top-down synthesis method for preparing functionalized graphene, which is usually carried out by a ball mill, as shown in Fig. 4.17(a).[20] This process utilizes metal balls traveling at high speed (significant kinetic energy) to unzip the graphitic layers (C—C bond cracking), cause a chemical reaction at the unzipped edges (C—X bond formation), and physically delaminate graphitic layers into graphene nanoplatelets (GnPs, <10 layers). Active carbon species such as carbon radicals and ions (cations and anions) are generated at the unzipped edges. Active carbon species generated *in situ* are reactive enough to promptly pick up appropriate reactants that are present (Figs. 4.17(b) and 4.17(c)). Finally, edge-selective functionalized graphene nanoplatelets (EFGnPs, Fig 4.17(d)) are produced.

In general, the ball milling method offers edge functionalization of graphite flakes, but the *sp*^{2} carbon framework is almost retained. This is

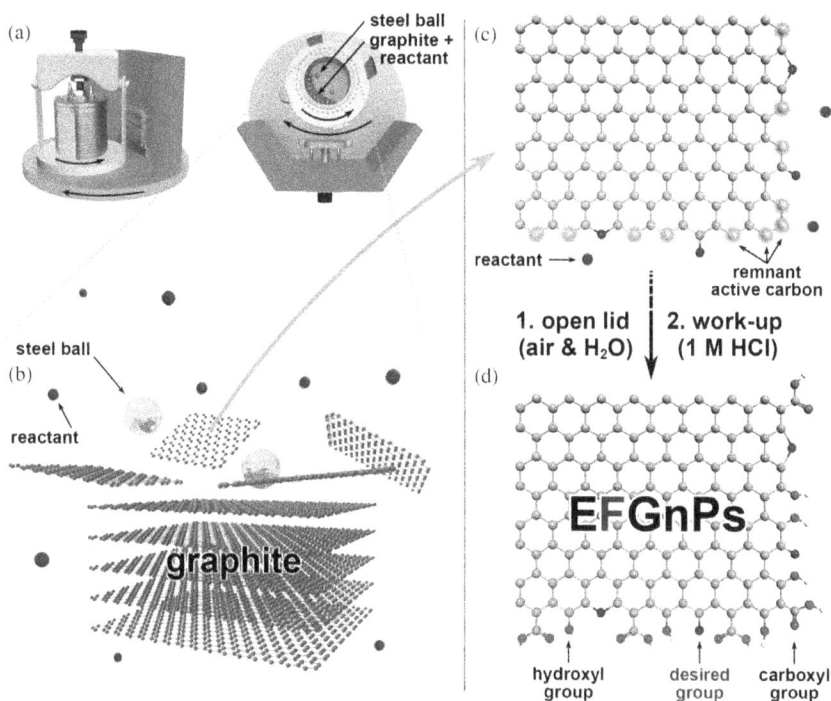

Figure 4.17. Schematic illustration of ball milling method[20]: (a) planetary ball milling machine and (b–d) schematic representation of the chemical reactions by ball milling graphite to produce edge-selective functionalized graphene nanoplatelets (EFGnPs).

because the edge carbon atoms are often dangling bonds or C—H bonds, which are more reactive than the carbon atoms in the 2D framework. The ball milling method does not use hazardous reagents (e.g., strong acids and carcinogenic reducing agents) in the synthesis process of functionalized graphene, and it is very suitable for large-scale production. On the other hand, the ball milling process can selectively introduce desired functional groups and heteroatoms (e.g., H_2O and HCl) at the edges of EFGnPs by changing reactants without additional procedures.

As shown in Figs. 4.17(c) and 4.17(d), when the reactants are air and water, functionalized graphene with hydroxyl and carboxyl groups at the edges can be obtained by ball milling of graphite. During ball milling of

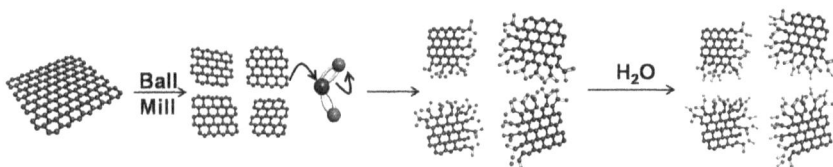

Figure 4.18. Schematic representation of physical cracking and edge carboxylation of graphite by ball milling in the presence of dry ice, and protonation through subsequent exposure to air moisture.[21]

graphite in the presence of dry ice, reactive carbon species (radicals, anions, and cations) generated by homolytic and heterolytic cleavages of the graphitic C—C bonds react with carbon dioxide (CO_2) to give edge-carboxylated graphene nanosheets, followed by protonation with moisture in the air.

4.2.2.5 *Microfluidization*

Microfluidization is a homogenization technique that is commonly used in food, pharmaceuticals, and daily chemical applications. In this synthesis method, high pressure (up to 207 MPa) is applied to a fluid, forcing it to pass through a microchannel (diameter, d <100 μm); thus, three effects, i.e., shear stress (shear rate ~ 10^8 s^{-1}), collision, and cavitation, are applied to the whole fluid volume. Under these effects, the materials in the fluid can be emulsified, homogenized, and decomposed. This technique can also be used to yield functionalized graphene from graphite: Shear stress is in charge of exfoliating graphite to produce single-layered or few-layered graphene, while the cavitation effect is able to cut and fragment graphene sheets, generating dimension- or structure-modified functionalized graphene.

As shown in Fig. 4.19, microfluidization can be employed to fabricate graphene quantum dots (GQDs) by taking graphite as the raw material. GQDs, sheets of graphene with less than 10 layers and lateral dimensions smaller than 100 nm, possess strong quantum confinement and edge effects. GQDs can be synthesized and fabricated by either a bottom-up or a top-down approach, and microfluidization is an efficient top-down synthesis strategy. Graphite–aqueous suspensions (Fig. 4.19(a)) are powered using a high-pressure pump (up to 30 kpsi) through microsized Z-shaped

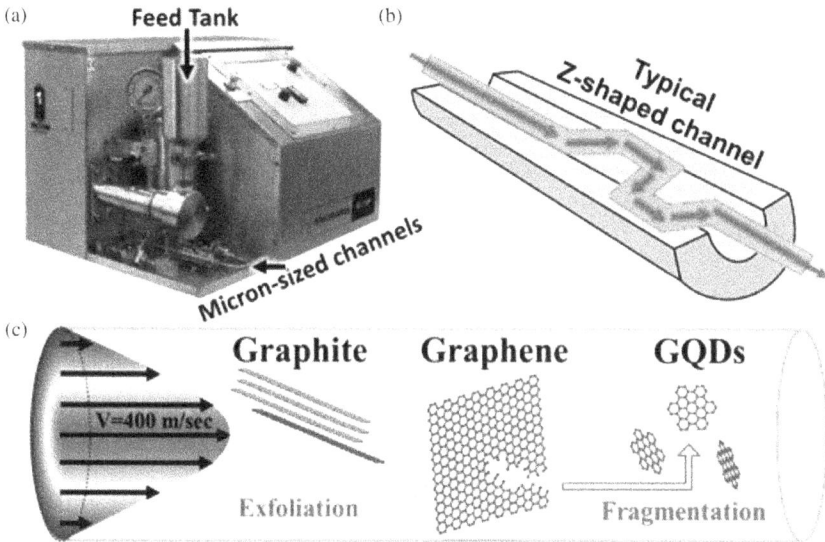

Figure 4.19. Synthesis of functionalized graphene by shear stress in the microfluidization method[22]: (a) photograph of typical microfluidizer; (b) schematic of Z-shaped channels with diameters ranging from 400 μm to 87 μm; and (c) typical flow profile within the channel with a maximal flow speed of 400 m/s, with the graphite flakes being exfoliated into graphene sheets and further fragmented into nanosized GQDs.

channels (Fig. 4.19(b)). As a result, the millimeter-sized graphite flakes are exfoliated into graphene sheets and are further fragmented into nanosized GQDs (Fig. 4.19(c)). GQDs synthesized in this technique are 2.7 ± 0.7 nm in diameter and 2–4 nm in thickness, namely, between two and four layers of graphene sheets.[22]

In the microfluidization process, in addition to shear stress, the cavitation effect can also be used to synthesize functionalized graphene. As shown in Fig. 4.20(a), a one-step preparation of graphene nanomeshes (GNMs) from pristine graphite flakes using a fluid-based method is reported.[23] In this method, bulk graphite particles are exfoliated into single- or few-layer graphene and are simultaneously physically punched by cavitation-induced micro jets to form pore structures. The critical part of the system is the nozzle, which is equipped with a variable cross-sectional flow channel for inducing cavitation and turbulence flow as schematized in Fig. 4.20(b).

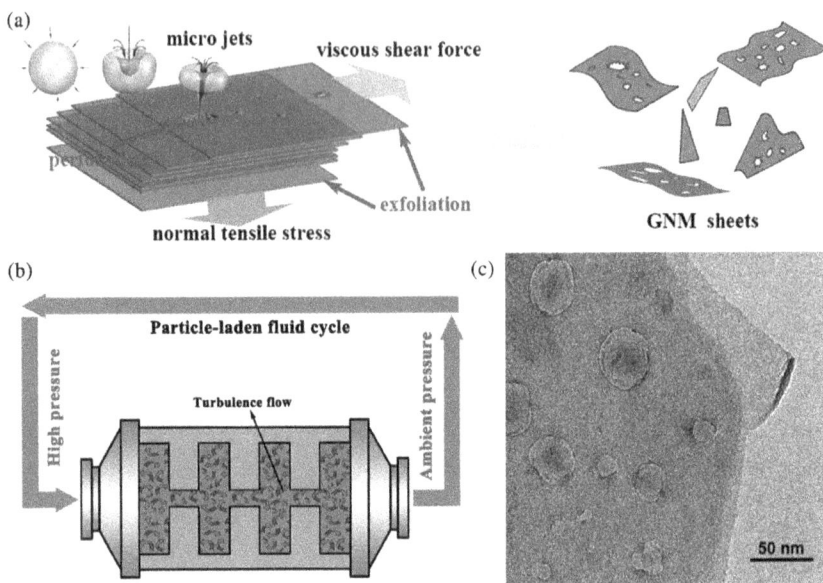

Figure 4.20. Preparation of functionalized graphene by cavitation effect using the microfluidization method[23]: (a) schematic illustration showing the working mechanism of the fluid-based method for preparing GNMs; (b) schematic of the fluid-based device; and (c) typical TEM image of as-produced GNMs.

The graphite dispersion passes through the nozzle under high pressure (30 MPa), and the fluid inside the nozzle forms a turbulent flow where the velocities distribute non-homogeneously. This will result in a velocity gradient between graphite flakes and the carrying fluid, thus leading to a viscous shear force that peels off thin graphene layers from the graphite flakes.

At the same time, cavitation occurs in the graphite dispersion passing through the nozzle and induces micro jets and compressive shock stress waves (Fig. 4.20(b)). These shock stress waves emitted by the implosion of cavitation bubbles can reach the magnitude of several MPa, which induces micro jets in the fluid. When the cavitation-induced micro jets exert on a tiny area in the sheet surface, they will cause a punching effect and leave holes or pits on the graphene flakes.

The interlayer binding force of layered graphite belongs to the van der Waals force and is relatively weak. Therefore, when exerting on the surface of the graphite flakes and propagating to the opposite side surface, the stress waves will reflect as normal tensile stress waves and drag graphene layers off the bulk graphite.

In addition, cavitation induced by the pressure difference and changes in geometry produces forces normal to the layers of graphite, which drag graphene layers off the bulk graphite. In summary, two requirements should be met, i.e., graphene exfoliation and perforation on graphene sheets, to prepare GNMs from pristine graphite flakes. The thickness of as-produced GNMs is 1–1.5 nm, which can be identified as single- or bilayer. TEM analysis shows micropores with diameters of 10–50 nm, and the short distances of ~50 nm between neighboring pores confirm the pore density of GNMs. Based on statistical analysis, the total area of the pores within 1 μm^2 of the GNM sheet can be estimated as ~0.15 μm^2 and the number of pores (pore density) as ~22 μm^{-2}.

4.3 Graphitic Non-covalently Functionalized Graphene Composites

In addition to the covalent functionalization of graphene, non-covalent functionalization of graphene composites can also be prepared by means of top-down approaches. As mentioned earlier, graphene contains conjugated sp^2 carbons and has an extended delocalized π electron system. Therefore, its conjugated framework is expected to have a variety of non-covalent interactions with other molecules, such as π-π interaction, van der Waals force interaction, and C—H...π interaction. In the process of top-down exfoliation of graphite, non-covalently functionalized graphene composites can be fabricated by simultaneous exfoliation and non-covalent functionalization of graphite by introducing appropriate functional molecules. This category of synthesis methods mainly includes liquid-phase exfoliation and the supercritical fluid (SCF) method.

4.3.1. *Liquid-phase exfoliation*

The liquid-phase exfoliation method is often used to synthesize high-quality single-layered or few-layered graphene sheets in large quantities (Fig. 4.21(a)). The synthesis process typically involves three steps: (i) Graphite is immersed in an appropriate liquid media (solvent or solution containing surfactant); (ii) with the help of stirring, ultrasound,

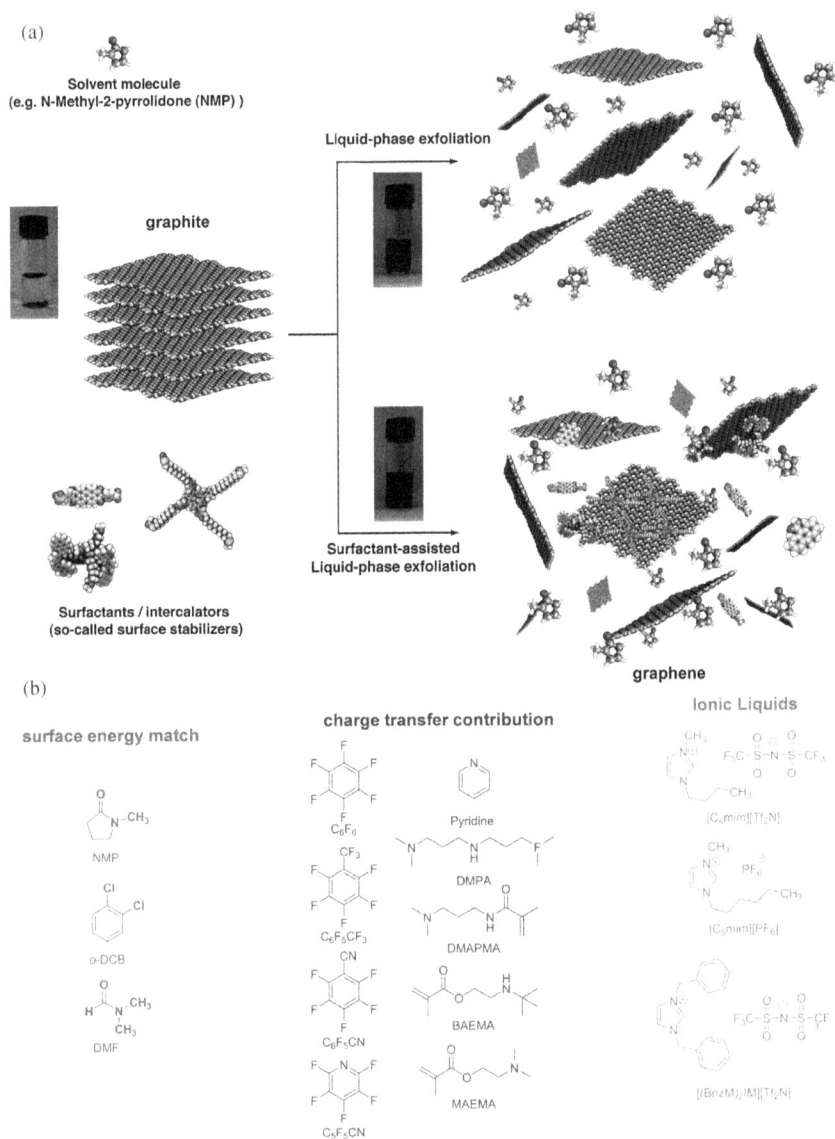

Figure 4.21. Synthesis of graphene nanosheets and non-covalently functionalized graphene composites by liquid-phase exfoliation: (a) schematic representation of the liquid-phase exfoliation process of graphite in the absence (top right) and presence (bottom right) of surfactant molecules[24] and (b) chemical structures of the solvents and ILs used in graphene exfoliation.[25]

microwave, or heating, the solvent and surfactant interact with the graphite, and the van der Waals force between sheets is broken and the graphite is exfoliated; and (iii) the unexfoliated graphite is removed by centrifugation to obtain single- or few-layered graphene.

It has been demonstrated that low-viscosity solvents, with a surface tension (γ) of 40–50 mJ/m², which is close to that of graphene, show the best performance in the exfoliation of graphite, e.g., N-methyl pyrrolidone (NMP, γ = 40 mJ/m²), N,N-dimethylformamide (DMF, γ = 37.1 mJ/m²), and ortho-dichlorobenzene (o-DCB, γ = 37 mJ/m²). Besides physical forces like surface tension, charge transfer phenomena also play an important role in liquid-phase exfoliation of graphene. As shown in Fig. 4.21(b), solvents with strong electronegativity, such as hexafluorobenzene (C_6F_6), octafluorotoluene ($C_6F_5CF_3$), and pentafluoropyridine (C_5F_5CN), are capable of withdrawing electrons from electron-rich graphene. Both the π-π interaction and charge transfer through π-π stacking are capable of stabilizing the exfoliated graphene in these electron-withdrawing solvents, leading to more efficient dispersion of graphene. In addition, the aromatic donors may also exfoliate graphite in the reverse manner, that is, charge transfer through π-π stacking from the solvent molecules to graphene, forcing the latter to act as an electron-withdrawing species. Significant examples are the direct exfoliation of graphite with DMPA [3,3'-iminobis(N,N-dimethyl-propylamine] and DMAPMA (N-[3-(dimethylamino)propyl] methacrylamide) as shown in Fig. 4.21(b). Furthermore, ILs have also been employed as dispersing and stabilizing media in graphene exfoliation.

In addition to the use of the aforementioned solvents for direct exfoliation of graphite, surfactants (Fig. 4.22) can be employed to stabilize exfoliated graphene in water or organic solvents (Fig. 4.21(a)). Adsorption of these surfactants onto the graphene surface occurs through π-π interaction and a charge transfer effect between the planar π-conjugated surfaces, thus reducing the surface free energy of the dispersion for exfoliation of graphene. It is worth noting that those molecules or moieties interacting strongly with graphene (Fig. 4.22) are still adsorbed on the resulting exfoliated graphene and interact with graphene to fabricate non-covalent composites, that is, non-covalently functionalized graphene composites.

As shown in Fig. 4.23, the N,N'-dimethyl-2,9-diazaperopyrenium dication (MP²⁺) with an extended π-conjugated structure can intercalate graphite and interact non-covalently with graphene sheets by π-π interaction.[26] Upon addition of graphite to an aqueous solution of MP·2Cl and

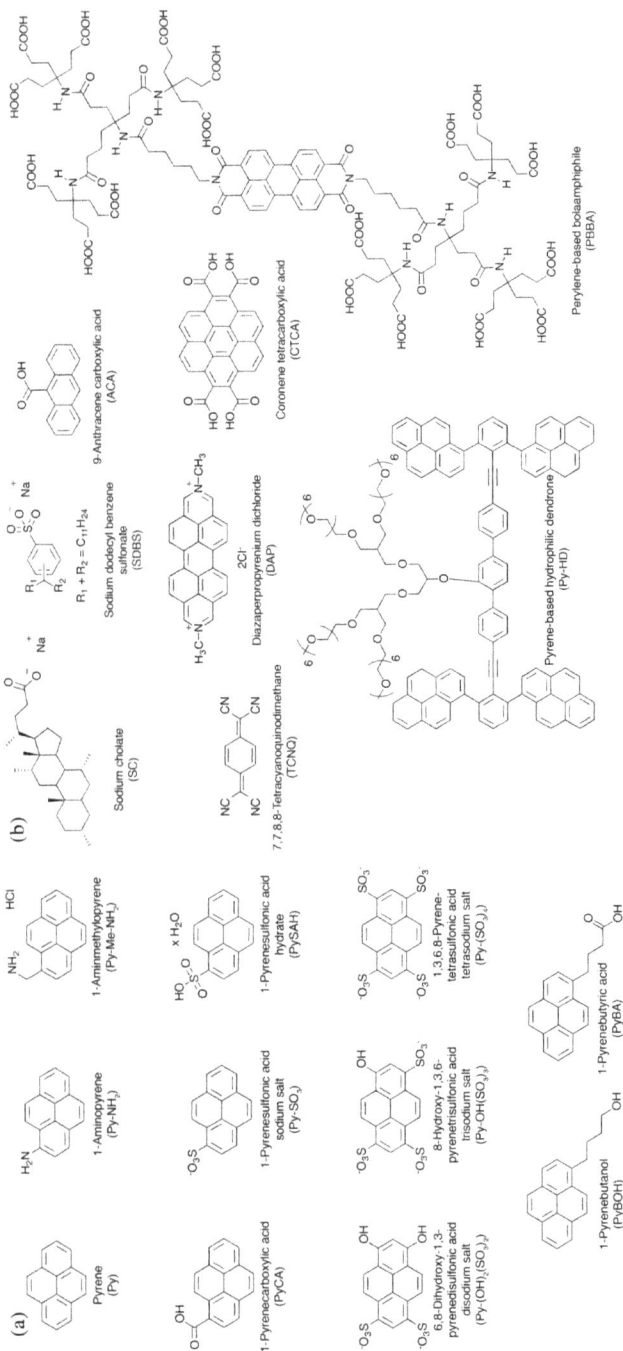

Figure 4.22. Chemical structure of organic molecules used as surfactants in the process of LPE of graphite toward graphene[24]: (a) pyrene and its water-soluble derivatives and (b) other water-soluble surfactants.

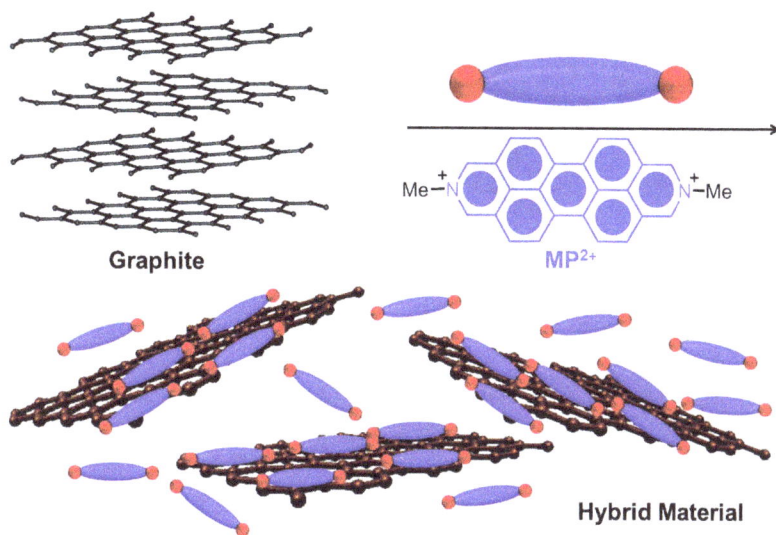

Figure 4.23. Schematic representation of exfoliation of graphite and stabilization of graphene in a solvent using the MP^{2+} dication.[26]

following sonication, MP^{2+} is adsorbed onto the graphene surface due to the π-π interaction between them, resulting in exfoliation of graphite flakes. Following the removal of larger graphitic materials, the MP·2Cl/ graphene composite dispersion in water is found to be stable for more than three weeks without any further aggregation. The electrostatic repulsion between the positively charged regions present in MP^{2+} minimizes the aggregation of graphene layers and helps sustain a stable dispersion of the graphene sheets in water. Meanwhile, the efficient fluorescence quenching of MP^{2+} dication is witnessed in the presence of the resulting functionalized graphene composite, which indicates that the composite acts as an electron acceptor and that fluorescence quenching takes place by energy transfer from MP^{2+} to graphene. In addition, MP hexafluorophosphate can be dissolved in DMF to obtain an organic solution of MP^{2+}. Upon addition of graphite to this organic solution of MP^{2+}, MP^{2+}/graphene noncovalently composites can also be obtained. Meanwhile, the dicationic nature of MP^{2+} aids in the dispersion of this composite by charge repulsion, thereby minimizing agglomeration in solution.

Most of the surfactants shown in Fig. 4.22 contain π-conjugated structures and can interact with graphene through π-π interaction to form

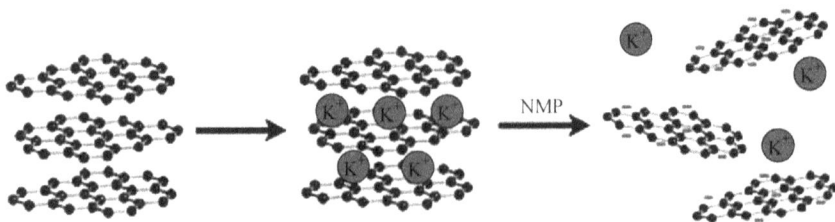

Figure 4.24. Schematic illustration of exfoliation of negatively charged graphene sheets using potassium GIC.[27]

composites. Therefore, these molecules are able to exfoliate graphite in organic or aqueous solutions, and highly dispersed and stable non-covalent functionalized graphene composites in the corresponding solutions are synthesized.

In addition to the non-covalent functionalization of graphene composites with surfactants, negatively charged functionalized graphene composites can also be prepared by exfoliating graphite through metal intercalation. The electronegativity of carbon atoms is moderate (2.55), so they can either accept electrons so as to be reduced or lose electrons so as to be oxidized. Therefore, graphite composed entirely of carbon atoms is able to withdraw electrons from alkali metals such as lithium, sodium, and potassium, whose electronegativity is lower than that of graphite, leading to negatively charged graphite, while the cations of reducing agent lose electrons and get positive charged and intercalated into graphite, after which GICs are finally obtained.

Commonly used reductive intercalation substances with lower electronegativity than carbon atoms are alkali metals, alkaline earth metals, and lanthanide metals. Studies have shown that lithium and calcium can be intercalated into graphite to produce LiC_6 and CaC_6 compounds, that is, six carbon atoms share one lithium ion or calcium ion, while potassium intercalation results in the KC_8 compound. This category of GICs can be dissolved in polar solvents, such as THF, NMP, and water.

Intercalation increases the interlayer spacing of graphite, and the graphite is easy to exfoliate due to the Coulombic repulsion of the graphene sheets. Therefore, GICs can be exfoliated to negatively charged graphene in a suitable solvent, that is, negatively charged functionalized graphene. As shown in Fig. 4.24, potassium GIC can be synthesized by mixing graphite powder with a potassium-naphthalene solution using

THF as the solvent; naphthalene functions as a charge transfer agent, which is used to enhance the ability of potassium to reduce intercalated graphite.[27] The GIC is mixed with NMP in a controlled inert atmosphere and stirred at room temperature for 24 hours; the GIC could be exfoliated to obtain negatively charged graphene sheets.

The negatively charged graphene sheets are air sensitive and can be oxidized back to the neutral state; thus, they can be directly deposited on a variety of substrates to make graphene-based functional devices. In addition, the negatively charged graphene sheets show high reactivity and can further react with electrophiles (such as diazonium salts, halogenated alkanes, halogenated aromatic hydrocarbons, and olefins) to produce covalently functionalized graphene (Fig. 4.24).

4.3.2. Supercritical fluid method

The SCF method is a technique based on SCFs. SCFs are any substances at a temperature and pressure above their critical point, which are very sensitive to temperature and pressure. SCFs, which possess gas-like diffusivity, liquid-like density, low viscosity, and surface tension, show better transport properties than conventional organic solvents. These features make SCFs superior solvents for rapid penetration as well as intercalation of the layered structure. When graphite is mixed with a SCF, the SCF can effectively penetrate between the layers of graphite. The graphite layers expand following the intercalation of solvents, and a sudden depressurization reverts the SCF to its subcritical state, causing large pressure gradients that push the graphite layers apart, thus forming graphene sheets of single to multiple layers depending on the degree of intercalation.

The SCF method facilitates the exfoliation and fragmentation of graphite to graphene flakes, as well as the functionalization of graphene and fabrication of non-covalently functionalized graphene composites.[28] As shown in Fig. 4.25, graphite and pyrene derivatives (1-pyrenamine, 1-pyrenecarboxylic acid, and 1-pyrenebutyric acid) are dispersed in DMF by bath sonication. Then, the mixture is transported to a stainless steel reactor, and carbon dioxide is charged into the reactor and maintained above its critical temperature and critical pressure. The mixture is stirred in SC CO_2 for 6 hours. Under the SCF experimental conditions, CO_2 permeates into the layers of graphite, which facilitates the penetration and intercalation of soluble pyrene derivatives into the interlayer of graphite. A sudden depressurization in the reactor leads to a significant expansion

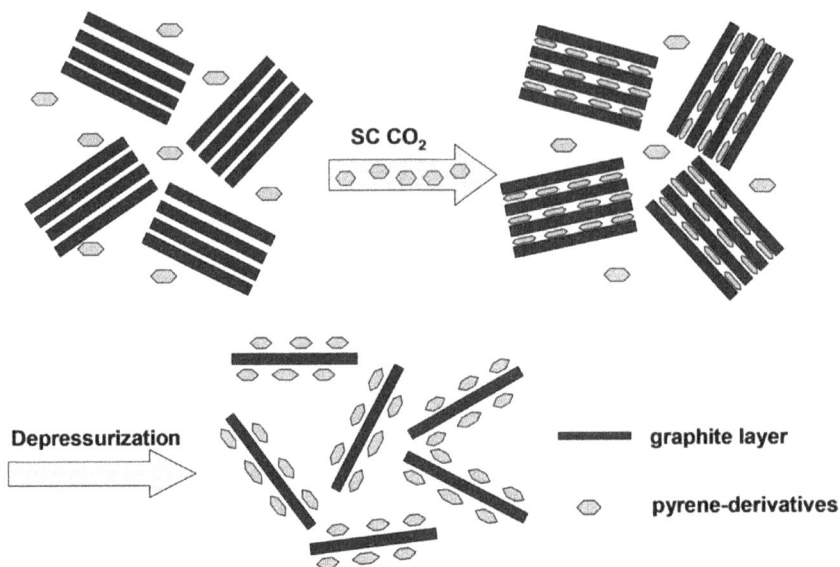

Figure 4.25.   Schematic illustration of exfoliation and non-covalent functionalization of graphene using pyrene derivatives with the assistance of SCF.[28]

of CO_2 in the graphite layers; thus, graphene is exfoliated from graphite. Meanwhile, pyrene derivatives can be firmly adsorbed on the surface of graphene layers because of strong π-π interactions, and non-covalent functionalized single- or few-layer graphene composites are obtained.

4.4 Non-Graphitic Functionalized Graphene

In the top-down synthesis of functionalized graphene, in addition to using graphite as the carbon source, CNTs, fullerenes, carbon black, carbon fibers, and even graphene can be utilized as carbon sources.

4.4.1 *Carbon nanotubes as carbon source*

CNTs are sp^2 nanocarbon materials with a 1D tubular structure, which are concentrically aligned tubes with one to several layers of carbon atoms arranged in hexagons. The distance between adjacent layers of carbon nanotube is identical (~0.34 nm) and the diameter is generally 2–20 nm.

It is well known that CNTs are considered to be GNRs rolled up into seamless tubes. Because of these structural features, a variety of GNRs or graphene nanosheets as well as other types of functionalized graphene with specific functions can be synthesized by unzipping CNTs via physical or chemical approaches.

4.4.1.1 *Chemical oxidation*

Strong oxidizers such as H_2SO_4 and $KMnO_4$ are able to oxidize carbon atoms on the CNTs to form oxygen-containing groups, while the C—C bonds are opened, producing GNRs and graphene nanosheets. As shown in Fig. 4.26, multi-walled carbon nanotubes (MWCNTs, with outer diameters of 40−80 nm and inner diameters of 15−20 nm) are suspended in concentrated H_2SO_4 and then treated with 500 wt.% $KMnO_4$ at room temperature for 1 h; the reaction mixture is stirred at room temperature for 1 h and then heated to 55−70°C for an additional 1 h. After filtration and washing, GNRs with oxidized edges are obtained; the width of these GNRs increases to >100 nm and they have linear edges with little pristine MWCNT side-wall structure remaining (Fig. 4.26(c)). The resulting GNRs are highly soluble in water, ethanol, and other polar organic solvents, which is due to distributed oxygen-containing functionalities such as carbonyls, carboxyls, and hydroxyls that have been shown to exist at the edges and the surface. However it is worth noting that when MWCNTs prepared by other methods than CVD are used as carbon sources and treated with the same oxidation, fewer nanoribbon-like structures are detected. This is indicative that the formation of GNRs is closely related to the raw materials used.

The mechanism of opening is based on the oxidation of alkenes by permanganate in acid (Fig. 4.26(b)). The proposed first step in this process is manganate ester formation (**2**, Fig. 4.26(b)) as the rate-determining step, and further oxidation is possible to afford the dione (**3**, Fig. 4.26(b)) in the dehydrating medium. Juxtaposition of the buttressing ketones distorts the β,γ-alkenes (red in **3**), making them more prone to the next attack by permanganate. As the process continues, the buttressing-induced strain on the β,γ-alkenes lessens because there is more space for carbonyl projection; however, the bond angle strain induced by the enlarging hole (or tear if originating from the end of the nanotube)makes the β,γ-alkenes (**4**, Fig. 4.26(b)) increasingly reactive. Hence, once an opening has been initiated, its further opening is enhanced relative to an unopened tube or an

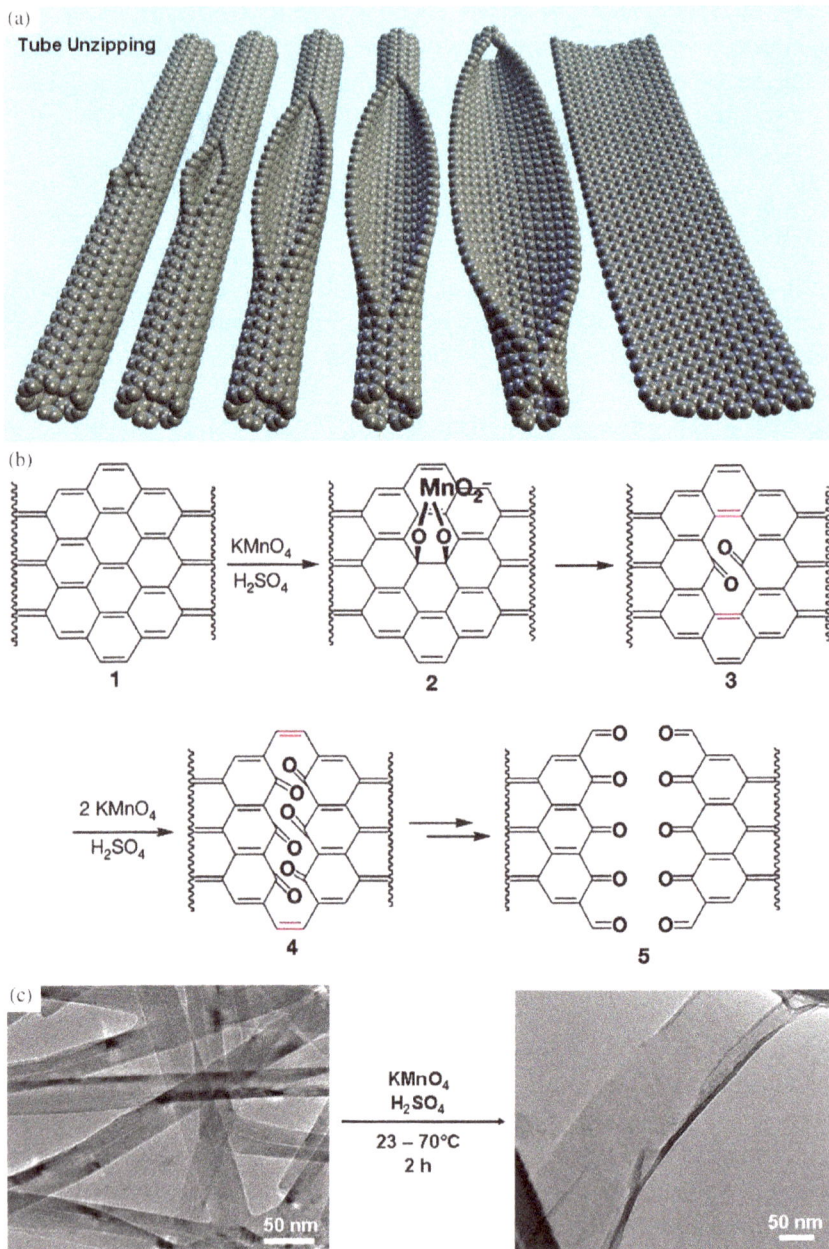

Figure 4.26. GNRs prepared by unzipping of CNTs as carbon source[29]: (a) representation of the gradual unzipping of one wall of a CNT to form a nanoribbon; (b) the proposed chemical mechanism of nanotube unzipping; and (c) TEM images depicting the transformation of MWCNTs (left) into oxidized nanoribbons (right).

uninitiated site on the same tube (**5**, Fig. 4.26(b)). Thus, neighboring carbon atoms are attacked by permanganate, contrasting with the random attack on non-neighboring carbon atoms. This also explains the preference for sequential bond cleavage over random opening and subsequent cutting. This could occur in a linear longitudinal cut or a spiraling manner, depending upon the initial site of attack and the chiral angle of the nanotube. Although depicted in Fig. 4.26(a) as occurring on the mid-section of the nanotube rather than at one end, the location of the initial attack is not known. Theoretical calculation indicates that the initial site is located in the middle of the carbon nanotube. The same unzipping process in single-walled carbon nanotubes (SWCNTs) is achieved to produce narrow nanoribbons, but their subsequent disentanglement is more difficult.

The ketones on the edge of GNRs are highly reactive and can be further converted to carboxylic acids, so they can be well dissolved in polar solvents. The edge-oxidized GNRs can be reduced with N_2H_4 to remove the oxygen-containing groups in the presence of ammonia, but residual defect sites are distributed at the irregular-shaped edges of the reduced GNRs.

4.4.1.2 *Gas-phase oxidation and liquid-phase sonication method*

Different from oxidation using strong oxidizers in solution, CNTs can be efficiently produced by unzipping gas-phase oxidized MWCNTs using mechanical sonication, and the obtained GNRs are of high quality without excessive oxidization (Fig. 4.27).[30] As shown in Fig. 4.27(a), the MWCNTs synthesized by arc discharge (Fig. 4.27(b)) are first calcined in air at 500°C to remove impurities and etch/oxidize MWCNTs at the defect sites and ends. Compared with the strong oxidative environment in the solution, this mild condition will not oxidize the pristine sidewalls of the nanotubes. The nanotubes are then dispersed in a 1,2-dichloroethane (DCE) organic solution of poly(*m*-phenylenevinylene-co-2,5-dioctoxy-*p*-phenylenevinylene) (PmPV) by sonication for 1 h. Under shear stress and the cavitation effect generated by sonication, the calcined CNTs are found to gradually unzip into nanoribbons (Figs. 4.27(c) and 4.27(d)). It is observed that the starting nanotubes with an average diameter of ~8 nm are unzipped to high-quality, smooth-edged graphene nanoribbons with widths in the range of 10–30 nm. The yield of nanoribbons is ~2% of the starting raw nanoribbons, significantly higher than previous methods capable of producing high-quality narrow nanoribbons.

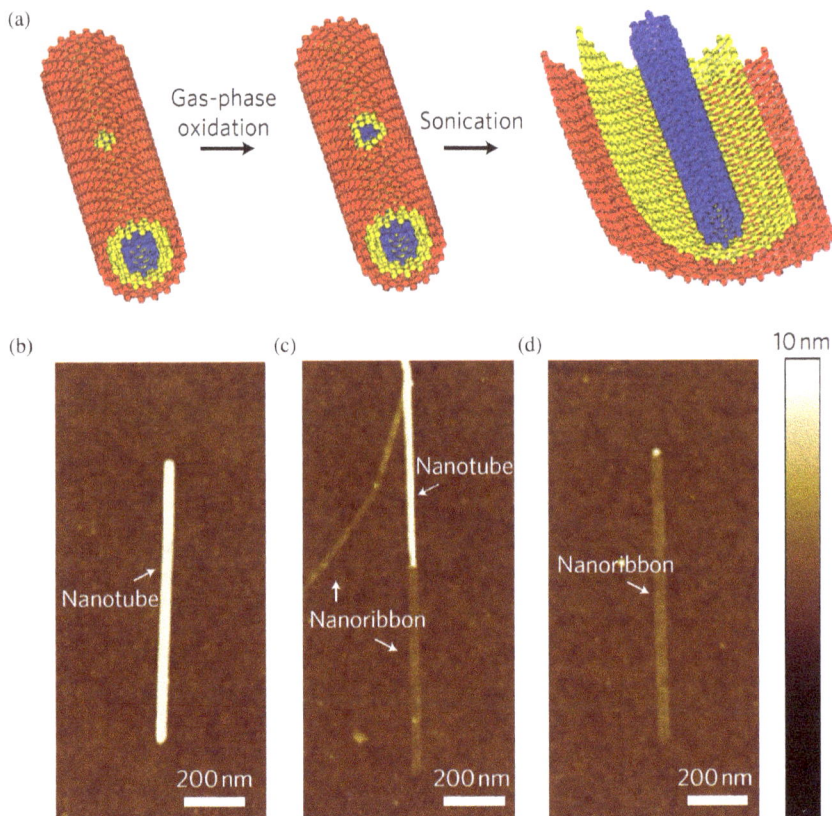

Figure 4.27. Unzipping of MWCNTs using a two-step method in gas and liquid phases[30]: (a) schematic of the unzipping processes and (b–d) AFM images of pristine, partially, and fully unzipped nanotubes, respectively.

4.4.1.3 *Catalytic hydrogenation of metal clusters*

In addition to the chemical unzipping of MWCNTs with strong oxidizers, the use of other chemical unzipping methods, such as catalytic hydrogenation of metal clusters, can cut open MWCNTs in order to produce GNRs. As shown in Fig. 4.28, transition metal nanoparticles (Ni or Co) can be used to catalyze MWCNTs and N-doped MWCNTs via catalytic hydrogenation, resulting in the unzipping of MWCNTs and the synthesis of GNRs.[31]

First, the transition metal nanoparticles are deposited on the CNTs, and this process can be achieved by two methods: (i) The MWCNTs are

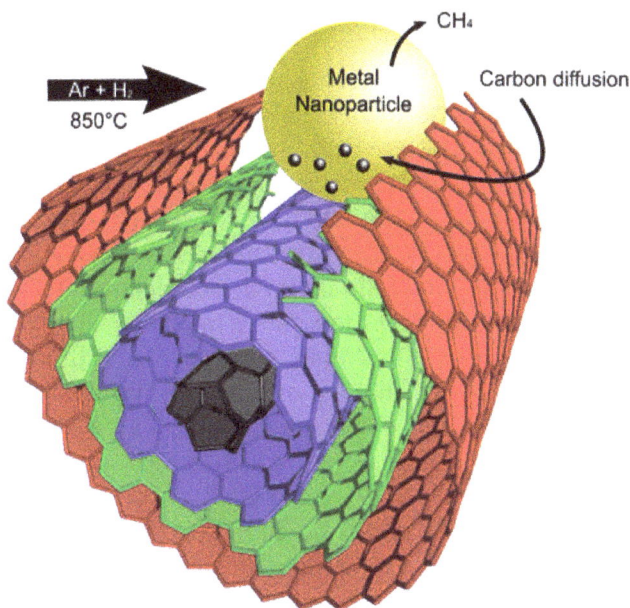

Figure 4.28. Schematic illustration of the synthesis of GNRs by unzipping MWCNTs via catalytic hydrogenation of metal clusters.[31]

introduced into the methanol solution of $CoCl_2$ (or NiCl), and this solution is dispersed ultrasonically and then dropped on a clean Si wafer. Subsequently, the substrate is heated to 500°C for 1 h in order to allow proper nucleation of Co nanoparticles on the surface of the tube. (ii) Alternatively, magnetron sputtering is used to deposit the Co nanoparticles on the Si wafer. The catalytic hydrogenation of MWCNTs with metal nanoparticles deposited on the surface is performed at 800°C for 30 min, the unzipping is achieved, and GNRs are obtained. During catalytic hydrogenation, carbon atoms are dissociated on the metal nanoparticles and, since the process takes place in an Ar−H atmosphere, dissociated carbon atoms further react with H_2 to form methane. As the reaction proceeds, the MWCNTs are continuously cut open. The cutting direction and depth are mostly determined by the nanoparticle size and the number of step edges where the nanoparticle nucleates. It should be noted that because the size of metal nanoparticles and their distribution on CNTs cannot be well controlled, the GNRs cut by this method are irregularly shaped and have poor uniformity.

4.4.1.4 *Metal intercalation and exfoliation method*

As mentioned earlier, the interlayer spacing of MWCNTs is about 0.34 nm. Therefore, similar to the alkali metal ions intercalated graphite, alkali metals such as Na and K can also be intercalated into MWCNTs; however, the stress generated by metal ion intercalation is not enough to cut open CNTs. MWCNTs can also be cut into GNRs if supplemented with cointercalating agents and appropriate post-treatments. As shown in Fig. 4.29, the uncapped CVD MWCNTs are cut in 3:1 concentrated $HNO_3:H_2SO_4$ (the cutting procedure also helps to remove catalytic metal particles and amorphous carbon).[32] Cut MWNTs are suspended in THF and added to liquid NH_3, followed by the addition of Li under magnetic stirring. The NH_3 is allowed to evaporate slowly under ambient air at room temperature, and the mixture is sonicated for 2 h in 10% HCl, filtered, and dried. Finally, it is calcined at 1000°C to obtain GNRs.

Defects on the walls, induced by cutting oxidizing acids, facilitate intercalation of $Li-NH_3$ and unwrapping and are the starting points for the detachment of graphene pieces. In addition, the defects on the opened tips also facilitate the insertion of $Li-NH_3$. Therefore, the intercalation of $Li-NH_3$ makes the CNTs partially exfoliated (Figs. 4.29(a) and 4.29(b)). During acid treatment, HCl reacts with Li ions and neutralizes NH_3, this exothermic reaction results in exfoliation, which further opens the nanotubes. Some species remain intercalated, possibly a combination of $Li-NH_3$ complexes and Li without NH_3, and their sudden removal during the thermal treatment results in further exfoliation. This completely opens many nanotubes; a smaller number of nanotubes remain partially exfoliated. Due to the heat treatment, the edges of the final graphene nanoribbons are oxidized to form oxygen-containing groups.

By using potassium intercalation to exfoliate MWCNTs, GNRs free of oxidized surfaces can be prepared.[33] In this process, potassium and MWCNTs, with a starting outside diameter of 40−80 nm and approximately 15−20 inner nanotube layers, are sealed in a glass tube, heated in a furnace at 250°C for 14 h, and quenched with ethanol as depicted in Fig. 4.30(b). Intromission of potassium atoms results in the formation of intercalation compounds within the interstices of MWCNT sidewalls (Fig. 4.30(a)) as commonly observed with other forms of graphitic carbon; however, in the case of the MWCNTs used here, the splitting could be further assisted by the generation of H_2 upon ethanolic quenching (Fig. 4.30(b)). Under sonication in chlorosulfonic acid, the split

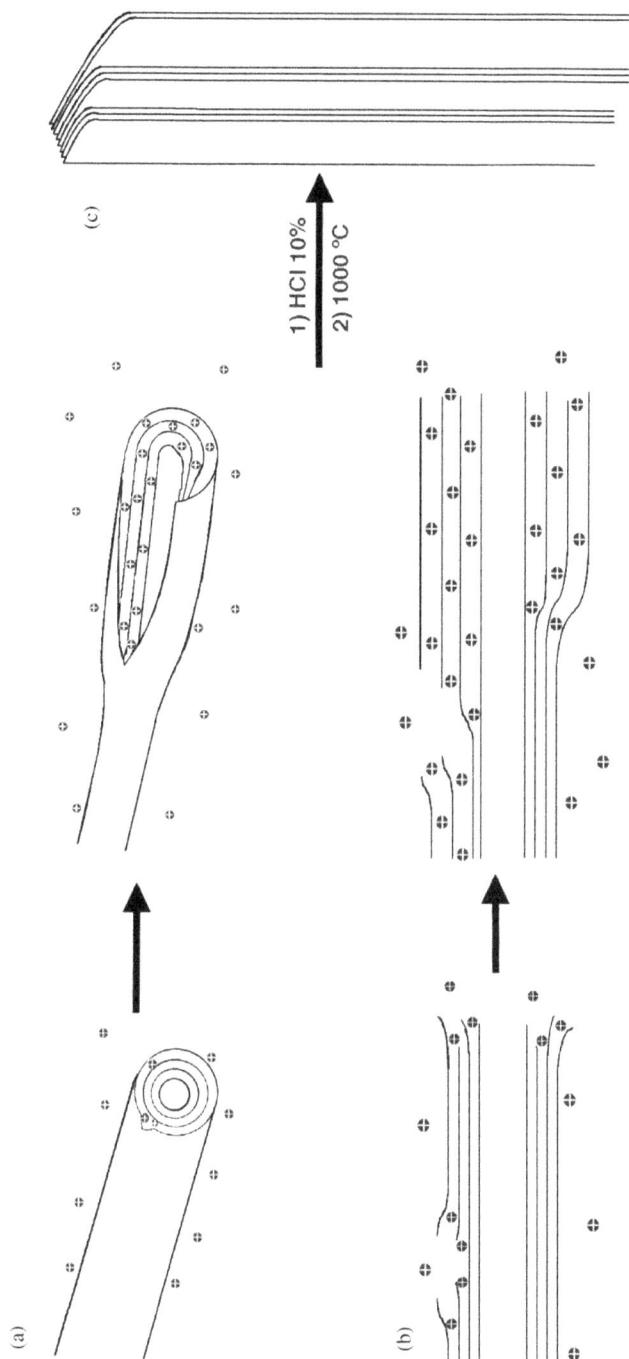

Figure 4.29. Proposed mechanism of metal intercalation and unwrapping of MWCNTs to synthesize GNRs[32]: (a) Li intercalation and partially unzipped MWCNTs; (b) cross-sectional view of MWCNTs during cutting; and (c) as-synthesized GNRs.

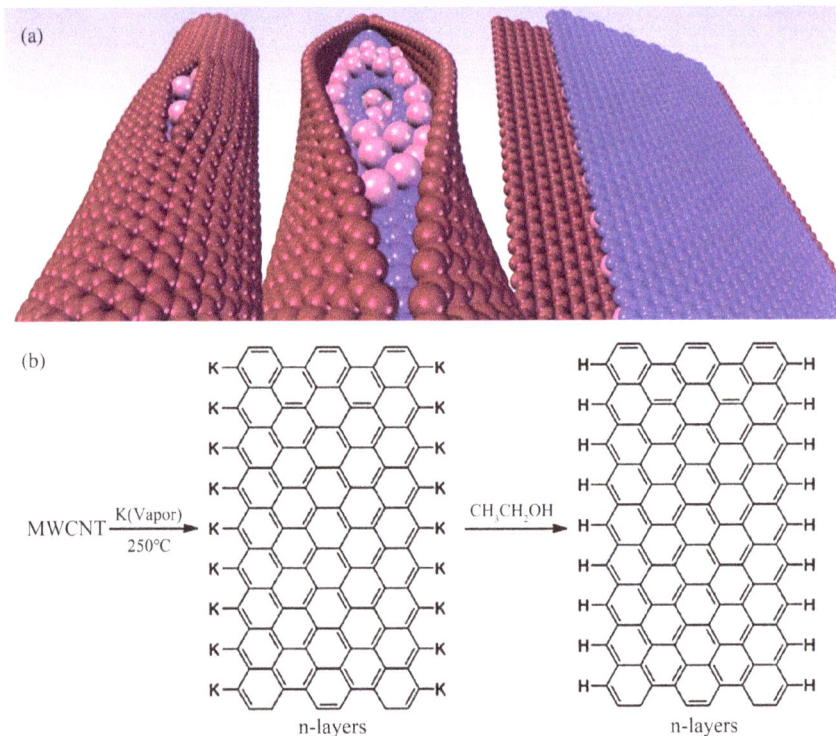

Figure 4.30.   GNRs synthesized by potassium splitting of MWCNTs[33]: (a) schematic of potassium intercalation and sequential splitting and (b) chemical schematic of the splitting processes.

MWCNTs are further exfoliated to form GNRs. The use of high-quality pristine CNTs as GNR precursors allows the production of material free from oxidative damage with conductivities paralleling the properties of the best samples of mechanically exfoliated graphene (measured on SiO_2 substrates). This simple, scalable, and inexpensive technique provides multigram quantities of stacked GNRs that can be further exfoliated with the aid of acid and mild bath sonication.

4.4.1.5 *Current sintering method*

In addition to the abovementioned chemical and mechanical approaches for cutting CNTs to prepare GNRs, the electrical method, by passing an electric current through a nanotube, can also be used to achieve

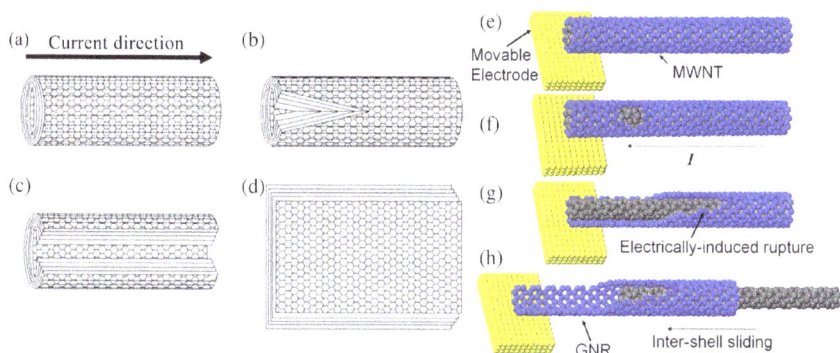

Figure 4.31. Schematic of synthesis of GNRs from MWCNTs by high DC pulses[34,35]: (a) MWCNT before current flowing; (b) the starting of tearing at the edge area; (c) breakage of sp^2 bonding along the current path; (d) transformed graphene layers; (e) MWCNT before the partial wall rupture; (f) electrical current induces rupture of the outer wall of the MWCNT; (g) partial outer-wall rupture of the MWNT results in a precursor GNR which is under the MWCNT inner core; and (h) inter-shell sliding between the GNR and the inner core results in a suspended, electrically contacted GNR.

unzipping, such as pulse current sintering (PCS). PCS is a field-activated sintering technique based on electric spark discharge, in which a high-energy, low-voltage sparking pulse current momentarily generates a spark plasma at high local temperatures. Because CNT is a cylindrical shape of 2D rolling sheets of carbon atoms and also electric current flows along the direction of CNT's surface, high DC pulse can convert CNTs into GNRs.

Graphene sheets are fabricated from MWCNTs with a high DC pulse through a pulsed current sintering process.[34] High DC pulses (1500–2500 A, <5 V) are applied on the MWCNTs synthesized under high temperature and high pressure. Because of the plasma created during PCS at high temperature with a high DC pulse, this energy will move forward directly from end to end over the surface of MWCNTs (Fig. 4.31(a)). The breaking of sp^2 carbon bonding starts at the tip of MWCNTs and follows the direction of the current (Fig. 4.31(b)). This phenomenon leads to the breakage of C—C bonds along the current path (Fig. 4.31(c)). The broken MWCNTs in a straight line are stretched away by tension on the curved surface of the MWCNTs and then transformed into graphene layers (Fig. 4.31(d)). The PCS technique has the advantages of short preparation time, high efficiency, and shows great potential for synthesis of GNRs from commercially CNTs on a large scale. However, the edge of GNRs prepared by this method is of an irregular shape.

In addition, a combination of electrical current and nanomanipulation technology is used to unzip MWCNTs to produce GNRs.[35] As shown in Fig. 4.31(e), arc-grown MWCNTs are attached to an aluminum wire using conductive epoxy, and the wire is then mounted to the stationary side of the holder. An etched tungsten probe is mounted to the opposite mobile side of the holder. Although the bare tungsten probe can itself serve as the mobile electrode, typically this electrode is first coated with a bundle of MWCNTs or an amorphous carbon–MWCNT composite to facilitate carbon–carbon contact between the mobile electrode and the MWNT to be unwrapped. The probe is moved such that the mobile electrode touches the tip of the MWCNT on the wire, creating carbon–MWCNT contact. Stable electrical and mechanical contact at the junction is established by annealing with a high current. When a current is applied between the movable electrode and the aluminum wire, the area of MWCNT–electrode contact has a high temperature due to contact resistance, which will cause the breaking of the carbon–carbon bond on the outer wall of the CNTs near the contact (Fig. 4.31(f)). With proper voltage bias control, only part of the MWCNT outer wall (upper portion in the schematic) is severed and, as shown in Fig. 4.31(g), a precursor GNR is created which clings to the remaining MWCNT inner core. The newly formed GNR can be easily removed from the MWCNT, or the sliding process can be terminated when a desired amount of GNR has slid off (Fig. 4.31(h)). The width of the GNR is about 45 nm, suggesting that about half (circumferentially) of the MWCNT (30 nm diameter) outermost shells are vaporized during the electrical unwrapping process. Although this method can accurately control the cutting process of CNTs and the size of the product graphene nanoribbons, the efficiency is low and it is not suitable for large-scale production. However, it offers the possibility of small-scale production of sophisticated devices.

4.4.1.6 *Plasma etching*

It is difficult to obtain GNRs with smooth edges and controllable widths at high yields using most of the methods mentioned earlier. However, these problems can be perfectly addressed by plasma etching. As shown in Fig. 4.32, MWCNTs are embedded in a poly(methyl methacrylate) (PMMA) layer as an etching mask, and GNRs with smooth edges and a narrow width distribution (10–20 nm) are obtained by plasma etching of MWCNTs.[36] First, the arc discharge MWCNTs are dispersed in

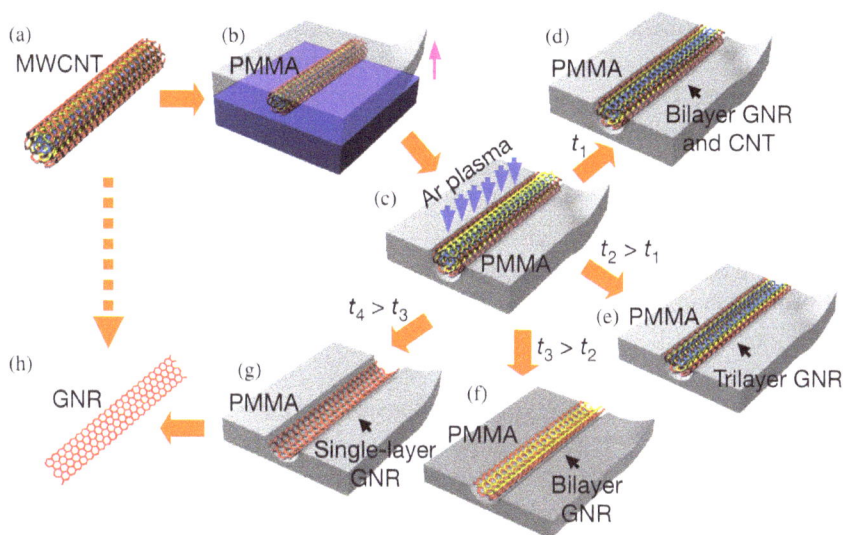

Figure 4.32. Schematic illustration of synthesis of GNRs by plasma etching of MWCNTs partly embedded in a polymer film.[36]

1% surfactant solution by brief sonication (Fig. 4.32(a)) and deposited onto a Si substrate, and a 300-nm-thick film of PMMA is spin coated on top of the MWCNTs. After baking, the PMMA–MWCNT film is peeled off in a KOH solution (Fig. 4.32(b)). MWCNTs embedded in the resulting PMMA film have a narrow strip of sidewall not covered by PMMA, owing to conformal PMMA coating on the substrate. The PMMA–MWCNT film is then exposed to 10 W Ar plasma for various times (Fig. 4.32(c)). Owing to protection by the PMMA, the top-side walls of MWCNTs are etched faster and removed by the plasma (Fig. 4.32(d–g)). Finally, the PMMA film is removed by washing to obtain GNRs (Fig. 4.32(h)). Monolayer, bilayer, and trilayer GNRs with inner CNT cores are produced depending on the diameter, number of layers of the starting MWCNT, and the etching time (Fig. 4.32(e–g)). By controlling the etching time, the inner CNT cores can also be retained, and a mixture of CNTs and bilayer GNRs can be obtained (Fig. 4.32(h)).

4.4.2 *Fullerenes as carbon source*

Like graphite and CNTs, fullerenes are a large class of allotropes of carbon. Fullerenes are a class of hollow molecules composed of a single layer of sp^2-hybridized carbon, and they are shaped like hollow spheres, ellipsoids, or tubes; thus graphene, which is a 2D construction block for carbon substances of all different dimensionalities, can be wrapped into 0D fullerene. Unlike graphene sheets, which only have six-membered carbon rings, fullerene molecules contain not only six-carbon rings but also five- and seven-membered rings. The best-known member of the fullerene's family is its C_{60} isoform which consists of 60 carbon atoms.

C_{60} molecules have well-defined dimensions and shapes; if C_{60} molecules are cut into slices by precise manipulation, precise dimension control of graphene sheets can be obtained. The molecular size of C_{60} is on the nanoscale (diameter ~7.1 Å), so fragmented graphene is also on the nanoscale, that is, GQDs, which represent a class of zero-dimensional carbon nanoparticles with typical dimensions of ca. <20 nm.

Like CNTs, C_{60} is also composed of sp^2-hybridized carbon; just like CNTs can be cut by strong oxidizers to produce GNRs, the C_{60} can also be cut into graphene nanosheets in a strong oxidizing environment, after which GQDs are obtained.[37] Fullerene C_{60} is treated with concentrated sulfuric acid, sodium nitrate, and potassium permanganate based on the modified Hummers method. In the presence of strong acid and oxidizers, the carbon–carbon bonds of C_{60} molecules are broken, and the sp^2-hybridized carbon atoms are oxidized to form oxygen-containing functional groups such as hydroxyl, carbonyl, and carboxyl moieties, which allow C_{60} to yield GQDs with oxidized edges. GQDs with well-defined diameters of 2–3 nm are produced, which remain fully dispersed in aqueous suspension due to the oxidized edges and exhibit exceptional luminescence properties, with a maximum intensity at 460 nm for a 340-nm excitation wavelength.

Surface-catalyzed decomposition of C_{60} on reactive transition metals can also be conducted to derive GQDs. As shown in Fig. 4.33, geometrically well-defined GQDs can be synthesized on a ruthenium surface using C_{60} as a precursor.[38] In this process, the strong C_{60}–Ru interaction induces the formation of surface vacancies in the Ru single crystal and a subsequent embedding of C_{60} molecules in the surface. It is found that the C_{60} molecule is lying with one six-membered ring parallel to the Ru(0001) substrate (left, Fig. 4.33(c)). Theoretical calculations show that the bottom-hemisphere carbon atoms of the C_{60} have stronger interactions with the Ru substrate, indicating the weakening of these C—C bonds. The C—C long bonds, which are between a hexagon and pentagon ring in C_{60} (labeled in red in Fig. 4.33(c)), have been extended by ~2.6% on average compared to the perfect C_{60} molecule, and the C—C short

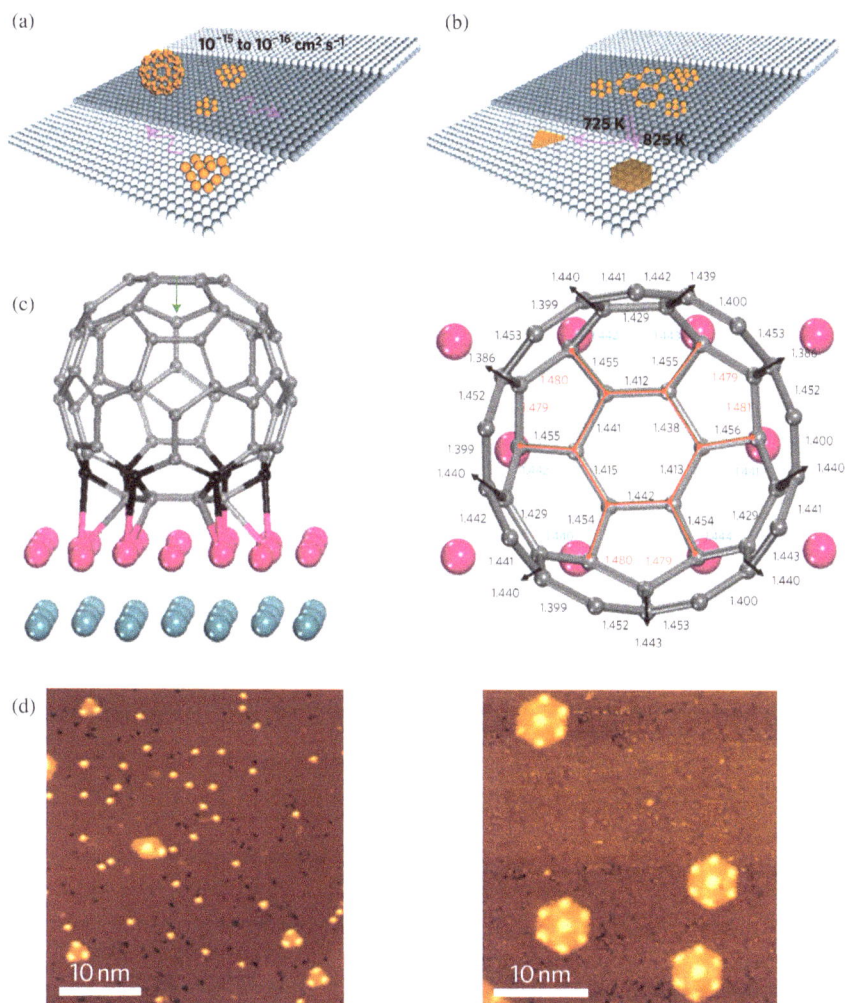

Figure 4.33. GQDs synthesized by metal surface-catalyzed decomposition[38]: (a) a majority of C_{60} molecules adsorb on the terrace and decompose to produce carbon clusters with restricted mobility; (b) temperature-dependent growth of GQDs with different equilibrium shapes from the aggregation of the surface-diffused carbon clusters; (c) C_{60} molecules adsorbed on the Ru surface and the bonds between them (left), C—C bond lengths of the bottom hemisphere of the C_{60}; and (d) corresponding STM images for the well-dispersed triangular and hexagonal equilibrium shaped graphene GQDs produced from C_{60}-derived carbon clusters.

bonds between two hexagon rings in C_{60} (labeled in blue in Fig. 4.33(c)) have been lengthened by ~3.0%. During high-temperature heat treatment, sufficient energy can be generated to rupture the fullerene cages into two unsymmetrical hemispheres. The surface-retained fragment derived from the bottom hemisphere of the ruptured C_{60} evolves eventually into the observed carbon clusters on the Ru surface, while the top hemisphere of the C_{60} cage may desorb into the gas phase. These carbon clusters undergo diffusion and aggregation to form GQDs. It is demonstrated that the mobility of these carbon clusters is variable, and GQDs of different shapes can be derived from the different mobility (Fig. 4.33(b)). Therefore, the equilibrium shape of GQDs can be tailored by optimizing the annealing temperature, such as triangular- and hexagonal-shaped (with lateral dimensions 2.7 nm and 5 nm, respectively) GQDs (Fig. 4.33(d)).

4.4.3 *Other carbon sources*

In addition to carbon allotropes such as CNTs and fullerenes, other carbon sources (coal, carbon black, and carbon fibers) can also be used as raw materials to synthesize functionalized graphene by top-down approaches.

4.4.3.1 *Coal as carbon source*

The unique coal structure has an advantage over pure sp^2-carbon allotropes (graphite, CNT, and fullerene) for producing GQDs. The structure of coal is complex, but the simplified composition contains angstrom or nanometer-sized crystalline carbon domains. An SEM image of bituminous coal is shown in Fig. 4.34(b).[39] These small-sized carbon crystalline domains contain defects that are linked by aliphatic amorphous carbon (Fig. 4.34(a)). GQDs can be obtained by treating coal with strong oxidizing agents. As shown in Fig. 4.34(c), bituminous coal is suspended in concentrated sulfuric acid and nitric acid and sonicated for 2 h. The reaction is then stirred and heated at 100°C or 120°C for 24 h. Then, GQDs can be obtained by filtration and dialysis. These GQDs show a hexagonal structure of 2.96 ± 0.96 nm diameter and 1.5–3 nm height, suggesting that there are two to four layers of GO-like structure. In addition, a few larger GQDs (>20 nm) are observed in the product. These large-sized GQDs are not fully cut and they are linked by amorphous carbon. All these GQDs

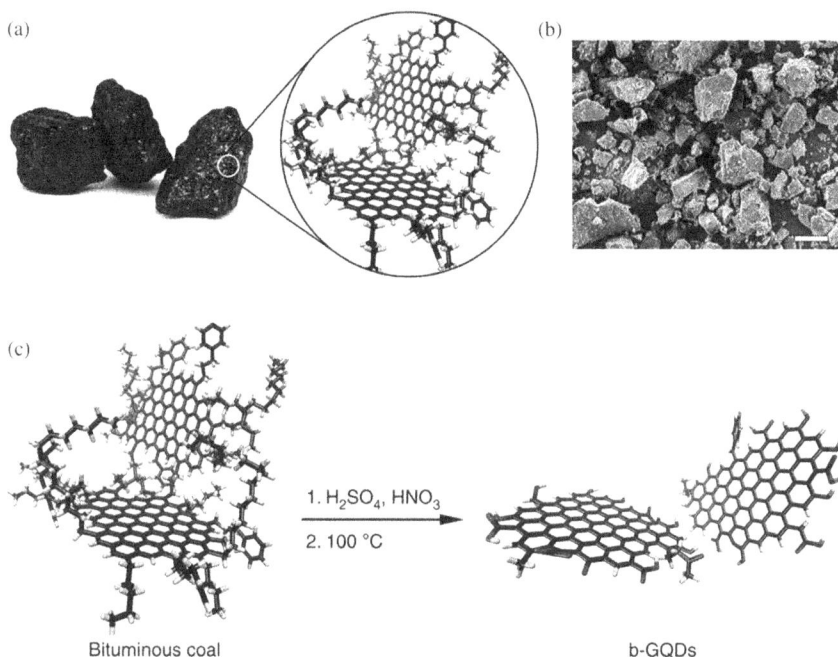

Figure 4.34. Synthesis of functionalized graphene by taking coal as a carbon source[39]: (a) macroscale image and simplified illustrative nanostructure of coal; (b) SEM image of bituminous coal; and (c) schematic illustration of GQDs synthesized by chemical oxidation of bituminous coal.

have hydrophilic bonds of C—O, C=O, and O—H on their edges, so they show high solubility in water.

The disordered configuration and small crystalline domains that are inherent in coal confer advantages over graphite such as easy dispersion, exfoliation, functionalization, and chemical cutting. This is because when graphite is oxidized to prepare GQDs under the same oxidative reaction, a lot of precipitates will be generated in the solution, while the solution of bituminous coal after the oxidation reaction is clear with little sediment.

Similarly, GQDs are also synthesized from coke and anthracite using the same method that is used for bituminous coal. The size and thickness of GQDs produced from bituminous coal are different from those synthesized from coke and anthracite, which is probably due to the different intrinsic morphologies of the starting coals. SEM analysis shows that ground bituminous coal and anthracite have irregular size and shape distributions but coke has a regular spherical shape. Bituminous coal has more carbon oxidation than anthracite and coke, and the presence of C—O, C=O, and O—H bonds is observed for bituminous coal, while C—O bonds are observed for anthracite. In addition, coke does not contain oxidized carbon since it is produced from devolatilization and carbonization of tars and pitches.

The morphologies of the GQDs are different depending on the different compositions and structure of coals from which they originate. Like bituminous coal, GQDs synthesized from anthracite and coke both show crystalline hexagonal structures and high solubility in water, but the GQDs obtained from anthracite exhibit a stacked structure of 29 ± 11 nm diameter. The GQDs prepared from coke have a uniform size of 5.8 ± 1.7 nm. In terms of size and shape, GQDs produced from bituminous coal are smaller and more uniform than those synthesized from coke and anthracite. This method has the prominent advantages of using a common carbon source, low cost, and high yield, and thus is of significance in the large-scale production of GQDs.

4.4.3.2 *Carbon black as carbon source*

GQDs can also be prepared from carbon black by refluxing it with concentrated nitric acid.[40] As shown in Fig. 4.35, XC-72 carbon black is formed as aggregates of spherical graphite particles with an approximate diameter of 30 nm. It is refluxed in concentrated nitric acid for 24 h, and after cooling to room temperature, the suspension is centrifuged for 10 min to obtain a supernatant and a sediment. The supernatant is heated at 200°C to evaporate the water and nitric acid, and GQDs1 are obtained. GQDs1 are washed with HCl and further adjusted to about pH 8 with ammonia water, and then GQDs2 are obtained. Both GQDs1 and GQDs2 contain abundant graphitic structures, carboxyl, and hydroxyl groups, suggesting excellent solubility in water; their average sizes are similar, about 15 nm and 18 nm, respectively. However, their thicknesses are different. The topographic heights of GQDs1 are mostly less than 0.7 nm, with an average height of about 0.5 nm, indicating that GQDs1 are mostly single layered. The topographic heights of GQDs2 are mainly between 1 and 3 nm, implying that they are multilayered (2−6 layers). This method shows prominent advantages of low cost, high yield, and a one-step facile process.

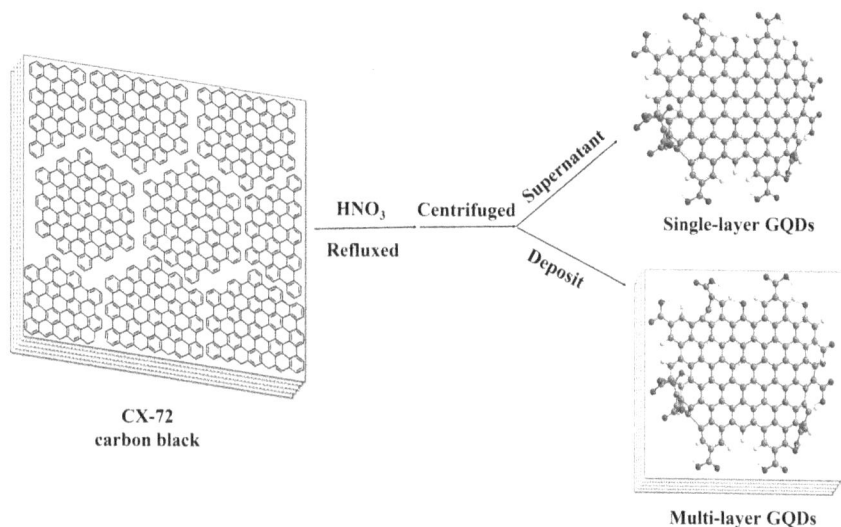

Figure 4.35.   Schematic illustration of the synthesis of GQDs from carbon black.[40]

4.4.3.3 *Carbon fiber as carbon source*

Carbon fiber is a fibrous carbon material composed of flake graphite microcrystals along the fiber axial direction; it is obtained from the carbonization and graphitization of organic polymer fiber. Therefore, carbon fibers can also be cut into GQDs chemically or physically. As shown in Fig. 4.36, pitch carbon fibers are added into a mixture of H_2SO_4 and HNO_3, and the solution is sonicated for 2 h and stirred for 24 h at high temperatures (80°C, 100°C, and 120°C).[41] The mixture is diluted and adjusted to pH 8, and the final product solution is further dialyzed to obtain GQDs. The sizes and heights of these GQDs are 1–4 nm and 0.4–2 nm, respectively, corresponding to 1–3 graphene layers. These GQDs prefer zigzag edges more than armchair ones, and the hydrophilic groups, including carbonyl, carboxyl, hydroxyl, and epoxy groups, are introduced to the edges during the oxidation, making them highly soluble in water and other polar organic solvents, such as DMF and DMSO. The interlayer spacing is 0.403 nm, which is broader than that of the original carbon fiber (0.364 nm), but smaller than that of GO. This result could be attributed to the oxygen-containing groups introduced in the exfoliation and oxidation of CF, which enhances the interlayer distance. However, the interlayer distance of GQDs is smaller than that of GO, which is due to GQDs being

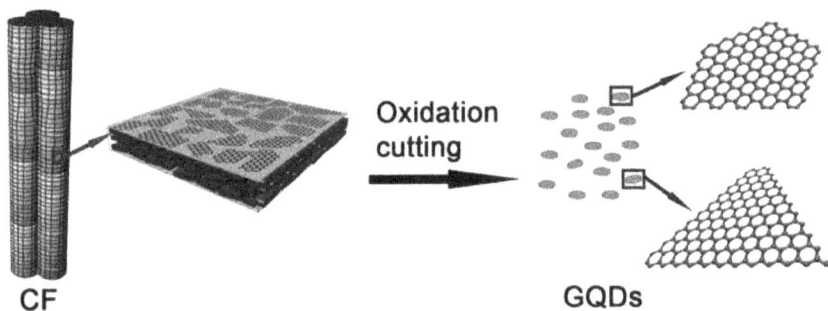

Figure 4.36.   Representation scheme of oxidation cutting of carbon fiber into GQDs.[41]

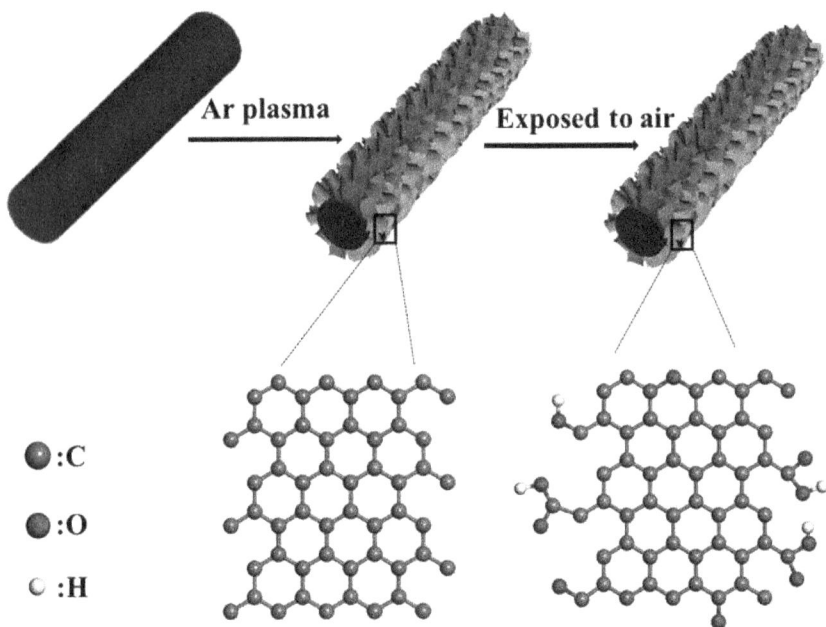

Figure 4.37.   Schematic of exfoliation of functionalized graphene from carbon fiber.[42]

oxidized only on the edges due to the very small size. The size of GQDs can be tuned by changing the reaction temperature in this method.

In addition, carbon fibers can be used as raw materials. The *in situ* exfoliation of functionalized graphene on the surface of carbon fibers is successfully conducted, and the resulting material is used as a self-standing electrocatalyst.[42] As shown in Fig. 4.37, Ar plasma is used to treat the carbon fiber surface to peel off the graphite microcrystals, generating edge active sites on the surface of carbon fibers, and edge-rich and oxygen-functionalized graphene is obtained. In this as-synthesized functionalized graphene, the edges are doped by oxygen and defects are distributed. These defects and oxygen-containing functional groups can serve as active sites in electrocatalytic reactions. At the same time, the plasma-etched carbon fibers also possess improved specific surface area, leading to enhanced electrocatalytic performances of the carbon fiber.

4.5 Summary

Materials such as graphite, CNT, fullerene, coal, and carbon fiber can be used to synthesize functionalized graphene via a top-down strategy. By means of various methods such as redox, electrochemical oxidation, arc discharge, and mechanochemical method, functionalized graphene can be synthesized by exfoliation as well as simultaneous doping, dimension control, and covalent functionalization of graphene sheets via breaking down the non-covalent or covalent bond in the carbon raw material.

A top-down strategy is highly suitable for mass production because of the abundant precursor materials and simple operation. This strategy has been receiving considerable interest from both academia and industry and has been considered a key enabler in widespread applications. However, many functionalization methods based on this strategy are not perfect. For example, most methods based on this strategy are not very precise in functionalization (such as controlling sites and groups of functionalization as well as structures of functionalized graphene), while some more accurate methods, such as plasma etching of CNTs, are not suitable for large-scale production. In addition, top-down approaches can also be used to prepare functionalized graphene composites. For example, while exfoliating graphite to prepare graphene, functionalized graphene composites, such as non-covalent functionalized graphene composites, can be fabricated through *in situ* approaches. The synthesis and applications of functionalized graphene composites will be discussed in Chapter 6.

Therefore, compared with bottom-up synthesis strategies, top-down protocols have both advantages and disadvantages, and the two can complement each other. In practical applications, proper selection of strategy is required.

References

1. Eigler, S. and Hirsch, A. (2014). Chemistry with graphene and graphene oxide—Challenges for synthetic chemists, *Angew. Chem. Int. Ed.*, 53(30), 7720–7738.
2. Casabianca, L. B., Shaibat, M. A., Cai, W. W., Park, S., Piner, R., Ruoff, R. S., and Ishii, Y. (2010). NMR-based structural modeling of graphite oxide using multidimensional ^{13}C solid-state NMR and ab initio chemical shift calculations, *J. Am. Chem. Soc.*, 132(16), 5672–5676.
3. Ren, W. and Cheng, H.-M. (2014). The global growth of graphene, *Nat. Nanotechnol.*, 9(10), 726–730.
4. Marcano, D. C., Kosynkin, D. V., Berlin, J. M., Sinitskii, A., Sun, Z., Slesarev, A., Alemany, L. B., Lu, W., and Tour, J. M. (2010). Improved synthesis of graphene oxide, *ACS Nano*, 4(8), 4806–4814.
5. Dimiev, A., Kosynkin, D. V., Alemany, L. B., Chaguine, P., and Tour, J. M. (2012). Pristine graphite oxide, *J. Am. Chem. Soc.*, 134(5), 2815–2822.
6. Zhang, L., Li, X., Huang, Y., Ma, Y., Wan, X., and Chen, Y. (2010). Controlled synthesis of few-layered graphene sheets on a large scale using chemical exfoliation, *Carbon*, 48(8), 2367–2371.
7. Yu, P., Lowe, S. E., Simon, G. P., and Zhong, Y. L. (2015). Electrochemical exfoliation of graphite and production of functional graphene, *Curr. Opin. Colloid Interface Sci.*, 20(5), 329–338.
8. Parvez, K., Li, R., Puniredd, S. R., Hernandez, Y., Hinkel, F., Wang, S., Feng, X., and Müllen, K. (2013). Electrochemically exfoliated graphene as solution-processable, highly conductive electrodes for organic electronics, *ACS Nano*, 7(4), 3598–3606.
9. Cao, J., He, P., Mohammed, M. A., Zhao, X., Young, R. J., Derby, B., Kinloch, I. A., and Dryfe, R. A. W. (2017). Two-step electrochemical intercalation and oxidation of graphite for the mass production of graphene oxide, *J. Am. Chem. Soc.*, 139(48), 17446–17456.
10. Pei, S., Wei, Q., Huang, K., Cheng, H.-M., and Ren, W. (2018). Green synthesis of graphene oxide by seconds timescale water electrolytic oxidation, *Nat. Commun.*, 9(1), 145.
11. Wang, H. S., Tian, S. Y., Yang, S. W., Wang, G., You, X. F., Xu, L. X., Li, Q. T., He, P., Ding, G. Q., Liu, Z., and Xie, X. M. (2018). Anode coverage for enhanced electrochemical oxidation: a green and efficient strategy towards water-dispersible graphene, *Green Chem.*, 20(6), 1306–1315.
12. Ejigu, A., Kinloch, I. A., and Dryfe, R. A. W. (2017). Single stage simultaneous electrochemical exfoliation and functionalization of graphene, *ACS Appl. Mater. Interfaces*, 9(1), 710–721.
13. Liu, N., Luo, F., Wu, H., Liu, Y., Zhang, C., and Chen, J. (2008). One-step ionic-liquid-assisted electrochemical synthesis of ionic-liquid-functionalized

graphene sheets directly from graphite, *Adv. Funct. Mater.*, 18(10), 1518–1525.

14. Saïdane, K., Razafinimanana, M., Lange, H., Huczko, A., Baltas, M., Gleizes, A., and Meunier, J. L. (2004). Fullerene synthesis in the graphite electrode arc process: local plasma characteristics and correlation with yield, *J. Phys. D: Appl. Phys.*, 37(2), 232.

15. Wu, X., Liu, Y., Yang, H., and Shi, Z. (2016). Large-scale synthesis of high-quality graphene sheets by an improved alternating current arc-discharge method, *RSC Adv.*, 6(95), 93119–93124.

16. Pham, T. V., Kim, J.-G., Jung, J. Y., Kim, J. H., Cho, H., Seo, T. H., Lee, H., Kim, N. D., and Kim, M. J. (2019). High areal capacitance of N-doped graphene synthesized by arc discharge, *Adv. Funct. Mater.*, 29(48), 1905511.

17. Nan, Y., Li, B., Zhang, X., and Song, X. (2018). Catalyst-free, tunable doping content of graphitic-N in arc-discharged graphene via gas and solid nitrogen sources and their formation mechanisms, *J. Nanopart. Res.*, 20(10), 274.

18. Kim, S., Song, Y., Takahashi, T., Oh, T. and Heller, M. J. (2015). An aqueous single reactor arc discharge process for the synthesis of graphene nanospheres, *Small*, 11(38), 5041–5046.

19. Kim, S., Song, Y., and Heller, M. J. (2017). Seamless aqueous arc discharge process for producing graphitic carbon nanostructures, *Carbon*, 120, 83–88.

20. Jeon, I.-Y., Bae, S.-Y., Seo, J.-M., and Baek, J.-B. (2015). Scalable production of edge-functionalized graphene nanoplatelets via mechanochemical ball-milling, *Adv. Funct. Mater.*, 25(45), 6961–6975.

21. Jeon, I.-Y., Shin, Y.-R., Sohn, G.-J., Choi, H.-J., Bae, S.-Y., Mahmood, J., Jung, S.-M., Seo, J.-M., Kim, M.-J., Wook Chang, D., Dai, L., and Baek, J.-B. (2012). Edge-carboxylated graphene nanosheets via ball milling, *PNAS*, 109(15), 5588–5593.

22. Buzaglo, M., Shtein, M., and Regev, O. (2016). Graphene quantum dots produced by microfluidization, *Chem. Mater.*, 28(1), 21–24.

23. Liang, S., Yi, M., Shen, Z., Liu, L., Zhang, X., and Ma, S. (2014). One-step green synthesis of graphene nanomesh by fluid-based method, *RSC Adv.*, 4(31), 16127–16131.

24. Ciesielski, A. and Samorì, P. (2014). Graphene via sonication assisted liquid-phase exfoliation, *Chem. Soc. Rev.*, 43(1), 381–398.

25. Bottari, G., Herranz, M. Á., Wibmer, L., Volland, M., Rodríguez-Pérez, L., Guldi, D. M., Hirsch, A., Martín, N., D'Souza, F., and Torres, T. (2017). Chemical functionalization and characterization of graphene-based materials, *Chem. Soc. Rev.*, 46(15), 4464–4500.

26. Sampath, S., Basuray, A. N., Hartlieb, K. J., Aytun, T., Stupp, S. I., and Stoddart, J. F. (2013). Direct exfoliation of graphite to graphene in aqueous media with diazaperopyrenium dications, *Adv. Mater.*, 25(19), 2740–2745.

27. Vallés, C., Drummond, C., Saadaoui, H., Furtado, C. A., He, M., Roubeau, O., Ortolani, L., Monthioux, M., and Pénicaud, A. (2008). Solutions of negatively charged graphene sheets and ribbons, *J. Am. Chem. Soc.*, 130(47), 15802–15804.

28. Li, L., Zheng, X., Wang, J., Sun, Q., and Xu, Q. (2013). Solvent-exfoliated and functionalized graphene with assistance of supercritical carbon dioxide, *ACS Sustain. Chem. Eng.*, 1(1), 144–151.

29. Kosynkin, D. V., Higginbotham, A. L., Sinitskii, A., Lomeda, J. R., Dimiev, A., Price, B. K., and Tour, J. M. (2009). Longitudinal unzipping of carbon nanotubes to form graphene nanoribbons, *Nature*, 458(7240), 872–876.

30. Jiao, L., Wang, X., Diankov, G., Wang, H., and Dai, H. (2010). Facile synthesis of high-quality graphene nanoribbons, *Nat. Nanotechnol.*, 5(5), 321–325.

31. Elías, A. L., Botello-Méndez, A. R., Meneses-Rodríguez, D., Jehová González, V., Ramírez-González, D., Ci, L., Muñoz-Sandoval, E., Ajayan, P. M., Terrones, H., and Terrones, M. (2010). Longitudinal cutting of pure and doped carbon nanotubes to form graphitic nanoribbons using metal clusters as nanoscalpels, *Nano Lett.*, 10(2), 366–372.

32. Cano-Márquez, A. G., Rodríguez-Macías, F. J., Campos-Delgado, J., Espinosa-González, C. G., Tristán-López, F., Ramírez-González, D., Cullen, D. A., Smith, D. J., Terrones, M., and Vega-Cantú, Y. I. (2009). Ex-MWNTs: Graphene sheets and ribbons produced by lithium intercalation and exfoliation of carbon nanotubes, *Nano Lett.*, 9(4), 1527–1533.

33. Kosynkin, D. V., Lu, W., Sinitskii, A., Pera, G., Sun, Z., and Tour, J. M. (2011). Highly conductive graphene nanoribbons by longitudinal splitting of carbon nanotubes using potassium vapor, *ACS Nano*, 5(2), 968–974.

34. Kim, W. S., Moon, S. Y., Bang, S. Y., Choi, B. G., Ham, H., Sekino, T., and Shim, K. B. (2009). Fabrication of graphene layers from multiwalled carbon nanotubes using high dc pulse, *Appl. Phys. Lett.*, 95(8), 083103.

35. Kim, K., Sussman, A., and Zettl, A. (2010). Graphene nanoribbons obtained by electrically unwrapping carbon nanotubes, *ACS Nano*, 4(3), 1362–1366.

36. Jiao, L., Zhang, L., Wang, X., Diankov, G., and Dai, H. (2009). Narrow graphene nanoribbons from carbon nanotubes, *Nature*, 458(7240), 877–880.

37. Chua, C. K., Sofer, Z., Šimek, P., Jankovský, O., Klímová, K., Bakardjieva, S., Hrdličková Kučková, Š., and Pumera, M. (2015). Synthesis of strongly fluorescent graphene quantum dots by cage-opening Buckminsterfullerene, *ACS Nano*, 9(3), 2548–2555.

38. Lu, J., Yeo, P. S. E., Gan, C. K., Wu, P., and Loh, K. P. (2011). Transforming C_{60} molecules into graphene quantum dots, *Nat. Nanotechnol.*, 6(4), 247–252.

39. Ye, R., Xiang, C., Lin, J., Peng, Z., Huang, K., Yan, Z., Cook, N. P., Samuel, E. L. G., Hwang, C.-C., Ruan, G., Ceriotti, G., Raji, A.-R. O., Martí, A. A., and Tour, J. M. (2013). Coal as an abundant source of graphene quantum dots, *Nat. Commun.*, 4(1), 2943.

40. Dong, Y., Chen, C., Zheng, X., Gao, L., Cui, Z., Yang, H., Guo, C., Chi, Y., and Li, C. M. (2012). One-step and high yield simultaneous preparation of single- and multi-layer graphene quantum dots from CX-72 carbon black, *J. Mater. Chem.*, 22(18), 8764–8766.

41. Peng, J., Gao, W., Gupta, B. K., Liu, Z., Romero-Aburto, R., Ge, L., Song, L., Alemany, L. B., Zhan, X., Gao, G., Vithayathil, S. A., Kaipparettu, B. A., Marti, A. A., Hayashi, T., Zhu, J.-J., and Ajayan, P. M. (2012). Graphene quantum dots derived from carbon fibers, *Nano Lett.*, 12(2), 844–849.

42. Liu, Z., Zhao, Z., Wang, Y., Dou, S., Yan, D., Liu, D., Xia, Z., and Wang, S. (2017). In situ exfoliated, edge-rich, oxygen-functionalized graphene from carbon fibers for oxygen electrocatalysis, *Adv. Mater.*, 29(18), 1606207.

Chapter 5

Functionalized Graphene Materials Modified by Controllable Functional Groups

5.1 Overview

In general, the dispersibility of graphene is poor, and van der Waals interactions and π-π stacking interactions between graphene sheets (GSs) tend to cause irreversible agglomeration in water and common organic solvents due to which their applications are greatly limited. In this regard, functionalization of graphene is required to improve the dispersibility and processability of graphene. Thus, the resulting functionalized graphene can be used in different applications across various fields. In addition, when graphene is functionalized, modifications with different functional groups can effectively tune the chemical composition and band structure of graphene, improve its inherent properties, and offer new characteristics, such as improvement of mechanical properties and introduction of unique magnetic and fluorescence properties. This is of great significance for both theoretical research of graphene and its practical applications.

Compared with graphene, graphene oxide (GO) shows the advantages of low cost, large-scale production, and easy processing, and a large number of oxygen-containing functional groups are distributed on the π-conjugated plane and edge of GO, such as epoxide, hydroxyl, carbonyl, carboxyl, and ester groups. These oxygen-containing functional groups can promote the dispersion of GO in solvents for ease of subsequent processing and use; these groups also offer the possibility of further modification and functionalization of GO by other functional groups.

The controllable modification of graphene and GO with functional groups is mainly divided into covalent and non-covalent modifications. Covalent functionalization of graphene (or GO) involves incorporating graphene with newly introduced groups in the form of covalent bonds to improve its performance. In contrast, non-covalent functionalization of graphene (or GO) results in the formation of a functionalized graphene composite by combining graphene and functional groups through π-π interaction and electrostatic interaction forces. From the perspective of the synthesis mechanism, both the covalent and non-covalent functionalization of graphene or GO can be categorized as bottom-up strategies. However, the controllable modification of graphene and GO with functional groups is an important research issue in the field of graphene functionalization, for which extensive research has been conducted and widespread applications have been presented. Therefore, in this chapter, the controllable modification of graphene and GO with functional groups will be reviewed. First, the synthesis of covalent functionalized graphene and GO is introduced, and then the non-covalent functionalization methods are described.

5.2   Covalent Functionalization of Graphene and GO

5.2.1   *Covalent functionalization of graphene*

Graphene is a 2D crystal composed of sp^2-hybridized carbon, which can be regarded as infinitely extended polycyclic aromatic hydrocarbon (PAH) (Fig. 5.1). The in-plane sp^2-orbitals form a σ bond, leaving a free p_z orbital perpendicular to the plane. The remaining orbital electron forms a large π bond, and this electron can move freely in the plane. Therefore, most of the reactions that occur in PAHs can also be applied to graphene.

By means of covalent reactions, various functional groups can be attached onto graphene sheets. The reactions commonly used for covalently functionalized graphene mainly include free radical addition, cycloaddition, nucleophilic addition, and substitution (Fig. 5.2).

The functionalization of graphene via the free radical addition approach has been achieved by both thermal and photochemical treatments. To this end, the common adducts or reactions utilized are

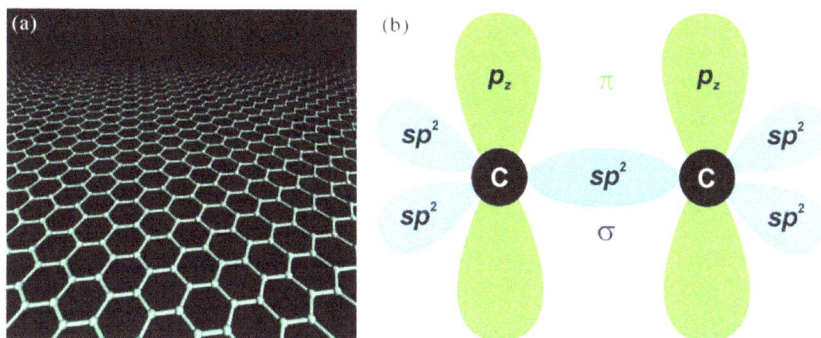

Figure 5.1. Structure and electron orbitals of graphene[1]: (a) pictorial representation of graphene sheet and (b) sp^2-hybridization in graphene.

aryl diazonium salts, peroxide, Bergman cyclization, and Kolbe electrosynthesis.

Cycloaddition includes [2 + 1] cycloaddition, [2 + 1] aziridine adduct reactions, [2 + 2] cycloaddition, [3 + 2] cycloaddition, and [4 + 2] cycloaddition.

Substitution has been successfully achieved via Friedel–Crafts acylation and hydrogen–lithium exchange methods.

5.2.1.1 *Free radical addition*

The functionalization of graphene via the free radical addition approach has been achieved by thermal or photochemical treatments, where the free radicals generated by adduct undergo addition reactions with sp^2-hybridized carbon on graphene. To this end, the common adducts are aryl diazonium salts and peroxides; thus, functional molecules with diazonium salts and peroxides can be incorporated with graphene via the free radical addition approach.

1. *Aryl diazonium salt* (*diazonium compound*)
Under neutral or alkaline conditions, diazonium salt or diazonium compound is reduced by single electron transfer to give an aryl radical which is accompanied by the release of N_2 gas. Then, this radical undergoes an adduct reaction with C=C bonds on graphene, resulting in a new C—C bond (Fig. 5.3(a)). During this process, graphene provides an electron to

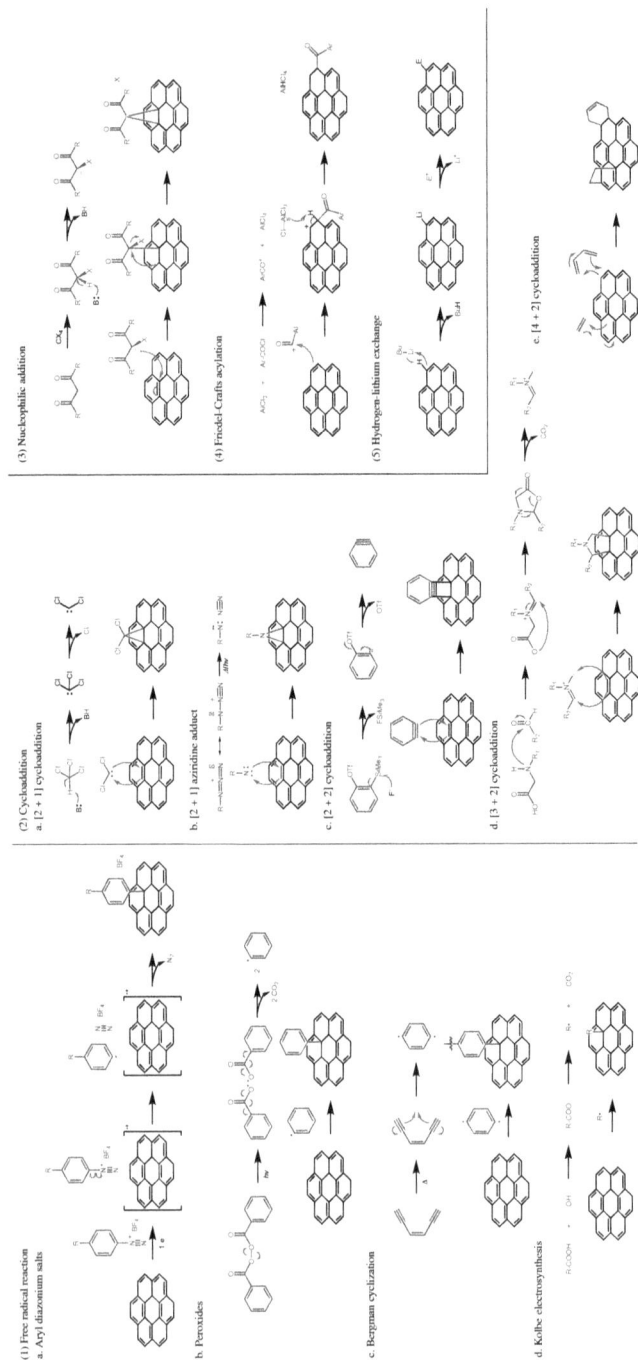

Figure 5.2.   Chemical reactions and mechanisms of the covalent functionalization of graphene.[2]

(a)

(b)

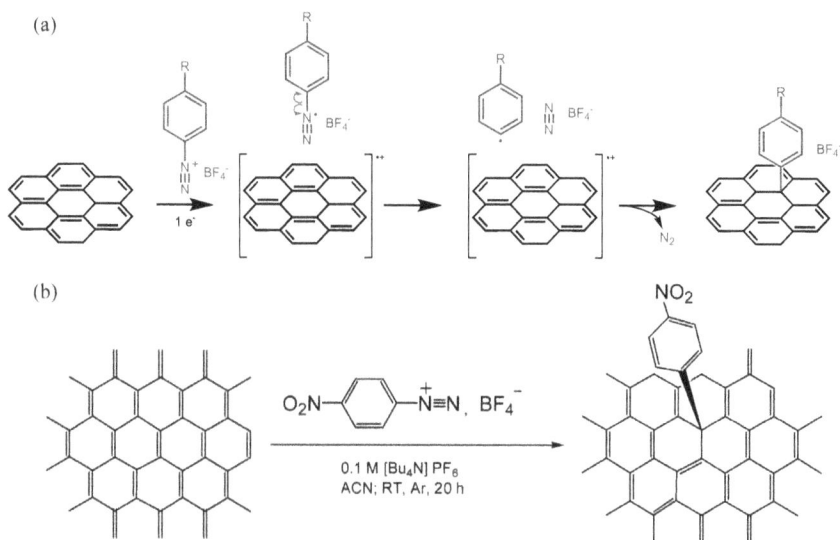

Figure 5.3. Mechanism and illustration of functionalization of graphene by aryl diazonium salts: (a) mechanism of free radical addition between diazonium salt and graphene[2] and (b) modification of nitrobenzene on the surface of graphene through this reaction.[3]

the diazonium salt, resulting in a radical aryl moiety that readily adds to the graphene sp^2 carbon network.

In 2009, Haddon and co-workers used the radical addition of diazonium salt to achieve surface modification of few-layer epitaxial graphene with nitrophenyl groups.[3] The synthesis method is shown in Fig. 5.3(b); epitaxial graphene is grown on SiC wafers as the substrate; in the presence of acetonitrile and tetrabutylammonium hexafluorophosphate ([Bu₄N]PF₆), 4-nitrophenyl diazonium (NPD) tetrafluoroborate undergoes free radical reaction with the graphene substrate to produce nitrophenyl-functionalized graphene. In 2010, Tour and co-workers studied the effect of 4-nitrophenyl modification on the electronic properties of GNRs.[5] The results showed that the covalent attachment of 4-nitrophenyl groups leads to a breakdown of the π-electron conjugated structure of graphene, resulting in the transition of the graphene carbon atoms from sp^2 to sp^3 hybridization, which in turn alters the conductivity. Studies have also shown that the conductivity of graphene can be tuned by controlling the grafting time of this radical reaction.

In addition, chemically reactive moieties can also be introduced via a radical reaction of diazonium salt, and then functional groups can be

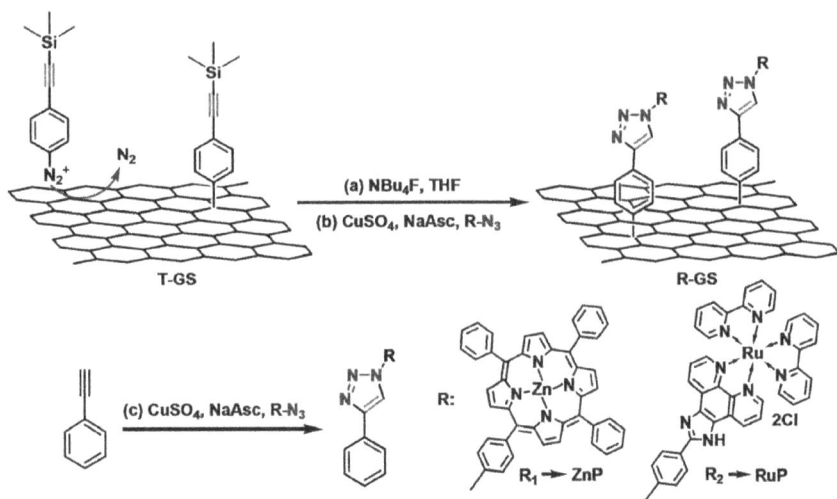

Figure 5.4.   Schematic illustration of the introduction of alkyne groups into graphene via a radical reaction of diazonium salt, followed by incorporation of functional groups via "click" chemistry.[4]

attached to graphene via these active moieties. For example, grafting with alkyne groups is carried out on the surface of graphene via a radical reaction of diazonium salt, and then functional groups are introduced into graphene through 1,3-dipolar cycloaddition of azides and alkynes reaction.[4] As shown in Fig. 5.4, the GSs are firstly modified by 4-(trimethylsilyl)ethynylaniline (noted as T-GS) via an aryl diazonium salt reaction, followed by an *in situ* deprotection step using tetrabutylammonium fluoride (NBu$_4$F), which removes the trimethylsilyl groups, giving phenylacetylene moiety-modified GSs with exposed alkyne groups. 1,3-dipolar cycloaddition between the deprotected T-GS and azide-terminated zinc-porphyrin (ZnP-N$_3$) as well as ruthenium-phenanthroline derivatives (RuP-N$_3$) is then performed to give zinc-porphyrin-triazole-GS (ZnP-GS) as well as ruthenium-phenanthroline-triazole-GS (RuP-GS), respectively. In addition, many different types of graphene-based nanocomposites can be obtained through 1,3-dipolar cycloaddition. For example, the attachment of 1,4-diketopyrrolo[3,4-c]pyrrole (DPP) onto the surface of graphene results in functionalized graphene with photoluminescence properties,[6] which is soluble in water when short-chain polyethylene glycol with carboxyl groups are grafted on the surface[7]; graphene functionalized by chain

(a)

(b)

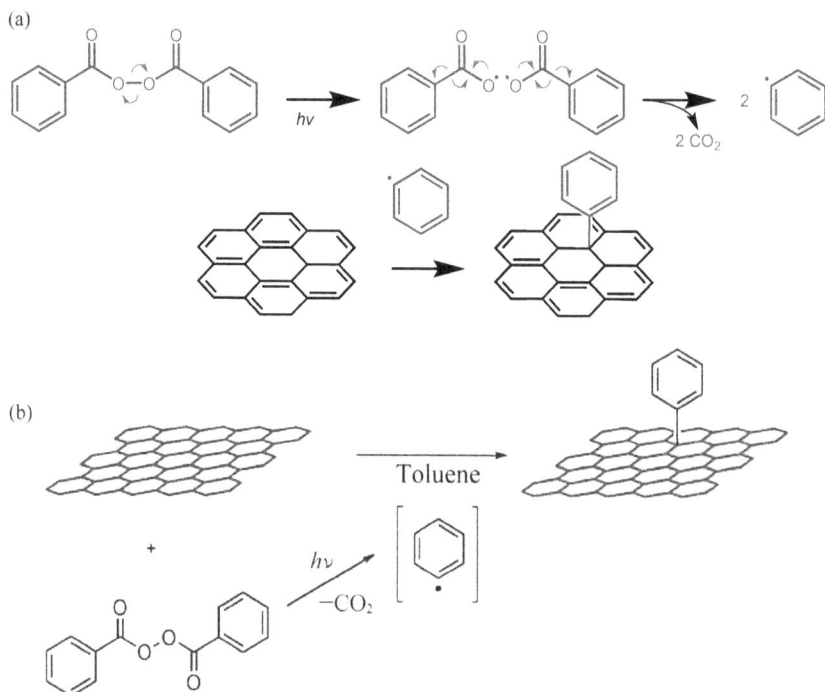

Figure 5.5. Mechanism and schematic of functionalization of graphene by peroxide: (a) mechanism of the (top) formation of phenyl radical from benzoyl peroxide and the (bottom) free radical addition of phenyl radical onto graphene[2] and (b) mechanism of grafting of a benzene ring on graphene.[9,10]

transfer reagents undergoes reversible addition-fragmentation chain transfer (RAFT) polymerization to produce a graphene–poly(N-isopropylacrylamide) molecular sieve.[8]

2. Peroxides

In addition to diazonium salts, benzoyl peroxide can also generate phenyl radicals. For example, benzoyl peroxide is used as a radical initiator under photochemical treatment to liberate two molecules of carbon dioxide, providing phenyl radical species, which are subsequently added onto the graphene carbon atoms to establish covalent linkages (Fig. 5.5(a)).[2] As shown in Fig. 5.5(b), the mechanically exfoliated graphene is placed on the SiO_2/Si substrate, and benzoyl peroxide undergoes a free radical addition reaction with graphene under laser irradiation.[9] The hot laser

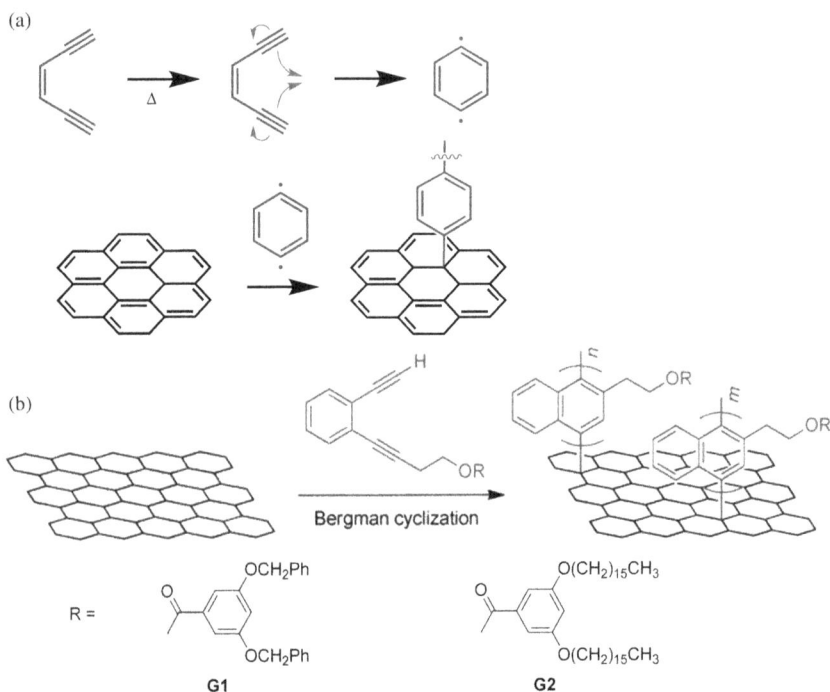

Figure 5.6.   Mechanism and schematic of functionalization of graphene via Bergman cyclization: (a) mechanism of formation of biradical from an enediyne moiety and the free radical addition of phenyl biradical onto graphene[2] and (b) mechanism of grafting an alkyl chain onto graphene.[11]

treatment initiates an electron transfer mechanism from the photoexcited graphene to the physisorbed benzoyl peroxide. The surface-adsorbed benzoyl peroxide accepts a hot electron, becomes photoexcited, and then decomposes to benzoate and benzoyloxyl radicals which are then converted to a phenyl radical by the elimination of CO_2. The produced phenyl radicals react with the sp^2-hybridized carbon atoms on the graphene plane to obtain benzene-derivatized graphene.

3. Bergman cyclization
The Bergman cyclization proceeds towards the formation of a six-membered ring. The precursor consists of an enediyne moiety which cycloaromatizes under high thermal treatment ($\sim$200°C) via a radical mechanism and the cyclization results in the formation of 1,4-benzenediyl biradical species. (Fig. 5.6(a)). The highly reactive biradical species can react with

sp^2-hybridized carbons on graphene to establish a covalent bond. Molecules with enediyne moieties can be functionalized onto the graphene surface through this reaction. As shown in Fig. 5.6(b), microcrystalline graphite is exfoliated to single-layer and few-layer graphene by ultrasonication in *N*-methylpyrrolidinone (NMP).[11] The collected homogeneous dispersion of exfoliated graphene in NMP is then degassed under high vacuum, followed by refluxing. To this hot suspension, enediyne G1 or G2 in NMP is added with a peristaltic pump and the reaction mixture is further heated for 12 h. Finally, alkyl chain functionalized graphene is obtained after cooling, filtration, washing, and drying. Due to the attachment of long alkyl chains on the surface, these two kinds of functionalized graphene can be dispersed in a variety of common organic solvents, such as NMP, *N,N*-dimethylformamide, tetrahydrofuran, dichloromethane, chloroform, ethyl acetate, and toluene.

4. *Kolbe electrosynthesis*

The Kolbe reaction involves the electrochemical oxidation of the carboxylate ions, which provides radical species via a subsequent decarboxylation step. The carboxylates are electrolyzed in a neutral or slightly acidic environment, and during this reaction, one carboxylate ion undergoes electrochemically oxidation to lose an electron, which allows the generation of a free radical and release of a CO_2 molecule. In general, the produced free radicals will proceed by means of dimerization reactions to provide highly active free radicals, which undergo free radical additions with sp^2-hybridized carbon on graphene; that is to say, the incorporation of functional groups on graphene is achieved (Fig. 5.7(a)). As shown in Fig. 5.7(b), the epitaxial graphene substrate serves as the working electrode, while the platinum wire and saturated calomel electrode are used as counter and reference electrodes, respectively.[12] The acetonitrile solution of tetrabutylammonium hexafluorophosphate (nBu$_4$NPF$_6$) is used as the electrolyte, and α-naphthylacetic acid and tetrabutylammonium hydroxide (nBu$_4$NOH) are added into this electrolyte for electrolysis. When the voltage is 0.93 V, α-naphthylacetic acid is oxidized and decarboxylated to generate α-anaphthylmethyl (α-NM) radicals. This rapidly leads to the covalent attachment of the α-NM functionality to the graphene lattice via C—C bond formation, and it creates a new sp^3 carbon center in place of sp^2 carbon atoms in the graphene lattice. This functionalization of graphene is reversible; that is, a higher potential (1.85 V) results in an electroerasing event where the grafted functionality is erased and α-NM functionality is oxidized, and the graphene surface recovers its original pristine state. Compared with

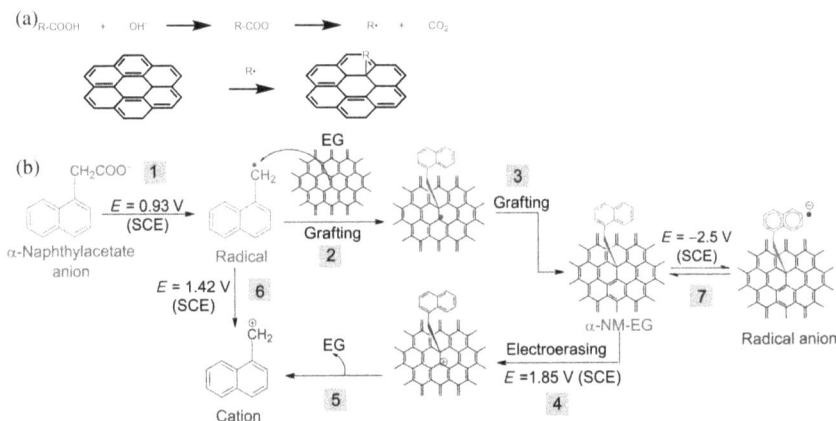

Figure 5.7. Mechanism and illustration of Kolbe electrosynthesis reaction to achieve graphene functionalization: (a) schematic of Kolbe reaction for the (top) formation of radical species and the (bottom) free radical addition of radical species onto graphene[2] and (b) mechanistic pathways associated with the grafting of α-naphthylmethyl (α-NM) groups to epitaxial graphene, SCE = standard calomel electrode.[12]

other covalent functionalization methods of graphene, the advantages of Kolbe electrooxidation are as follows: (i) The reaction is reversible; (ii) the grafted functionalities can be electrochemically erased; (iii) α-NM groups are found to offer well-ordered structures on graphene surfaces, and thus the resulting graphene derivative is anticipated to exhibit interesting magnetic and electronic behaviors; and (iv) the simplicity, versatility, and efficiency of the reaction makes the covalent binding of a wide variety of arylmethyl groups possible with appropriate substituents on the phenyl rings.

5.2.1.2 *Cycloaddition*

Cycloaddition usually involves interactions between dienophiles and C—C bonds in graphene. Functional groups with dienophiles are able to react with graphene via [2 + 1], [2 + 2], [3 + 2], and [4 + 2] cycloadditions, resulting in three-, four-, five-, and six-membered rings, respectively.

1. [2 + 1] *cycloaddition*
The [2 + 1] cycloaddition reaction between dienophiles and graphene mainly includes cycloaddition for the formation of cyclopropane or aziridine adducts. As shown in Fig. 5.8(a), singlet dichlorocarbene is

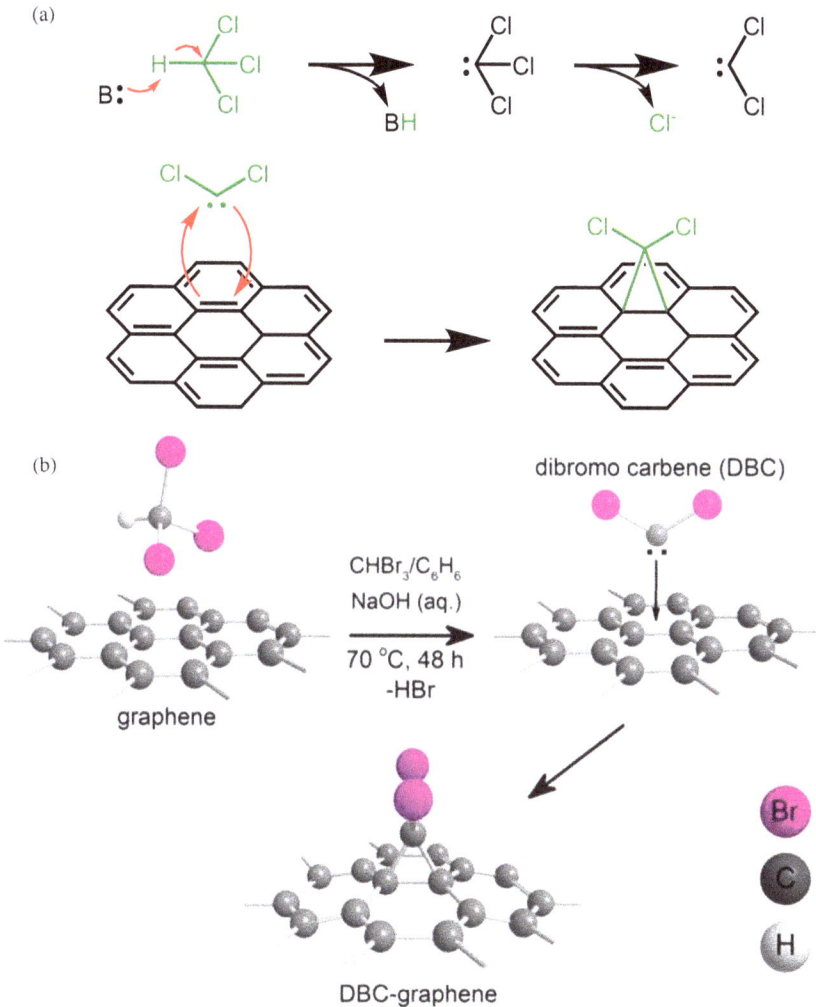

Figure 5.8. Mechanism and illustration of functionalization of graphene via cyclopropa-nation: (a) mechanism of the (top) formation of dichorocarbene with chloroform and base and (bottom) cyclopropanation of graphene with dichlorocarbene[2] and (b) the grafting of dibromocyclopropyl groups on the surface of graphene by [2 + 1] cycloaddition reaction.[13]

generated from a mixture of chloroform in a strong base (NaOH). Since dichlorocarbene is highly active, it would readily react with C=C bonds in the graphene in a concerted manner, resulting in a cyclopropane struc-ture, after which surface-modified graphene with dichloromethyl groups

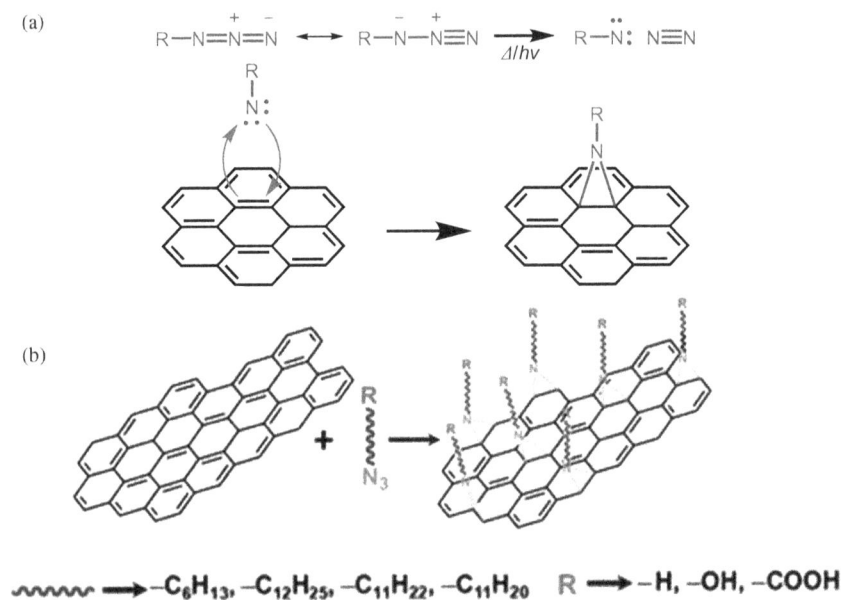

Figure 5.9.   Mechanism and illustration of the formation of nitrene from the decomposition of azide and cycloaddition of nitrene onto graphene: (a) mechanism of the (top) formation of nitrene from the decomposition of azide and (bottom) [2 + 1] cycloaddition of nitrene onto graphene[2] and (b) schematic of the covalent functionalization of graphene with various alkylazides by nitrene [2 + 1] cycloaddition.[14]

is obtained. Similarly, dibromomethyl groups can also be grafted on the surface of graphene via [2 + 1] cycloaddition (Fig. 5.8(b)).[13] In this research, bromoform and trihexylamine are added to a suspension of solution-exfoliated graphene in toluene. Then, the suspension is added drop-wise to a stirred aqueous solution of NaOH and allowed to stir for 48 h at 70°C. After filtration, washing, and drying, surface-modified graphene with dibromomethyl is obtained.

2. [2 + 1] *aziridine adduct*

Nitrene is a highly reactive intermediate, which is able to undergo a [2 + 1] cycloaddition reaction on the graphene sp^2 carbon network to provide an aziridine adduct. As shown in Fig. 5.9(a), the nitrene intermediate is usually generated from thermal or photodecomposition of an azide group. Due to the high activity of nitrene, it will react with graphene rapidly to generate aziridine; thus, the functional groups are modified on graphene.

Different alkyl chains with azide groups at the terminals can form and generate aziridine groups on the surface and edges of graphene via [2 + 1] cycloaddition, and hexyl, dodecyl, hydroxyundecyl, and carboxyundecyl groups can be grafted onto graphene, as shown in Fig. 5.9(b).[14] The modified functionalized graphene exhibits enhanced dispersibility in toluene and acetone, and the degree of functionalization depends on the amount of nitrene added in the reaction mixture.

Through this method, graphene with specific functions can be obtained by reacting it with different functional molecules containing azide groups. For example, the addition of azidotrimethylsilane onto epitaxial graphene is shown to have a profound impact on varying the band gaps of the graphene hybrid materials.[15]

3. [2 + 2] *cycloaddition*

In chemistry, an aryne is an uncharged reactive intermediate derived from an aromatic system by the removal of two ortho substituents. Aryne is highly reactive and can undergo [2 + 2] cycloaddition with a graphene sp^2 carbon network. As shown in Fig. 5.10(a), the presence of fluoride ion induces a desilylation step of *o*-trimethylsilyl-phenyl triflate to form aryne, which proceeds by means of the elimination of the triflate and trimethylsilyl group. The generated aryne rapidly undergoes [2 + 2] cycloaddition with graphene, and the benzene ring is grafted onto graphene. The functionalized graphene modified with a benzene ring shows stable dispersion in DMF, 1,2-dichlorobenzene, ethanol, chloroform, and water without obvious precipitation for weeks.

It should be noted that fluoride ion is required to catalyze the formation of aryne from *o*-trimethylsilyl-phenyl triflate; however, fluoride ion may lead to residual by-products.[17] Therefore, aryne can be obtained by thermal decomposition of the corresponding anhydrides under microwave irradiation. As shown in Fig. 5.10(b), different anhydrides and exfoliated few-layer graphene are thoroughly mixed in a mortar. Different homogenous powder mixtures are obtained at different cycles of microwave irradiation. In each cycle, 200 watts of microwave power are applied until 250°C is reached in about 5 seconds. After 5 cycles, the corresponding few-layer graphene modified by different aromatic moieties is obtained. These anhydrides are able to absorb microwave irradiation with high efficiency; therefore, they play two roles in the reaction process: as a reagent and as a microwave-absorbing matrix which allows high temperatures to be reached in short times under solvent-free conditions.[16]

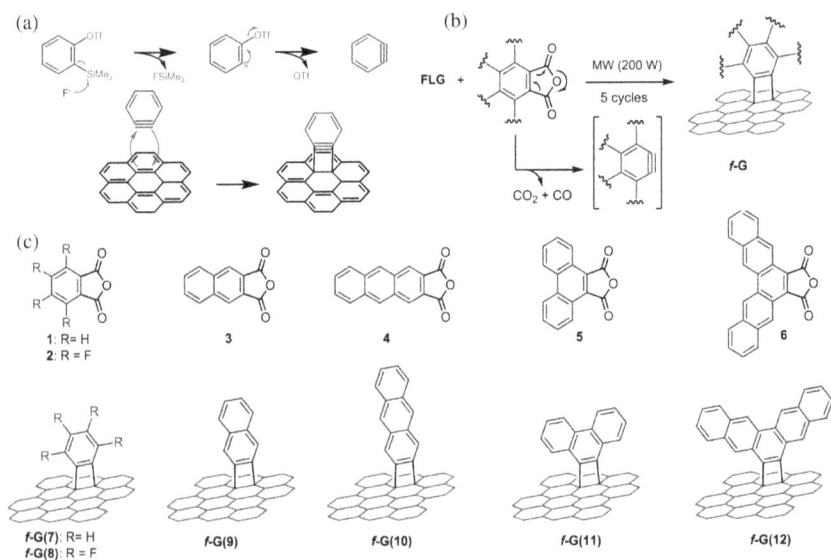

Figure 5.10. Mechanism and illustration of functionalization of graphene by [2 + 2] cycloaddition: (a) modification of benzene rings on the surface of graphene by [2 + 2] cycloaddition[2] and (b–c) modification of different aromatic rings on the surface of graphene by [2 + 2] cycloaddition.[16]

4. [3 + 2] cycloaddition

Azomethine ylides are a special class of carbonyl ylides formed by carbene association with carbonyl oxygen of an amide, and they are commonly used dienophiles, which undergo 1,3-dipolar [3 + 2] cycloaddition for the functionalization of graphene. As shown in Fig. 5.11(a), the amino acid reacts with aldehyde to form an azomethine ylide, which undergoes a [3 + 2] cycloaddition reaction with graphene to construct an azapentane five-membered ring on graphene. Functional groups containing aldehyde groups, such as tetraphenyl-porphyrin, with the assistance of sarcosine, can form amethymine ylides, and they undergo 1,3-dipolar [3 + 2] cycloaddition with rGO, and the nonlinear optical (NLO) performance is improved when the rGO is modified with tetraphenyl-porphyrin (Fig. 5.11(b)).[18]

5. [4 + 2] cycloaddition

The [4 + 2] cycloaddition reaction, better known as Diels–Alder cycloaddition, is the most famous pericyclic reaction in organic chemistry. The

(a)

(b)

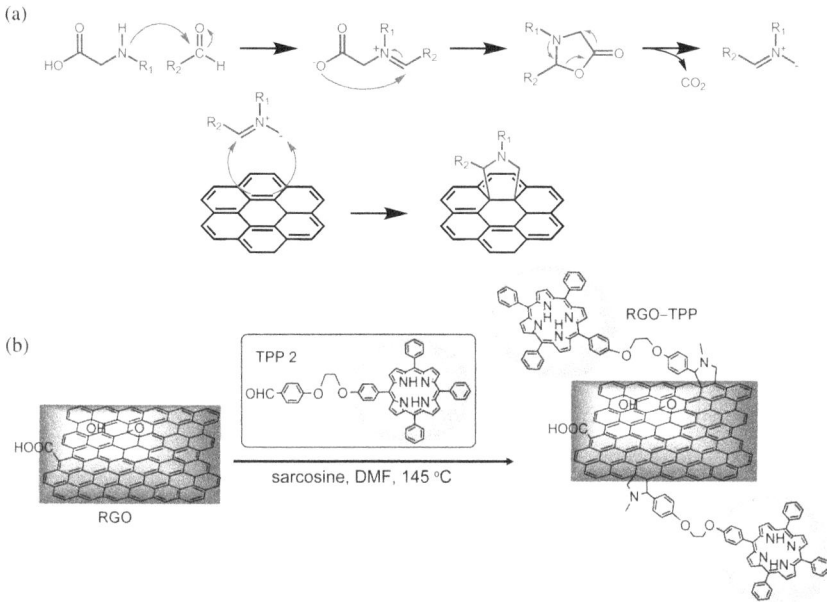

Figure 5.11. Mechanism and illustration of functionalization of graphene by [3 + 2] cycloaddition reaction: (a) mechanism of the (top) formation of azomethine ylide from amino acid and (bottom) 1,3-dipolar [3 + 2] cycloaddition of graphene with azomethine ylide[2] and (b) functionalization of graphene with tetraphenyl-porphyrin by [3 + 2] cycloaddition reaction.[18]

reaction proceeds due to the overlap between the highest occupied molecular orbital (HOMO) of the conjugated diene and the lowest unoccupied molecular orbital (LUMO) of the dienophile via heat treatment to form a six-membered ring, and thus, this reaction is also commonly applied for the functionalization of graphene. In the [4 + 2] cycloaddition reaction, the sp^2-hybridized carbon on graphene can act as a conjugated diene to react with a dienophile or as a dienophile to react with other conjugated dienes (Fig. 5.12(a)). As shown in Fig. 5.12(b), the graphene epitaxially grown on SiC is immersed for several tens of hours in a toluene solution containing maleimide derivatives to deposit these molecules onto graphene.[20] The results show that maleimide derivatives are grafted onto graphene through the Diels–Alder cycloaddition reactions. The conjugated structure at the covalently functionalized sites on graphene is destroyed, and sp^2-hybridized carbon atoms are transformed into

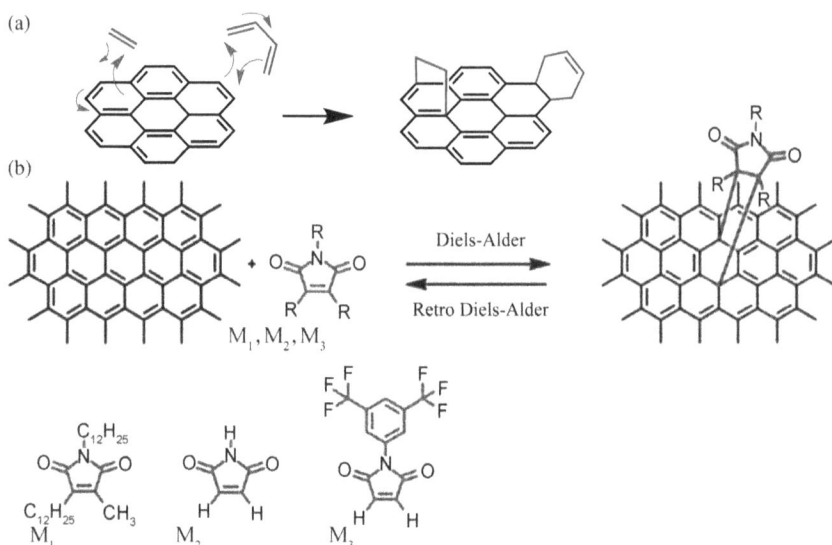

Figure 5.12. Mechanism and illustration of functionalization of graphene by [4 + 2] cycloaddition reaction: (a) mechanism of [4 + 2] cycloaddition (Diels–Alder reaction) on graphene[2] and (b) functionalization of graphene through [4 + 2] cycloaddition (Diels–Alder reaction).[19,20]

sp^3-hybridized ones. Due to the covalent functionalization of maleimide derivatives, a tendency for the opening of a gap is evidenced.

In another study, graphene was functionalized with cyclopentadienyl-capped poly(ethylene glycol) monomethyl ether through a Diels–Alder cycloaddition reaction (Fig. 5.13) without any catalyst, and the resulting functionalized graphene showed improved dispersion properties in various solvents, such as DMSO, DMF, NMP, THF, ethylene glycol, ethanol, water, acetone, and chloroform.[21]

5.2.1.3 *Nucleophilic addition*

Nucleophilic addition, also known as the Bingel reaction, originated from the cyclopropanation chemistry of fullerene. It utilizes a halide derivative of the diethyl malonate moiety in the presence of a base such as 1,8-diazabicyclo[5.4.0]undec-7-ene or sodium hydride (Fig. 5.14(a)). Nucleophilic addition has ever since found its usefulness in graphene chemistry given the ease of its reaction conditions. The base abstracts a

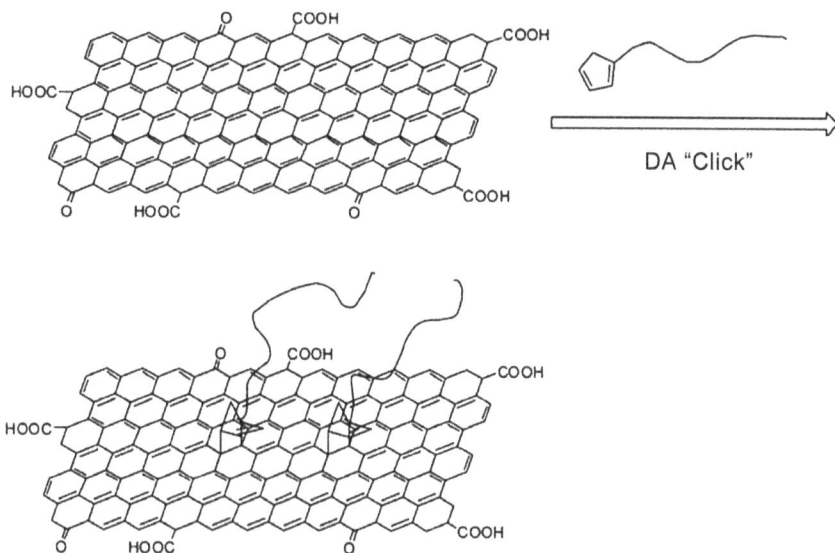

Figure 5.13. Schematic illustration of functionalization of graphene with cyclopentadienyl-capped polymer through a Diels–Alder [4 + 2] "click" reaction.[21]

proton from halide-malonate to provide an enolate which subsequently nucleophilically attacks a C=C bond on the graphene carbon framework. The resulting carbanion undergoes a subsequent nucleophilic substitution, which displaces the halide atom to provide a cyclopropane adduct via intramolecular ring closure.

In addition to using a strong base as a catalyst, dimethyl sulfoxide (DMSO) and sodium carbonate (Na_2CO_3) can also be employed to catalyze this reaction at room temperature. As shown in Figs. 5.14(b) and 5.14(c), a nucleophilic attack by dimsyl anion on the brominated α-carbon atom and the concomitant bromine departure lead to a dimethylsulfoxonium intermediate.[22] The dimethylsulfoxonium intermediate could easily strip protons under weak alkaline conditions to generate a carbanion intermediate because dimethylsulfoxonium has a stronger electron withdrawal ability than bromine. Next, the carbanion intermediate reacts with carbon atoms of C=C bonds on graphene to obtain intermediate graphene carbanion. This carbanion on graphene reacts with the α-carbon of diethyl malonate to provide a cyclopropanation product that undergoes ring closure with the intramolecular displacement of DMSO.

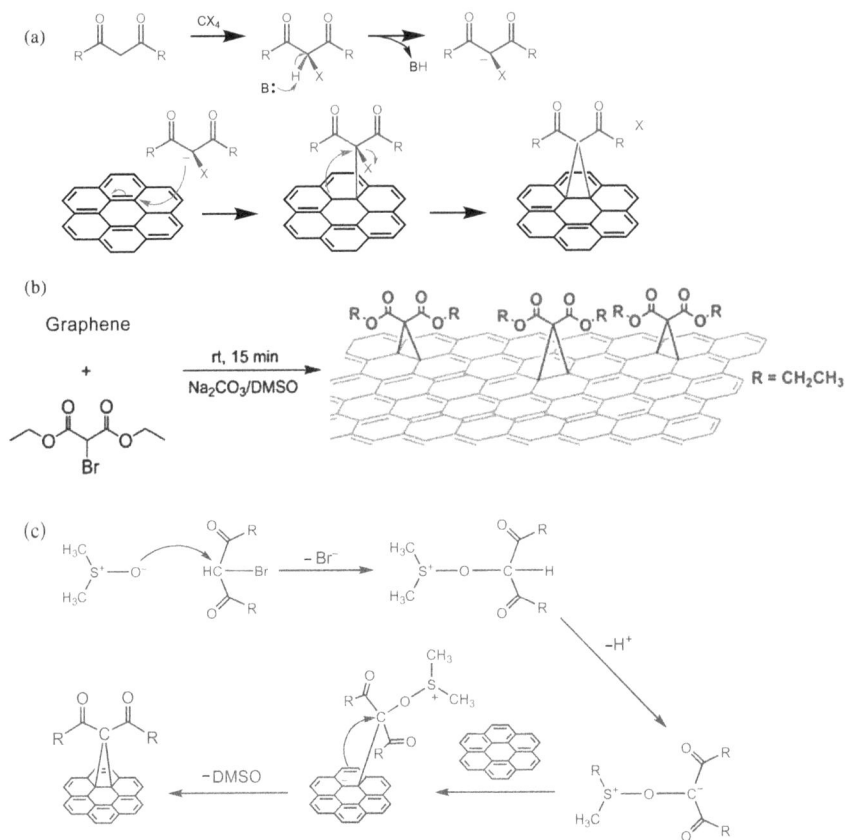

Figure 5.14. Mechanism and illustration of functionalization of graphene through nucleophilic addition: (a) mechanism of the (top) formation of halide derivative of enolate and (bottom) addition of halide derivative of enolate onto graphene[2] and (b–c) mechanism and illustration of DMSO and sodium carbonate catalyzed nucleophilic addition of diethyl bromomalonate to functionalized graphene.[22]

5.2.1.4 *Friedel–Crafts acylation*

Friedel–Crafts acylation is an electrophilic substitution reaction to introduce aryl ketone groups onto aromatic rings (Fig. 5.15(a)). A typical Friedel–Crafts acylation reaction goes through an intermediate of an acyl anion catalyzed by the presence of a Lewis acid such as aluminum chloride and phosphorus pentoxide. The acyl anion electrophilically attacks the graphene C=C bond to graft a ketone derivative onto the graphene.

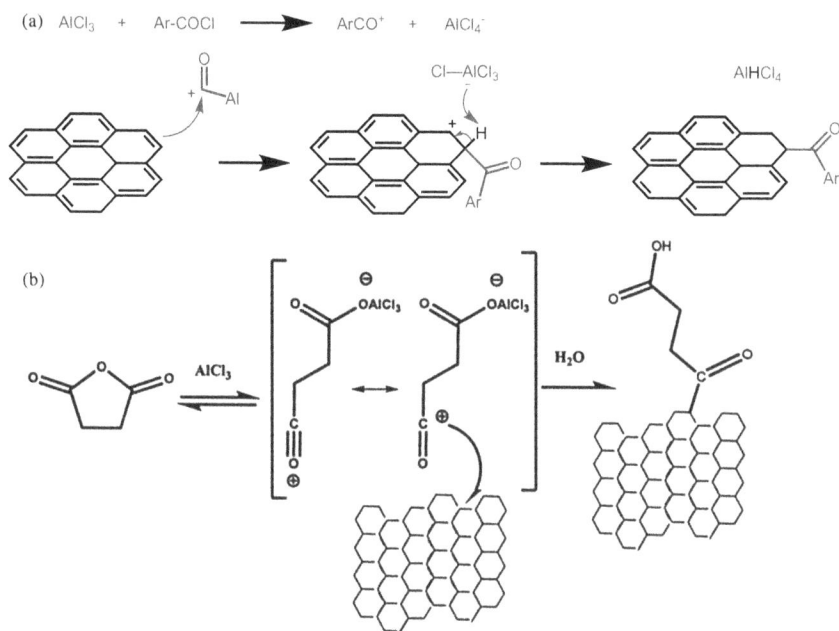

Figure 5.15. (a) Mechanism[2] and (b) synthetic route[23] of the functionalization of graphene via Friedel–Crafts acylation.

It is found that most derivatization occurs at the edges of the graphene. As shown in Fig. 5.15(b), the graphene grown on Si/SiO_2 by the CVD method is immersed in a cyclohexanone solution of succinic acid, and then aluminum trichloride is added.[23] After the mixture is allowed to stand at 60°C for 24 h, it is washed and dried to obtain functionalized graphene whose edges are derivatized with succinic acid.

5.2.1.5 *Hydrogen–lithium exchange*

Under the action of strong hydrogen-abstracting reagents (such as *n*-butyllithium), the hydrogen on the aromatics can be replaced by metals to produce organic compounds containing carbon-metal bonds, such as Ar-Li. After metallization, aromatics exhibit more reactivity and nucleophilicity. In the presence of electrophiles, the metalated aromatics thus react readily to form a covalent bond, resulting in the extension of the aromatic ring (Fig. 5.16(a)). This reaction can also be used for the

(a)

(b)

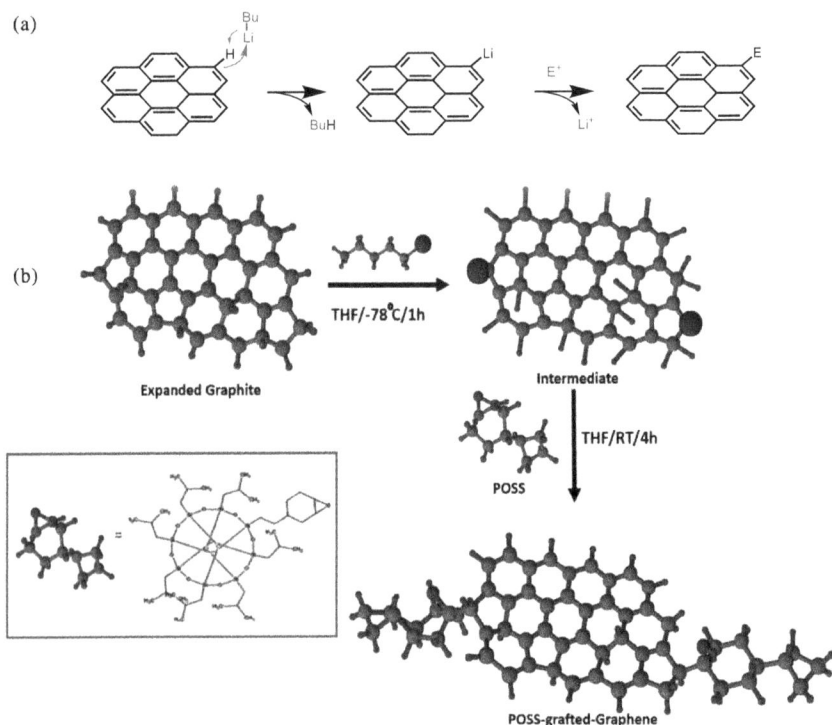

Figure 5.16.   Mechanism and illustration of functionalization of graphene by hydrogen–lithium exchange reaction: (a) mechanism of the hydrogen–lithium exchange leading to functionalized graphene[2] and (b) POSS grafting on the graphene surface based on this reaction.[24]

functionalization of graphene, such as the decoration of polyoligomeric silsesquioxane (POSS) on the surface of graphene (Fig. 5.16(b)).[24] Graphite in anhydrous THF is kept at 78°C under stirring, and nitrogen atmosphere is maintained in the reaction vessel, after which *n*-butyllithium is introduced. After the mixture is stirred for 1 h, epoxycyclohexylethyl-POSS is introduced. The reaction mixture is stirred for 4 h at room temperature, and the POSS-modified graphene is obtained after filtration, washing, and drying. During this process, treatment of *n*-butyllithium with graphene exclusively scavenges protons from the defect sites, and this anionic center is made to react with an epoxide ring containing POSS moiety, resulting in covalent bond formation.

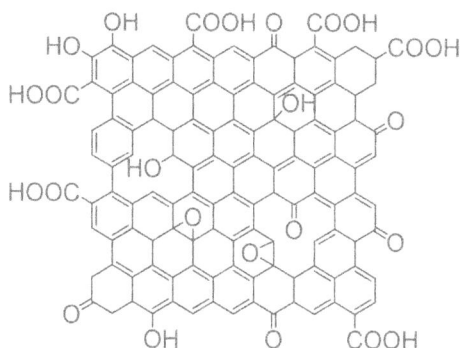

Figure 5.17.   Schematic representation of GO.

5.2.2 *Covalent functionalization of GO*

GO has a 2D aromatic structure similar to graphene, but its chemical properties are more active than those of graphene, which can potentially provide great opportunities for further modifications. In general, GO prepared by Hummers' method has a large number of oxygen-containing functional groups on the surface and edges. On the basal planes, there are both hydroxyl and epoxide groups; the edges can include carboxyl, carbonyl, phenol, lactone, and quinone groups (Fig. 5.17). The presence of these oxygen-containing functional groups damages the conjugated structure of graphene, resulting in the loss of its inherent properties, such as a significant decrease in electrical conductivity; however, it provides a large number of reactive sites for the functionalization of graphene. In this context, as epoxide, hydroxyl, carboxyl, and other moieties are abundant on the surface and edge of GO, most functionalization strategies target these functional groups.

5.2.2.1 *Epoxide opening*

In the covalent functionalization of GO, epoxide groups generally react with nucleophiles under alkaline conditions to form covalent bonds via ring opening. In the presence of alkaline, a nucleophile (such as amine and hydroxyl) is usually negatively charged. A nucleophilic attack at the electron-deficient carbon atom of an epoxide, occurring from the backside of the GO sheet, releases the ring strain and forms a new bond between the

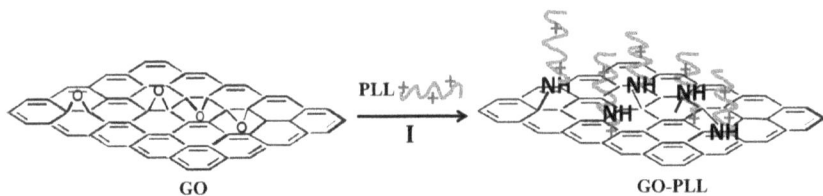

Figure 5.18. Preparation of peptide chain functionalized GO via nucleophilic ring-opening of epoxides.[25]

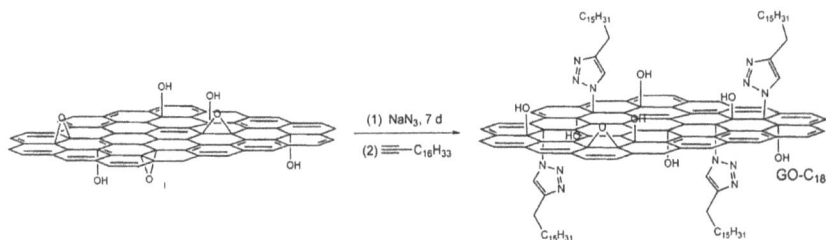

Figure 5.19. Illustration of functionalized GO via ring opening of epoxide groups to attach azides followed by azide click chemistry.[26]

nucleophile and the carbon atom of GO, in addition to the formation of a new hydroxyl group on the basal plane of GO. Amines, thiols, and the azide anion (N_3) are examples of nucleophiles that are highly reactive toward epoxides. As shown in Fig. 5.18, an amino-terminated peptide chain can be used as a nucleophile to attack the epoxides on GO in an alkaline solution, thereby resulting in the formation of a covalent bond and functionalization of a peptide chain onto the surface of GO.[25] In this study, GO is ultrasonically dispersed in deionized water, followed by the introduction of poly-L-lysine (PLL) and sodium hydroxide. Next, the pH of the mixture is adjusted to 9, and the mixture is stirred at 70°C for 24 h. Finally, PLL-modified GO (GO-PLL) is obtained by centrifugal washing.

The epoxides of GO can also be covalently bonded to the active groups, and then the functional molecules can be grafted onto the surface of GO through a reaction of the reactive groups with the functional molecules. As shown in Fig. 5.19, GO is dispersed by ultrasonication in a 1:1 solution of water/acetonitrile, NaN_3 is added, and the mixture is heated to reflux; then, an azide derivative of GO (GO–N_3) is obtained.[26] Next, a mixture of GO–N_3 and 1-octadecyne is stirred at room temperature, in which azide and alkyne groups undergo 1,3-dipolar cycloaddition

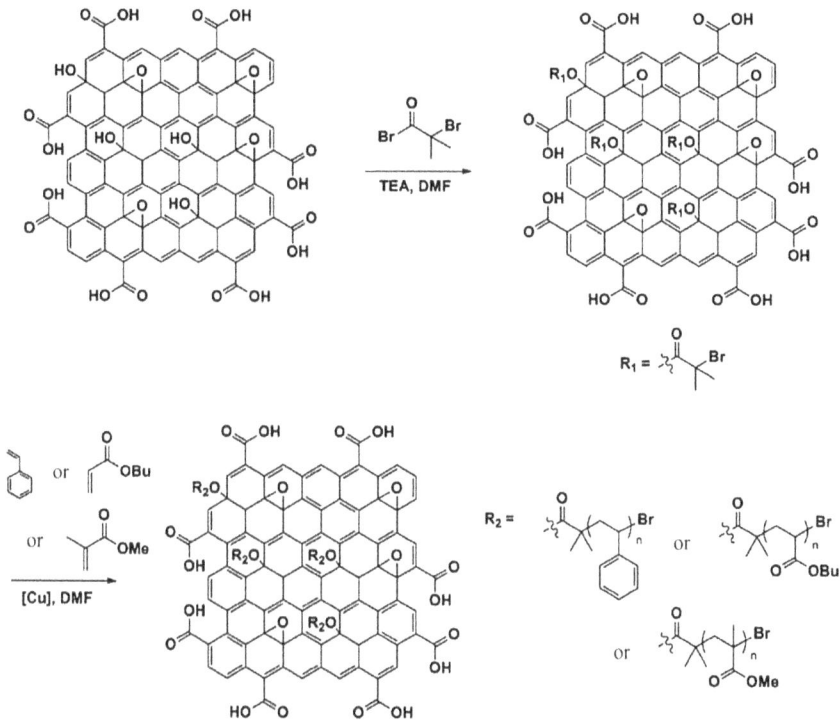

Figure 5.20. Synthesis of functionalized GO via attachment of an ATRP initiator on the hydroxyl group followed by polymerization of styrene, butyl acrylate, or methyl methacrylate.[27]

reactions (click reaction), after which functionalized GO (GO–C_{18}) is prepared. Due to the long alkyl chains on the surface, GO–C_{18} exhibits excellent solubility in various organic solvents.

5.2.2.2 *Hydroxyl group derivatization*

The hydroxyl groups on GO mainly react with electrophiles to form ether bonds via substitution reactions and also react with isocyanates to form urethane (carbamate) bonds. Through these two reactions, GO can be functionalized by grafting the functional molecules onto the hydroxyl moieties of GO.

As shown in Fig. 5.20, the attachment of α-bromoisobutyryl bromide onto GO is achieved by a substitution reaction of hydroxyl groups.[27] Next, a variety of different polymers are grafted onto the surface of GO through

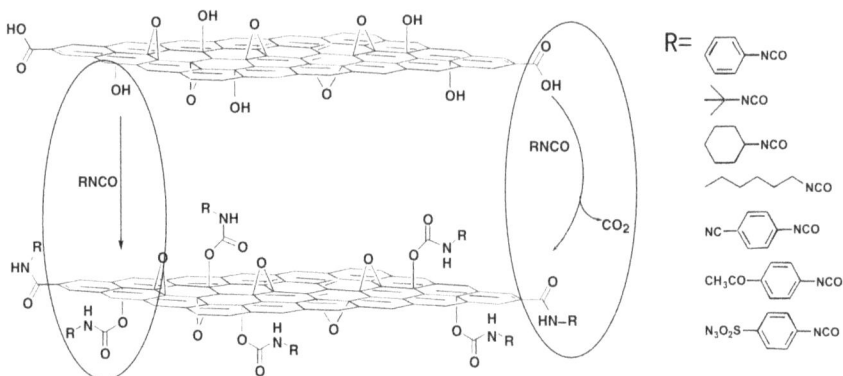

Figure 5.21.   Functionalization of GO by reaction between hydroxyl and isocyanate.[28]

the covalent attachment of an initiator (α-bromoisobutyryl bromide) followed by the polymerization of styrene, methyl methacrylate, or butyl acrylate using atom transfer radical polymerization (ATRP).

The use of isocyanates to react with hydroxyl groups is another strategy to prepare covalently functionalized GO. As shown in Fig. 5.21, organic isocyanates with different monomers are added to GO dispersion in DMF, and the mixture is stirred in a nitrogen atmosphere for 24 h.[28] The slurry reaction mixture is poured into methylene chloride to coagulate the product. Finally, different types of functionalized GO are prepared by filtration and washing. Due to the polar groups on the surface, the functionalized GO readily forms stable colloidal dispersions in polar aprotic solvents. It should be noted that isocyanates also react with carboxyl groups, releasing CO_2 to form amides, but the reactivity of the carboxyl group is lower than that of the hydroxyl group.

In addition, an electrosynthesis approach can be used to oxidize the catechol to *o*-quinone, which then reacts with hydroxyl groups on partially reduced GO (rGO) via a nucleophilic addition reaction, resulting in a GO/rGO composite.[29] This composite can be used as an electrode material for supercapacitors and exhibits excellent performance.

5.2.2.3 Carboxyl derivatization

The carboxyl derivatization strategies of GO mainly include the following: (i) amidation of carboxyl (or acyl chloride converted from carboxyl)

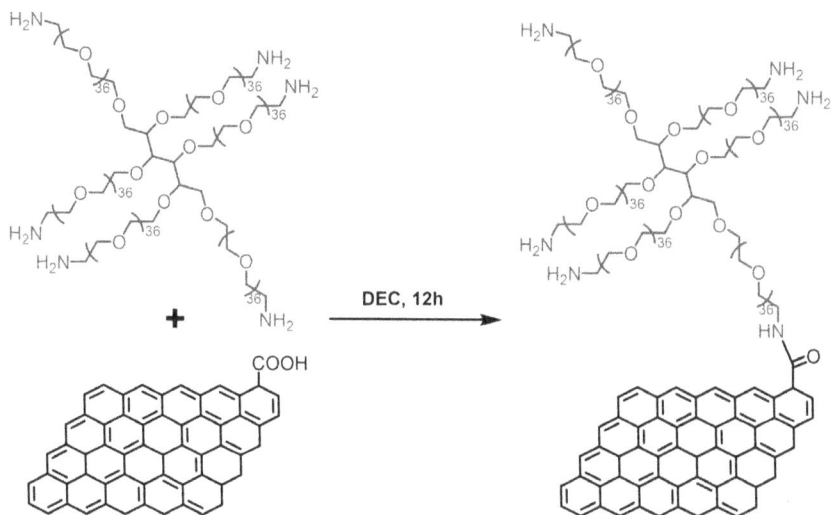

Figure 5.22. Schematic of functionalization of GO through amidation of carboxyl groups.

with amino, (ii) reaction of carboxyl with isocyanate, (iii) condensation of carboxyl and hydroxyl, and (iv) Friedel–Crafts acylation of carboxyl.

1. Amidation
As shown in Fig. 5.22, the carboxylic acid groups on GO are activated by N-(3-Dimethylaminopropyl-N'-ethylcarbodiimide hydrochloride (EDC) and react with a 6-arm polyethylene glycol (PEG)-amine to form amide bonds, and PEG is grafted onto the surface of GO.[7] The resulting functionalized GO exhibits excellent stability in all biological solutions tested including serum and cell medium, so it is expected to be used for the delivery of water-insoluble drugs *in vivo*. For example, a camptothecin (CPT) analog SN38, which is very insoluble in water, can be immobilized on the surface of the PEGylated GO by a non-covalent van der Waals interaction, thus rendering excellent stability in physiological solution.

In general, carboxylic acid groups on GO can be activated by thionyl chloride ($SOCl_2$) to produce the reactive intermediate, acyl chloride, for further condensation reactions with aminos. For example, $SOCl_2$ is employed to acyl chlorinate the carboxyl acid groups on the edge of GO.[30] Then, the resulting acyl chloride in GO reacts with stearylamine to obtain functionalized GO that could be dispersed in THF, carbon

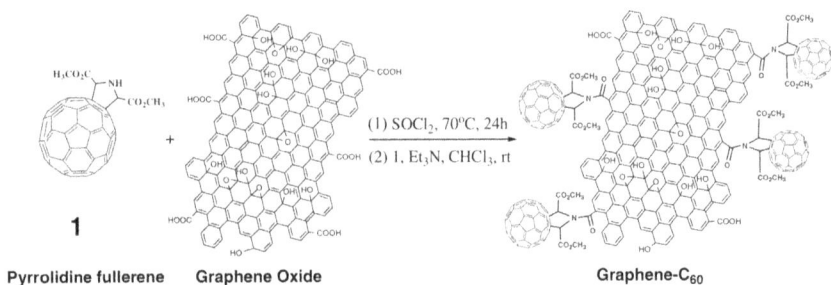

Pyrrolidine fullerene   **Graphene Oxide**   **Graphene-C₆₀**

Figure 5.23. Hybridization of GO with fullerene via a condensation reaction between carboxyl groups and amino groups.[30]

tetrachloride, and dichloroethane. Due to the higher reactivity of acyl chlorides, GO can be usually acyl chlorinated prior to further functionalization. As shown in Fig. 5.23, with the help of triethylamine, graphene/fullerene composites are synthesized by GO treated with SOCl₂ and pyrrolidine derivatives of fullerene in chloroform at room temperature via condensation reaction.

2. Reaction with isocyanate
As mentioned earlier, carboxyl groups can also react with isocyanates to form amides (Fig. 5.21).[28] However, isocyanates also react with hydroxyl groups on GO, making this strategy non-chemoselective.

3. Condensation reaction with hydroxyl groups
The carboxylic acids on GO can also condense with the hydroxyl groups of functional molecules to form ester groups, and the functional groups can be grafted onto GO. As shown in Fig. 5.24, the polyvinyl alcohol (PVA) is grafted to GO via a condensation reaction between the carboxylic acid groups on the GO and the hydroxyl groups at the end of the PVA. This reaction can be completed in two different synthetic strategies. The first one involves a condensation reaction of the carboxylic groups on GO and the hydroxyl groups on PVA under the catalysis of N,N-dicyclohexylcarbo-diimide (DCC) and 4-dimethylaminopyridine (DMAP), while the second one goes through acyl chloride derivative of GO carboxylic group to condense with PVA. The functionalized GO with PVA is soluble in DMSO and water with the aid of heat.

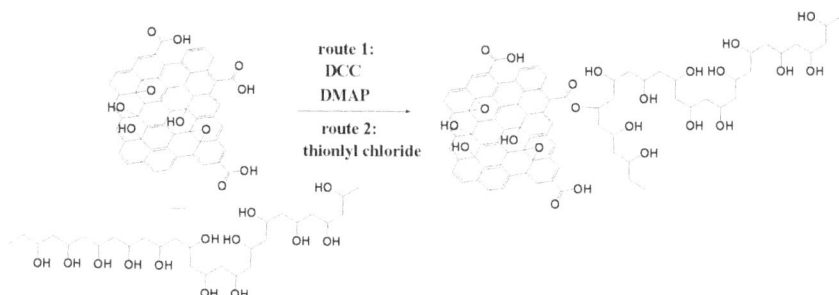

Figure 5.24. Functionalization of carboxylic groups and hydroxyl groups on GO with PVA via a condensation reaction.[31]

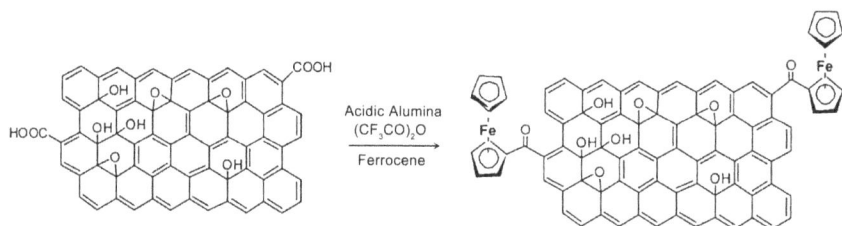

Figure 5.25. Functionalization of GO via Friedel–Crafts acylation.[32]

4. Friedel–Crafts acylation

The carboxylic group on GO can also be used as an acylating agent to react with other aromatic hydrocarbons via Friedel–Crafts acylation, thereby grafting other aromatic hydrocarbons on GO. As shown in Fig. 5.25, under the co-catalysis of acidic alumina and trifluoroacetic anhydride, the GO is covalently functionalized with ferrocene by a Friedel–Crafts acylation reaction.[32] Functionalized GO with ferrocene exhibits a unique magnetic behavior.

5.3 Non-covalent Functionalization of Graphene and GO

In addition to covalent functionalization, non-covalent functionalized graphene composites can also be synthesized by non-covalent functionalization of graphene and GO. The basal plane of graphene consists of 2D conjugated sp^2-hybridized carbon atoms, so other functional molecules can be easily bonded through π-π interaction to achieve functionalization; GO not only contains conjugated planes but also has hydroxyl, epoxide, and carboxyl groups. Therefore, GO can be functionalized not only through the π-π interaction but also through hydrogen bonding between the hydroxyl group and other functional molecules and the electrostatic interaction between the carboxyl group and other molecules. In addition, graphene and rGO exhibit a hydrophobic nature, so they can also be functionalized through hydrophobic interactions. The major advantages of non-covalent functionalization are simple operation, mild conditions, and non-destructive nature, which can preserve the intrinsic properties of graphene to the greatest extent.

5.3.1 π-π *interactions*

A delocalized conjugated system is established by sp^2-hybridized carbon atoms in graphene; thus, π-π interactions can be involved between graphene and functional molecules with the same π system. There are three limiting interaction geometries between two aromatic rings utilizing the benzene dimer as an example: face-to-face, slipped, and C−H$\cdots\pi$ (i.e., edge-to-face) arrangements.[33,34] In face-to-face stacking, the centers of the two aromatic rings are completely overlapping and the planes of the two aromatic rings are completely parallel; in slipped stacking, the centers of the two aromatic rings do not overlap and the planes of the two aromatic rings partially overlap; and finally, C−H$\cdots\pi$ interaction, also known as edge-to-face interaction, occurs when the edge of one aromatic ring (such as a benzene ring) is perpendicular to the plane of another one and the hydrogen atom on the edge of one aromatic ring interacts with the plane of another one. In functionalized graphene or GO composites modified by other molecules through π-π interactions, the π-π interaction between functional molecules and graphene or GO mainly occurs in these three ways. For example, a water-soluble graphene-phthalocyanine hybrid material is fabricated and can be applied for photothermal therapy (PTT) and photodynamic therapy (PDT).[35]

As shown in Fig. 5.26, graphene is added to a Copper (II) phthalocyanine−3,4′,4″,4‴−tetrasulfonic acid tetrasodium salt (TSCuPc) aqueous solution, after which sonication at 140 W power output is performed for 3 h. A final water-soluble functionalized graphene composite material is

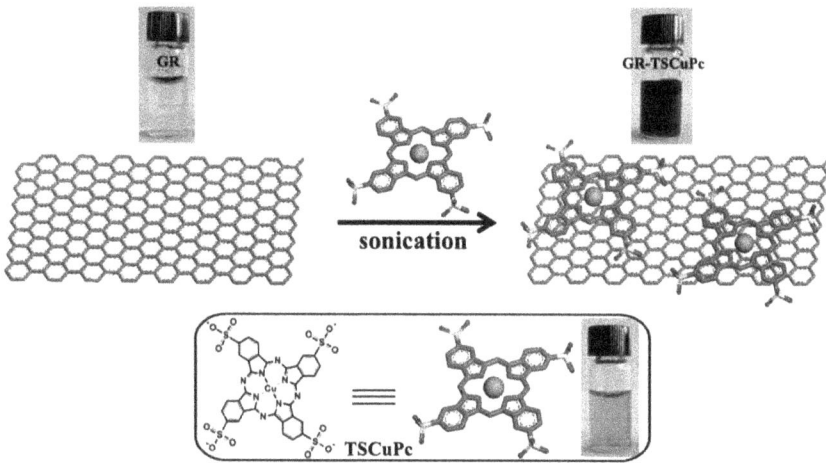

Figure 5.26. Schematic illustration of the synthesis of water-soluble phthalocyanine salt functionalized graphene via non-covalent π-π interactions.[35]

obtained after filtration, washing, and drying.[35] In this functionalized graphene composite, the water-soluble phthalocyanine salt is coated on the skeleton of pristine graphene via non-covalent π-π interactions.

Similarly, functionalization of GO can also be achieved by hybridization with functional molecules through π-π interactions. As shown in Fig. 5.27, GO is hybridized with hypocrellin via π-π interaction to prepare functionalized GO composites for PDT.[37] Hypocrellins, including hypocrellin A (HA) and hypocrellin B (HB), present high photodynamic activity toward many kinds of tumor cell lines. However, the clinical use of natural hypocrellins is severely limited by their poor water solubility. Hypocrellins tend to aggregate in blood plasma and block vascular networks after intravenous injection. If they are loaded on GO with excellent water solubility, the synthesized functionalized graphene composite material can be well dispersed in water and can be used for PDT. Hypocrellins are dissolved in water, using traces of DMSO as the latent solvent, and mixed with GO aqueous suspension at room temperature for 24 h with slow stirring. Then, the sample is ultracentrifuged, washed, and dried to obtain a hybrid of hypocrellins and GO. The results indicate that the large π-conjugated structure of GO can result in π-π stacking interaction with the benzene ring portion of hypocrellins.

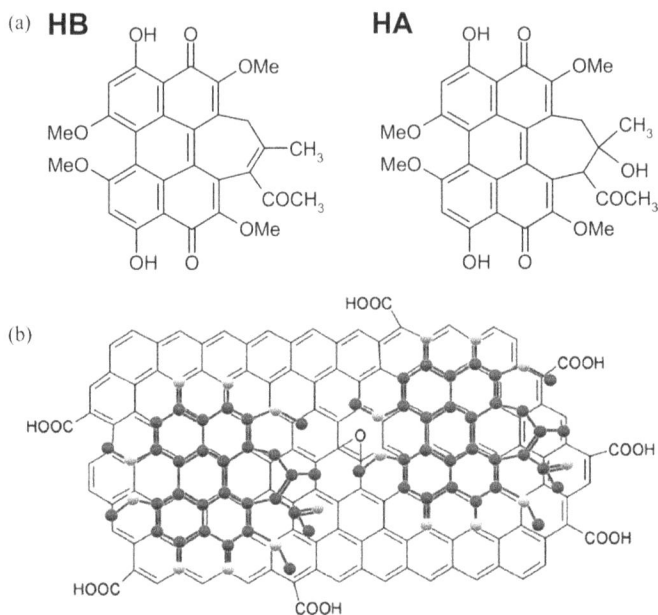

Figure 5.27.   Schematic representation of functionalization of GO with hypocrellins through π-π interaction.[37]

5.3.2 *Hydrogen bonding*

In chemistry, a hydrogen bond is a primarily electrostatic force of attraction between a hydrogen atom, which is covalently bound to a more electronegative "donor" atom or group, and another electronegative atom bearing a lone pair of electrons — the hydrogen bond acceptor. A large number of oxygen-containing groups such as epoxide, hydroxyl, and carboxyl groups are distributed on the surface of GO. These groups can act as both donors and acceptors of hydrogen bonds. Therefore, GO can be functionalized with other functional molecules by bonding with these oxygen-containing functional groups.[36]

As shown in Fig. 5.28, the hydroxyl groups of hydrophilic poly(vinyl alcohol) (PVA) can serve as either hydrogen-bond donors or acceptors, while the ester groups of hydrophobic poly(methyl methacrylate) (PMMA) can only serve as hydrogen-bond acceptors.[38] Therefore, although they have different hydrophilic/hydrophobic properties, they can be both used to synthesize nanocomposites by functionalizing GO via hydrogen

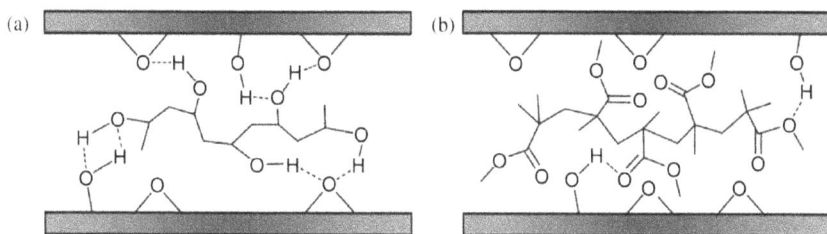

Figure 5.28. Schematic representation of functionalization of GO by (a) poly(vinyl alcohol) (PVA) and (b) poly(methyl methacrylate) (PMMA) via hydrogen bonding.[38]

bonding. The tensile strength and Young's modulus of PVA-functionalized GO are significantly improved, while the mechanical properties of PMMA-functionalized GO are also significantly improved. Hydrogen bonding plays a critical role in the mechanical properties of functionalized graphene composites. As another example, pH-responsive composites are prepared by functionalizing GO with PVA via hydrogen bonding.[39] In addition, polymers containing other hydrogen-bond donors/acceptors, such as polyaniline, polyamide, and polyurethane, can also be used to functionalize GO.[36]

5.3.3 *Electrostatic interaction*

GO can be stably dispersed in water thanks to electrostatic repulsions between the sheets with negatively charged oxygen-containing groups. Positively charged ions can be introduced to functionalize the GO via electrostatic interactions. As shown in Fig. 5.29, chitosan can functionalize GO through electrostatic interaction for the delivery of CpG oligodeoxynucleotides (ODNs).[40] Chitosan is dissolved in aqueous acetic acid and mixed with an aqueous dispersion of GO. After the mixture is ultrasonicated for 20 min, it is then stirred at room temperature for 2 h. Finally, chitosan-functionalized GO is prepared by centrifugation and washing. In this functionalized GO, chitosan is attached to the surface of GO through electrostatic interaction. Compared with GO, functionalized GO composites possess a smaller size, densely positively charged surface, and lower cytotoxicity and can be used as efficient nanocarriers for the delivery of CpG ODNs.

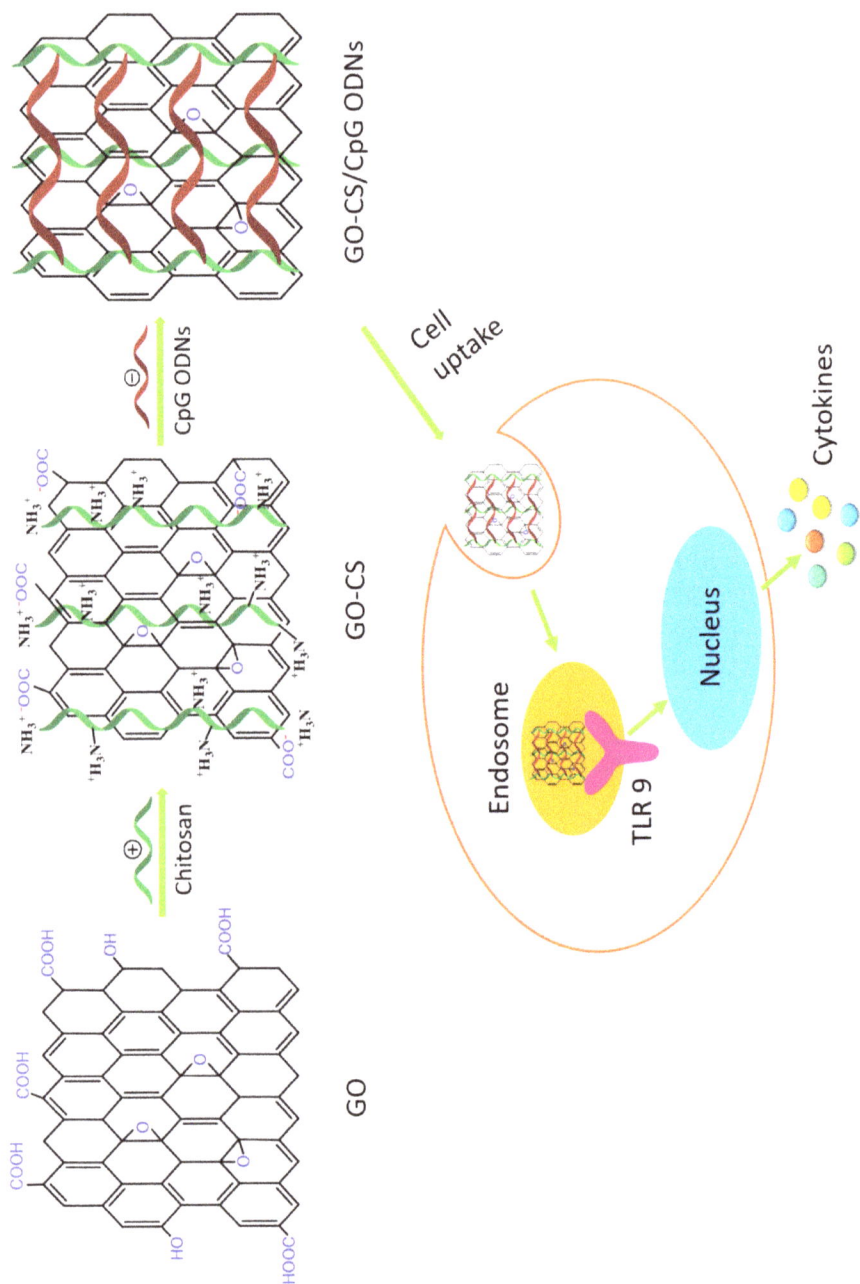

Figure 5.29.   Schematic illustration of chitosan-functionalized GO via electrostatic interaction as a carrier for intracellular delivery of CpG ODNs.[40]

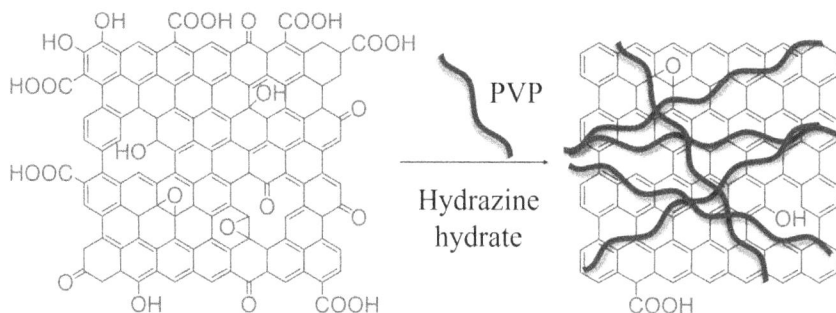

Figure 5.30. Schematic illustration of functionalization of rGO with poly(N-vinyl-2-pyrrolidone) via hydrophobic interaction.

5.3.4 *Hydrophobic interaction*

The sp^2-hybridized carbon atoms of graphene and rGO give rise to a hydrophobic basal plane. This hydrophobic conjugated plane is able to interact with neutral functional molecules to functionalize graphene or rGO. As shown in Fig. 5.30, water-soluble poly(*N*-vinyl-2-pyrrolidone) (PVP) can be used to functionalize rGO via hydrophobic interactions, and the product can be stably dispersed in water.[41] The PVP is mixed with aqueous GO solution, in which hydrazine hydrate is added, and the temperature is maintained at 80°C for 24 h. The GO solution is initially a bright brown color but gets darker as the reduction to graphene solution progresses. After the reaction, PVP-functionalized rGO is obtained. Due to the presence of water-soluble PVP, the prepared aqueous functionalized rGO solution is stable for several months without any aggregation or precipitation.

5.4 Summary

Controllable modifications with functional groups can effectively functionalize graphene and GO, allowing them to be easily processed and endowing them with additional properties to meet the needs of various applications. Controllable functionalization of graphene and GO with functional groups can be achieved through covalent and non-covalent approaches.

Covalently functionalized graphene and GO feature stable structures, multiple functionalities, and a wide variety of functionalization methods.

However, it is worth mentioning that the covalent functionalization of graphene compromises the sp^2 structure of graphene lattice, thus resulting in defects such as sp^3-hybridized carbon atoms. In this regard, it will jeopardize the intrinsic properties of graphene, such as electron mobility and high electric conductivity, resulting in a deterioration of its performance in applications. Therefore, in the process of functionalizing graphene, the route should be carefully optimized to avoid damaging the conjugate plane of graphene as much as possible. On the contrary, the non-covalent functionalization of GO is less problematic in this regard. The non-covalent functionalization of graphene and GO usually does not disrupt the extended π-conjugation and preserves their intrinsic properties, but the structure of non-covalently functionalized graphene materials is not very stable and it can easily dissociate by environmental influences. Therefore, it is necessary to choose a non-covalent approach according to the application scenario.

References

1. Tuček, J., Błoński, P., Ugolotti, J., Swain, A. K., Enoki, T., and Zbořil, R. (2018). Emerging chemical strategies for imprinting magnetism in graphene and related 2D materials for spintronic and biomedical applications, *Chem. Soc. Rev.*, 47(11), 3899–3990.
2. Chua, C. K. and Pumera, M. (2013). Covalent chemistry on graphene, *Chem. Soc. Rev.*, 42(8), 3222–3233.
3. Bekyarova, E., Itkis, M. E., Ramesh, P., Berger, C., Sprinkle, M., de Heer, W. A., and Haddon, R. C. (2009). Chemical modification of epitaxial graphene: Spontaneous grafting of aryl groups, *J. Am. Chem. Soc.*, 131(4), 1336–1337.
4. Wang, H.-X., Zhou, K.-G., Xie, Y.-L., Zeng, J., Chai, N.-N., Li, J., and Zhang, H.-L. (2011). Photoactive graphene sheets prepared by "click" chemistry, *Chem. Commun.*, 47(20), 5747–5749.
5. Sinitskii, A., Dimiev, A., Corley, D. A., Fursina, A. A., Kosynkin, D. V., and Tour, J. M. (2010). Kinetics of diazonium functionalization of chemically converted graphene nanoribbons, *ACS Nano*, 4(4), 1949–1954.
6. Lin, H., Xu, Z., Zhang, L., Yang, X., Ju, Q., Xue, L., Zhou, J., Zhuo, S., and Wu, Y. (2017). Diketopyrrolopyrrole derivative functionalized graphene for high performance visible-light photodetectors, *New J. Chem.*, 41(11), 4302–4307.
7. Liu, Z., Robinson, J. T., Sun, X., and Dai, H. (2008). PEGylated nanographene oxide for delivery of water-insoluble cancer drugs, *J. Am. Chem. Soc.*, 130(33), 10876–10877.

8. Pan, Y., Bao, H., Sahoo, N. G., Wu, T., and Li, L. (2011). Water-soluble poly(N-isopropylacrylamide)–graphene sheets synthesized via click chemistry for drug delivery, *Adv. Funct. Mater.*, 21(14), 2754–2763.

9. Liu, H., Ryu, S., Chen, Z., Steigerwald, M. L., Nuckolls, C., and Brus, L. E. (2009). Photochemical reactivity of graphene, *J. Am. Chem. Soc.*, 131(47), 17099–17101.

10. Criado, A., Melchionna, M., Marchesan, S., and Prato, M. (2015). The covalent functionalization of graphene on substrates, *Angew. Chem.*, 54(37), 10734–10750.

11. Ma, X., Li, F., Wang, Y., and Hu, A. (2012). Functionalization of pristine graphene with conjugated polymers through diradical addition and propagation, *Chem. Asian. J.*, 7(11), 2547–2550.

12. Sarkar, S., Bekyarova, E., and Haddon, R. C. (2012). Reversible grafting of α-Naphthylmethyl radicals to epitaxial graphene, *Angew. Chem.*, 51(20), 4901–4904.

13. Sainsbury, T., Passarelli, M., Naftaly, M., Gnaniah, S., Spencer, S. J., and Pollard, A. J. (2016). Covalent carbene functionalization of graphene: Toward chemical band-gap manipulation, *ACS Appl. Mater. Interfaces*, 8(7), 4870–4877.

14. Vadukumpully, S., Gupta, J., Zhang, Y., Xu, G. Q., and Valiyaveettil, S. (2011). Functionalization of surfactant wrapped graphene nanosheets with alkylazides for enhanced dispersibility, *Nanoscale*, 3(1), 303–308.

15. Choi, J., Kim, K.-j., Kim, B., Lee, H., and Kim, S. (2009). Covalent functionalization of epitaxial graphene by azidotrimethylsilane, *J. Phys. Chem. C*, 113(22), 9433–9435.

16. Sulleiro, M. V., Quiroga, S., Peña, D., Pérez, D., Guitián, E., Criado, A., and Prato, M. (2018). Microwave-induced covalent functionalization of few-layer graphene with arynes under solvent-free conditions, *Chem. Commun.*, 54(17), 2086–2089.

17. Zhong, X., Jin, J., Li, S., Niu, Z., Hu, W., Li, R., and Ma, J. (2010). Aryne cycloaddition: Highly efficient chemical modification of graphene, *Chem. Commun.*, 46(39), 7340–7342.

18. Wang, A., Yu, W., Xiao, Z., Song, Y., Long, L., Cifuentes, M. P., Humphrey, M. G., and Zhang, C. (2015). A 1,3-dipolar cycloaddition protocol to porphyrin-functionalized reduced graphene oxide with a push-pull motif, *Nano Res.*, 8(3), 870–886.

19. Daukiya, L. (2016). *Epitaxial Graphene Funtionalization: Covalent Grafting of Molecules, Terbium Interalation and Defect Engineering*, Doctor Thesis, Université de Haute Alsace.

20. Daukiya, L., Mattioli, C., Aubel, D., Hajjar-Garreau, S., Vonau, F., Denys, E., Reiter, G., Fransson, J., Perrin, E., Bocquet, M.-L., Bena, C., Gourdon, A., and Simon, L. (2017). Covalent functionalization by cycloaddition reactions of pristine defect-free graphene, *ACS Nano*, 11(1), 627–634.

21. Yuan, J., Chen, G., Weng, W., and Xu, Y. (2012). One-step functionalization of graphene with cyclopentadienyl-capped macromolecules via Diels–Alder "click" chemistry, *J. Mater. Chem.*, 22(16), 7929–7936.

22. Jin, B., Shen, J., Peng, R., Chen, C., Zhang, Q., Wang, X., and Chu, S. (2015). DMSO: An efficient catalyst for the cyclopropanation of C_{60}, C_{70}, SWNTs, and graphene through the bingel reaction, *Ind. Eng. Chem. Res.*, 54(11), 2879–2885.

23. Quiles-Díaz, S., Martínez, G., Gómez-Fatou, M. A., Ellis, G. J., and Salavagione, H. J. (2016). Anhydride-based chemistry on graphene for advanced polymeric materials, *RSC Adv.*, 6(43), 36656–36660.

24. Mondal, T., Bhowmick, A. K., and Krishnamoorti, R. (2014). Butyl lithium assisted direct grafting of polyoligomeric silsesquioxane onto graphene, *RSC Adv.*, 4(17), 8649–8656.

25. Ren, L., Zhang, Y., Cui, C., Bi, Y., and Ge, X. (2017). Functionalized graphene oxide for anti-VEGF siRNA delivery: preparation, characterization and evaluation in vitro and in vivo, *RSC Adv.*, 7(33), 20553–20566.

26. Salvio, R., Krabbenborg, S., Naber, W. J. M., Velders, A. H., Reinhoudt, D. N., and van der Wiel, W. G. (2009). The formation of large-area conducting graphene-like platelets, *Chem. Eur. J.*, 15(33), 8235–8240.

27. Lee, S. H., Dreyer, D. R., An, J., Velamakanni, A., Piner, R. D., Park, S., Zhu, Y., Kim, S. O., Bielawski, C. W., and Ruoff, R. S. (2010). Polymer brushes via controlled, surface-initiated atom transfer radical polymerization (ATRP) from graphene oxide, *Macromol. Rapid Commun.*, 31(3), 281–288.

28. Stankovich, S., Piner, R. D., Nguyen, S. T., and Ruoff, R. S. (2006). Synthesis and exfoliation of isocyanate-treated graphene oxide nanoplatelets, *Carbon*, 44(15), 3342–3347.

29. Jokar, E., Shahrokhian, S., and zad, A. I. (2014). Electrochemical functionalization of graphene nanosheets with catechol derivatives as an effective method for preparation of highly performance supercapacitors, *Electrochim. Acta*, 147, 136–142.

30. Zhang, X., Huang, Y., Wang, Y., Ma, Y., Liu, Z., and Chen, Y. (2009). Synthesis and characterization of a graphene–C_{60} hybrid material, *Carbon*, 47(1), 334–337.

31. Salavagione, H. J., Gómez, M. A., and Martínez, G. (2009). Polymeric modification of graphene through esterification of graphite oxide and poly(vinyl alcohol), *Macromolecules*, 42(17), 6331–6334.

32. Avinash, M. B., Subrahmanyam, K. S., Sundarayya, Y., and Govindaraju, T. (2010). Covalent modification and exfoliation of graphene oxide using ferrocene, *Nanoscale*, 2(9), 1762–1766.

33. Hunter, C. A., Lawson, K. R., Perkins, J., and Urch, C. J. (2001). Aromatic interactions, *J. Chem. Soc., Perkin Trans. 2*, (5), 651–669.

34. Wheeler, S. E. (2011). Local nature of substituent effects in stacking interactions, *J. Am. Chem. Soc.*, 133(26), 10262–10274.
35. Jiang, B.-P., Hu, L.-F., Wang, D.-J., Ji, S.-C., Shen, X.-C., and Liang, H. (2014). Graphene loading water-soluble phthalocyanine for dual-modality photothermal/photodynamic therapy via a one-step method, *J. Mater. Chem. B*, 2(41), 7141–7148.
36. Georgakilas, V., Tiwari, J. N., Kemp, K. C., Perman, J. A., Bourlinos, A. B., Kim, K. S., and Zboril, R. (2016). Noncovalent functionalization of graphene and graphene oxide for energy materials, biosensing, catalytic, and biomedical applications, *Chem. Rev.*, 116(9), 5464–5519.
37. Zhou, L., Jiang, H., Wei, S., Ge, X., Zhou, J., and Shen, J. (2012). High-efficiency loading of hypocrellin B on graphene oxide for photodynamic therapy, *Carbon*, 50(15), 5594–5604.
38. Putz, K. W., Compton, O. C., Palmeri, M. J., Nguyen, S. T., and Brinson, L. C. (2010). High-nanofiller-content graphene oxide–polymer nanocomposites via vacuum-assisted self-assembly, *Adv. Funct. Mater.*, 20(19), 3322–3329.
39. Bai, H., Li, C., Wang, X., and Shi, G. (2010). A pH-sensitive graphene oxide composite hydrogel, *Chem. Commun.*, 46(14), 2376–2378.
40. Zhang, H., Yan, T., Xu, S., Feng, S., Huang, D., Fujita, M., and Gao, X.-D. (2017). Graphene oxide-chitosan nanocomposites for intracellular delivery of immunostimulatory CpG oligodeoxynucleotides, *Mater. Sci. Eng.: C*, 73, 144–151.
41. Yoon, S. and In, I. (2011). Role of poly(N-vinyl-2-pyrrolidone) as stabilizer for dispersion of graphene via hydrophobic interaction, *J. Mater. Sci.*, 46(5), 1316–1321.

Chapter 6

Applications of Functionalized Graphene Materials

The modification of graphene with various functional groups with specific functions can endow graphene with rich and diverse properties so that graphene-based materials can meet the application requirements of different fields. The functional groups can be attached to graphene, GO, and rGO via covalent and non-covalent approaches (Fig. 6.1). The organic covalent functionalization of graphene includes two general routes: (a) the formation of covalent bonds between free radicals or dienophiles and C=C bonds of pristine graphene and (b) the formation of covalent bonds between organic functional groups and the oxygen groups of GO. The non-covalent interactions between functional groups and graphene/GO include hydrogen bonding, π-π interactions, hydrophobic interactions, and electrostatic interactions.

Functionalization is the best way to achieve the best performance out of graphene or GO and it reveals a great number of potential applications. In order to expand the number of potential applications of graphene-based materials, (functionalized) graphene can be hybridized with other functional materials to obtain functionalized graphene composites with multiple functionalities. Various functional materials (such as polymer, biological materials, metals, inorganic compounds, or organic materials) can be hybridized with (functionalized) graphene (graphene sheet, GO, rGO, and doped graphene) through one or more interactions to form composites with specific functions. As shown in Fig. 6.2, artificial nacre nanocomposites form a type of ordered layered structure hybridized by

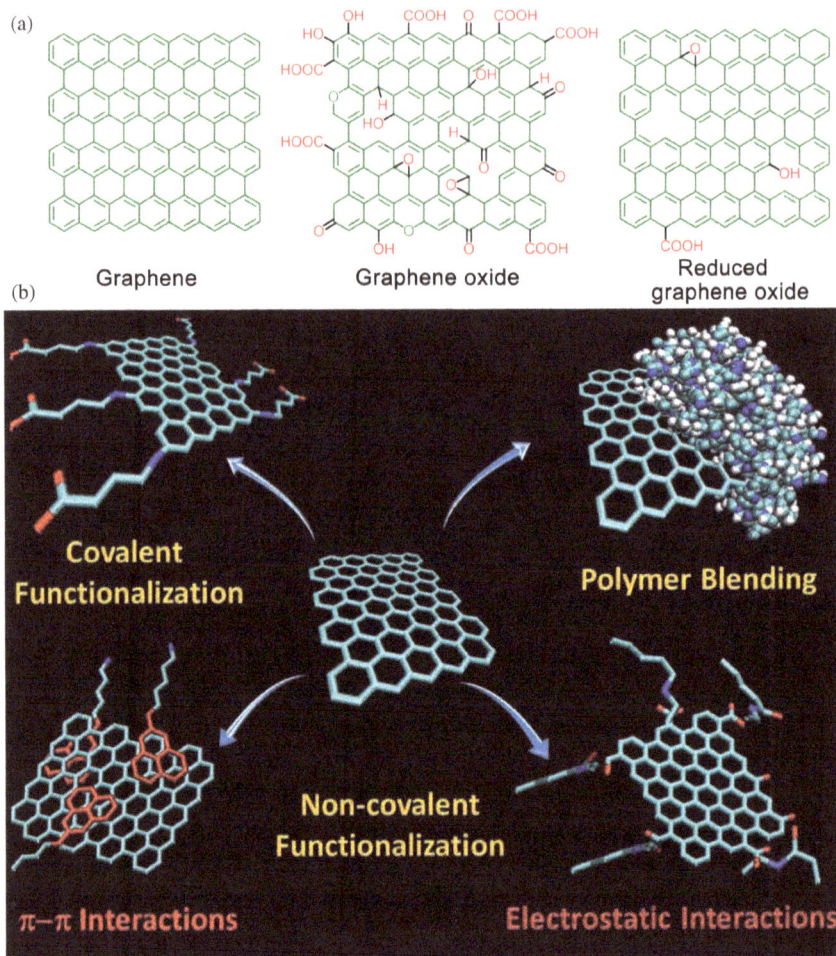

Figure 6.1. Schematic illustration of commonly used functionalized graphene and its further functionalized forms: (a) structure of graphene, GO, and rGO and (b) their functionalization via different interactions.[1]

(functionalized) graphene with a variety of functional materials. In such nanocomposites, graphene materials interact with functional molecules through more than one kind of interaction to form a nacre-like orderly layered structure. Functionalized graphene composites, which possess multifunctionalities, are also applicable in many fields. This chapter focuses on the applications of functionalized graphene materials

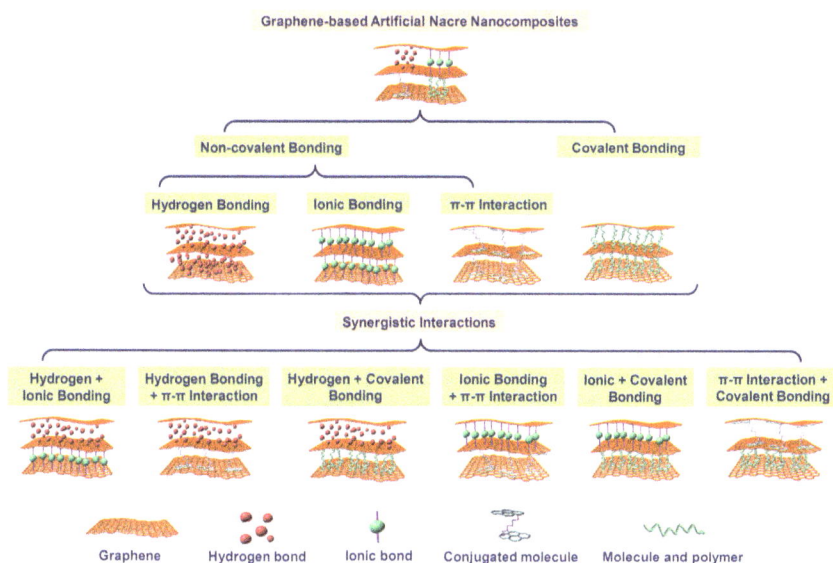

Figure 6.2.   Example of the interactions design for functionalized graphene composites: graphene-based artificial nacre nanocomposites.[2]

(functionalized graphene and functionalized graphene composite materials) in the fields of solar energy conversion, energy storage, biomedicine, environmental science, catalysis, electromagnetic shielding, and seawater desalination. The structures, fundamental operation mechanisms, and crucial issues of performance improvement of key devices in each field are introduced, and the functions and operation mechanisms of functionalized graphene materials in devices are discussed.

6.1 Solar Energy Conversion

Solar energy is a renewable and clean energy source. The radiation from the sun to the earth is immense, and the average energy received by the earth per square meter is about 1360 W. The effective use of solar energy can increase the share of clean energy in the total energy used by society and reduce the pollution generated during the utilization of other energy sources. Solar energy can be converted into electrical energy and chemical energy through photovoltaic devices and photocatalytic reactors. These two solar energy conversion devices both utilize electron–hole pairs (excitons) generated by photon-absorbing semiconductors or conjugated polymers. The electrons and holes flow to the external circuit to generate electrical energy, or the electrons and holes flow to the active surface and react with the corresponding molecules to generate chemical energy.

In order to effectively utilize solar energy, it is necessary to improve the efficiency of solar light collection and device conversion so as to improve the efficiency of solar conversion. The following three major factors affect energy conversion efficiency: (i) the generation of excitons (electron–hole pairs); (ii) the separation of electrons and holes; and (iii) the separation and transfer of electrons and holes, that is, electrons and holes migrating from photon absorbers to external circuits or active surfaces. The generation of excitons depends on the optical and electronic properties of the photon absorber. A good photon absorber is able to absorb more sunlight and generate a sufficient number of excitons. After the excitons are generated, if they are not extracted in time and migrated to the external circuit or the active surface, they will recombine in the device. Therefore, the structure and composition of the devices need to be optimized to efficiently extract and transport electron–hole pairs.

Due to its unique photoelectric and mechanical properties, graphene has shown great application potential in the field of solar energy conversion. Single-layer graphene exhibits excellent light transmittance (about 97.7% at 550 nm) and electrical conductivity (about 10^8 S/m); as a transparent electrode, it is an ideal material that can increase the solar conversion efficiency of devices. The Fermi level and Dirac point of single-layer graphene coincide, and it is a semi-metal with zero band gap [Fig. 6.3(a)];

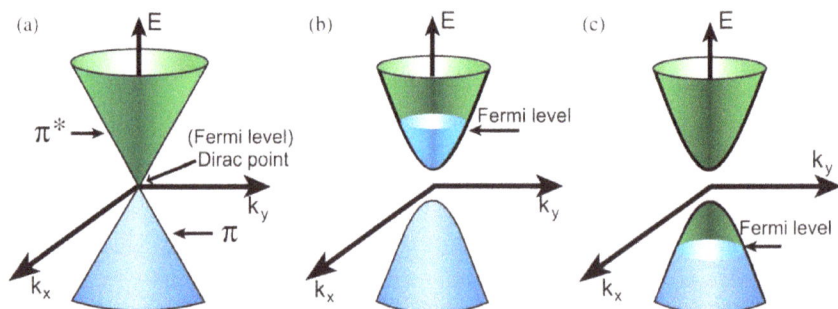

Figure 6.3. Band structure of graphene[3]: (a) approximation of the low-energy band structure of pristine graphene with two cones touching at a Dirac point; energy band structure of (b) *n*-type and (c) *p*-type graphene with a bandgap.

however, when graphene is functionalized, *n*-type graphene with a band gap (the Fermi level is at the bottom of the conduction band, which is higher than the Dirac point) and *p*-type graphene (the Fermi level is at the top of the valence band, which is lower than the Dirac point) can be obtained, which are shown in Figs. 6.3(b) and 6.3(c), respectively. Graphene and functionalized graphene materials with unparalleled properties can meet a variety of requirements for photovoltaic and photocatalytic applications and have been extensively used for solar energy conversion, for example in organic solar cells (OSCs), dye-sensitized solar cells, quantum dot solar cells, and photocatalytic cells.

6.1.1 *Organic solar cells*

Solar cells, also called photovoltaic cells, form a kind of device that converts solar energy into electricity by absorbing sunlight. At present, large-scale commercial solar cells are made of monocrystalline, polycrystalline, or amorphous silicon, and their conversion efficiency is close to 20%. Compared with silicon solar cells, OSCs hold tremendous potential in the photovoltaic market owing to a number of advantageous features, including their utilization of efficient solution processes, low manufacturing requirements, and flexible nature. With the development of fundamental theory, material technology, and device preparation technology, the conversion efficiency of OSCs is constantly improving. In 2018, OSCs with a lamellae structure prepared by Chen's group at Nankai University achieved a record energy conversion efficiency (17.3%).[4] Currently, OSCs comprise the following layers: cathode, hole transport layer, donor material, acceptor material, electron transport layer, and anode. The fabrication of OSCs can be divided into three types: single-layer Schottky, bilayer heterojunction, and bulk heterojunction (Fig. 6.4). Among them,

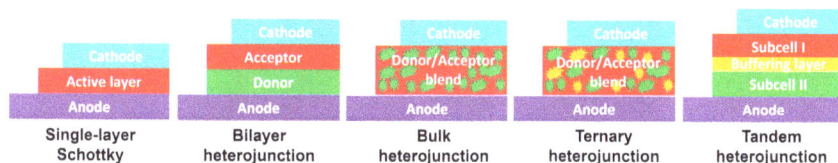

Figure 6.4. Schematic of the fabrication of OSCs.

bulk heterojunctions facilitate the separation and diffusion of excitons to the interface and achieve high conversion efficiency compared with planar heterojunctions, and they occupy a dominant position in current research. In order to further improve the performance of OSCs, researchers have developed ternary heterojunctions (i.e., a ternary blend of two acceptors and one donor) and tandem heterojunctions (i.e., two cells are connected in series) on the basis of bulk heterojunctions.

As shown in Fig. 6.5, when the OSC device is irradiated with light, the donor material absorbs light and photoexcitation occurs to generate excitons (electron–hole pairs). Exciton at the interface will separate to create free electrons and holes due to the energy level difference. At the donor/acceptor interface, electrons are extracted by the acceptor; at the donor/anode interface, the holes are filled by electrons from the anode, that is, the holes are transferred to the anode. Free carriers will be created that result in the separation of the electron and hole, and they accumulate on the corresponding electrodes and migrate to the external circuit. In order to enhance the transport properties of holes and electrons to the cathode and anode, respectively, a hole transport layer is usually deposited between the donor and the anode, and an electron transport layer is placed between the acceptor and the cathode. Graphene has been studied extensively in OSCs in an effort to increase charge transport and overall device efficiency, and it can be used as a transparent electrode (anode and cathode), acceptor, hole transport layer, and electron transport layer.

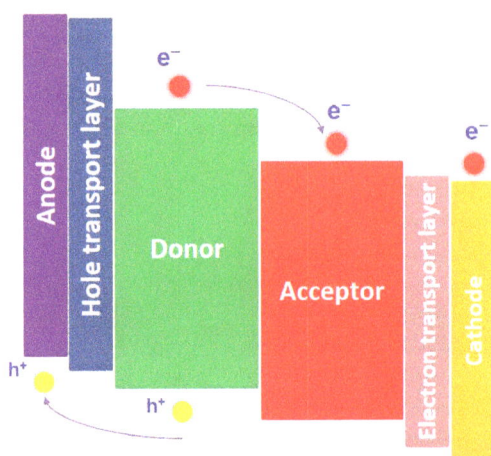

Figure 6.5.   Energy level diagram demonstrating the operation of OSCs.

6.1.1.1 *Anodes*

The transparent electrodes are inevitable in modern OSCs, and an ideal transparent electrode is required to have the following characteristics: (i) high optical transparency (>80%) that does not affect the absorption of incident light by active materials, (ii) high conductivity (area resistance < 100 Ω/sq), and (iii) appropriate work function to match the HOMO of the donor or the LUMO of the acceptor to lower the energy barrier for hole or electron conduction. The work function of pristine graphene is ~4.5 eV, which cannot match the HOMO of many donor-type organic semiconductors (~5.0 eV, the work function of the anode must be close to the HOMO of the donor). This introduces a large hole–injection barrier between the graphene and the donor materials, which hinders the conduction of holes from the acceptor to the graphene anode. Therefore, it is necessary to have *p*-type doping in graphene in its applications as an anode. In addition, compared with the commercial anode material ITO, the electrical conductivity of graphene material is relatively low. In order to improve the performance of devices based on graphene anodes, it is also necessary to improve the conductivity of graphene. In order to improve the work function and conductivity of graphene, *p*-type doping of graphene can be carried out by doping with electron acceptors. 7,7,8,8-tetracyanoquinodimethane (TCNQ) is a commonly used electron acceptor, which has a strong electron-withdrawing ability and is able to withdraw electrons from graphene to provide *p*-type doping. Single-layer graphene prepared by the CVD method is used to fabricate a stacked graphene/TCNQ/graphene/TCNQ/graphene composite (G/TCNQ/G/TCNQ/G) by a layer-by-layer molecular doping process (Fig. 6.6).[5] In this composite, TCNQ embedded

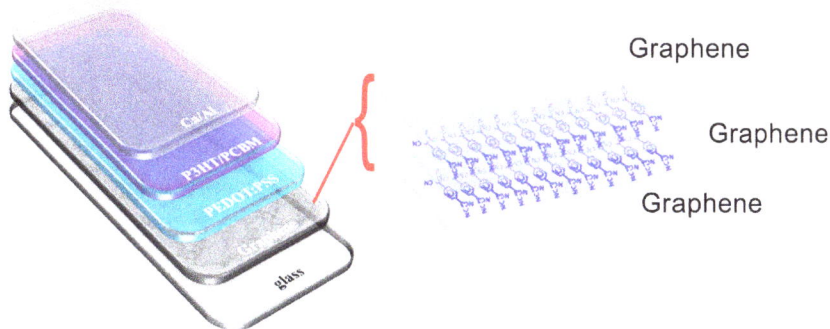

Figure 6.6. Scheme for OSC device with TCNQ-doped graphene as an anode and illustration of anode structure.[5]

between two adjacent graphene layers plays a role in the *p*-type doping of graphene, thereby increasing the hole carrier concentration of graphene. After doping, the work function of single-layer graphene prepared by this CVD method increases from 5.0 eV to 5.2 eV. These values match well with the work function of the commonly used hole transporter (PEDOT:PSS), suggesting that the work function of the G/TCNQ stacked films is tunable and suitable for anode applications. At the same time, the sheet resistance of single-layer graphene decreases from 839 Ω/sq to 278 Ω/sq, indicating significantly improved conductivity of the graphene composite. Thanks to the improvement of work function and conductivity after doping, the energy conversion efficiency of OSCs using the G/TCNQ/G/TCNQ/G composite as anode is significantly higher than that of single-layer graphene-based devices (2.58% vs 0.45%).

6.1.1.2 *Cathodes*

A graphene cathode in an OPV needs both high conductivity and an appropriate work function that should be close to the LUMO of the *n*-type organic semiconductor in the device to form an Ohmic contact for the collection of electrons. However, due to physically absorbed oxygen and water during the graphene fabrication and transfer process, pristine graphene often shows *p*-type doping with a work function (>4.5 eV) higher than the LUMO levels of normally used organic acceptors, such as PCBM (LUMO: ~4.2 eV). Thus, *n*-type doping is necessary for graphene cathodes to decrease their work function, for example, a single-layer graphene cathode modified by the Al-TiO$_2$ composite (TiO$_2$-Al-G).[6] When thin Al nanoclusters (~0.5 nm) are evaporated onto the single-layer graphene prepared by the CVD method, the work function decreases from 4.6 eV to 4.1 eV, which matches the LUMO of the acceptor, that is, [6,6]-phenyl-C61-butyric acid methyl ester (PCBM, 4.2 eV). In addition, the surface wettability of single-layer graphene is greatly improved after depositing aluminum nanoclusters: The contact angle is reduced from 95.7° to 48.0°; subsequently, the TiO$_2$ can be readily deposited onto the Al-G composite as an electron transport layer, which in turn effectively enhances the electron transport of the graphene composite cathode. The energy conversion efficiency of the OSCs using TiO$_2$-Al-G cathode is 2.58%, reaching 75% performance of control devices using the ITO cathode and identical Al-TiO$_2$ composite interface layer.

6.1.1.3 *Acceptors*

In OSCs, the acceptor material needs to have an appropriate LUMO level to match the HOMO level of the donor, thus offering a higher open-circuit voltage (V_{OC}). Meanwhile, the contact area of the acceptor and donor material needs to be as large as possible, and the acceptor is expected to adsorb light energy, so as to improve the light absorption capacity of the device. To date, the most efficient OSCs are based on the bulk heterojunction structure employing fullerene derivatives such as PCBM as electron acceptors. Over the past several years, although fullerene derivatives have been extensively used as electron-acceptor materials for OSCs, their various intrinsic limitations, such as poor absorption in the visible-light region, a difficult-to-adjust molecular structure, and morphological instability, have impeded further development of OSCs. To circumvent this constraint, considerable progress has been achieved recently due to the development of non-fullerene acceptors (NFAs) for high-performance non-fullerene OSCs, as NFAs have high absorption coefficients and suitable frontier molecular orbital energy levels that facilitate both the harvesting of solar photons and charge separation.

Functionalized graphene materials have emerged as new electron acceptors owing to their high electron mobility and large surface area for donor/acceptor interfaces. In 2008, Liu *et al.* first fabricated OPVs by

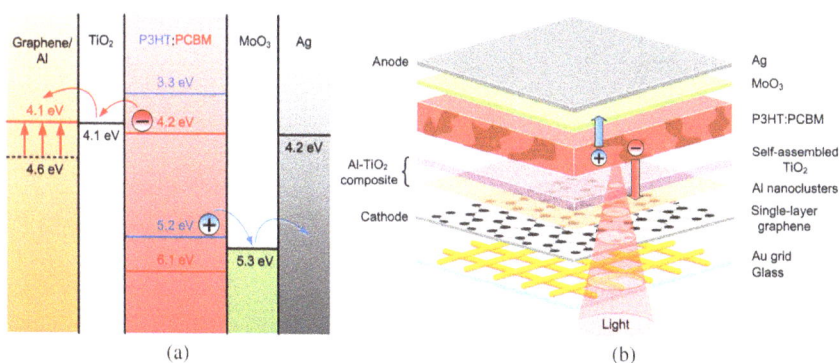

Figure 6.7.   (a) Flat-band energy diagram of OSCs using the graphene cathode modified by the Al-TiO$_2$ composite and (b) device structure of inverted OSCs using the graphene cathode shown in (a).[6]

using phenyl isocyanate-functionalized GO as an acceptor (Fig. 6.7).[7] Abundant hydroxyl groups on graphene synthesized by the Hummers' method are used to react with phenyl isocyanate to obtain a phenyl-functionalized surface, so as to improve compatibility with the donor material, that is, poly(3-octylthiophene) (P3OT). The highly delocalized 2D π-electrons in these graphene composites interact strongly with the π-electrons correlated with the lattice of the conjugated polymer skeleton. Spectral studies show remarkably reduced photoluminescence and consequently efficient charge transfer along the interface between phenyl isocyanate-functionalized GO and P3OT. This indicates that phenyl isocyanate-functionalized GO can act as an acceptor material to receive the photoexcited electrons from P3OT. In addition, the graphene composite has absorption capacity in the near-infrared (NIR) region (650–2000 nm), suggesting the enhanced light-absorbing ability of the device. The energy conversion efficiency of the OSCs with a bulk heterojunction structure using this graphene composite as the acceptor is 1.4% (Fig. 6.8). Although the conversion efficiency is moderate, this pioneering work indicates that graphene shows great potential as acceptor material for

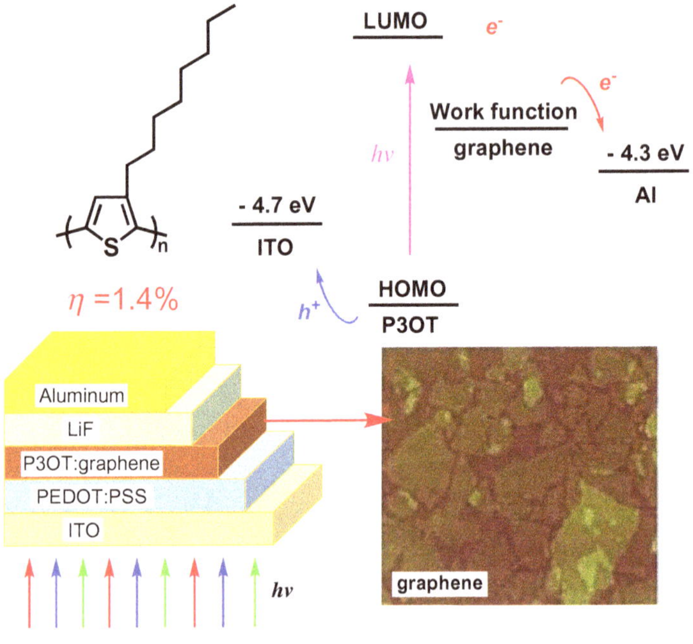

Figure 6.8. Schematic of structure and energy level diagram of OSCs using phenyl isocyanate functionalized GO as acceptor material.[7]

OSCs. In terms of OSC optimization and material development, devices with higher energy conversion efficiency have been fabricated; for example, OSCs with bulk heterojunction using 3,5-dinitrobenzoyl-functionalized graphene nanosheets as an efficient electron-cascade acceptor material exhibit an energy conversion efficiency of 6.59%.[8]

6.1.1.4 *Hole transport layers*

In OSCs, hole transport layers are used to improve the efficiency of donor materials in transporting holes to the anode. In the present study, the use of a hole transport layer is inevitable in high-efficiency OSCs, especially in bulk heterojunction cells. An ideal hole transport layer must be a *p*-type material with a wide band gap with the characteristics of high transparency, high conductivity, high stability, and chemical inertness. In OSCs, the hole transport layer can play the following roles: (i) adjusting the energetic barrier height between the active layer and electrode; (ii) transporting holes while blocking electrons; (iii) enabling the optimum morphology of the active layer; (iv) prohibiting a reaction between the active layer and electrode; and (v) acting as an optical spacer. At present, the most commonly used hole transport layer is the PEDOT:PSS polymer composite, which is coated on the surface of ITO by the solution method; it can match the energy level between the ITO and the donor. However, it is a highly acidic composite material that can erode the ITO anode and result in poor stability of the OSCs. To address this issue, new materials are needed to replace PEDOT:PSS as the hole transport layers in OSCs.

Graphene materials, especially GO-based materials, are found to be able to replace PEDOT:PSS as hole transport layer materials. GO is normally deposited by a neutral aqueous solution, which can avoid erosion of the ITO that occurs when PEDOT:PSS is coated. Moreover, the GO hole transport layers used in OSCs are much thinner and more transparent than their PEDOT:PSS counterparts, which is beneficial in improving the performance and reducing the volume of devices. However, it should be noted that due to the large electrical resistance of GO materials, the performance of OSCs with GO as hole transport layers is very sensitive to the thickness of the GO layer. When the thickness of the GO hole transport layers is greater than 3 nm, the performance of OSCs will rapidly decrease due to the high resistance of the GO layers. Therefore, in order to improve the conductivity of the GO hole transport layer, GO is commonly hybridized or derivatized with other conductive materials, such as GO/SWCNT

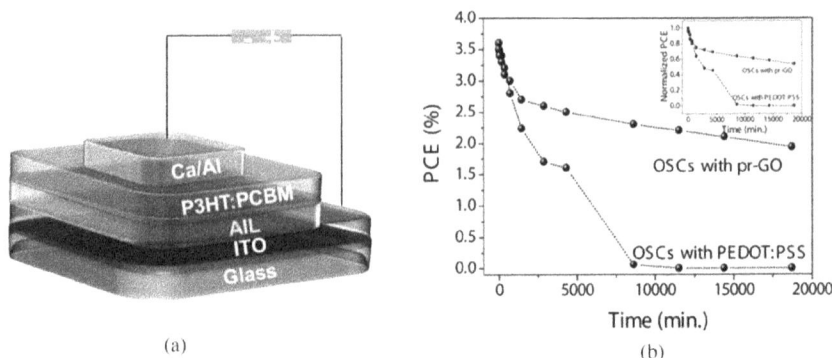

Figure 6.9. OSCs using rGO as the hole transport layer[9]: (a) device structure and (b) comparison of energy conversion efficiency (with OSCs using PEDOT:PSS as the hole transport layer).

composites, GO–OSO3H, or GO/PEDOT:PSS composites. In addition, GO can be reduced to rGO by chemical reduction or heat treatment to enhance its carrier conduction efficiency. For example, rGO is prepared by using a p-toluenesulfonyl hydrazide (p-TosNHNH$_2$) reductant, which is used as an anode interfacial layer (AIL), that is, the hole transport layer of OSCs [Fig. 6.9(a)].[9] As shown in Fig. 6.9(b), the overall photovoltaic characteristics of the cell with the rGO are greatly improved when compared to those of the cell with PEDOT:PSS, including the fill factor (FF), short-circuit current density (JSC), open-circuit voltage (VOC), average power conversion efficiency (PCE), and lifetime of the device. Furthermore, the work function of GO can be greatly increased to match the HOMO energy level of the donor material (typically greater than 5 eV) through derivatization, such as O$_2$ plasma treatment[10] or photochemical chlorination.[11]

6.1.1.5 *Electron transport layers*

In OSCs, an electron transport layer is often placed between the acceptor material and the cathode to improve the efficiency of the acceptor to transport electrons to the cathode, block the transport of holes to the cathode, and adjust the energy between the acceptor and the cathode. At present, the commonly used materials for electron transport layers are metal oxides, such as TiO$_x$ and ZnO$_x$. The commonly used methods to prepare metal oxide as an electron transport layer include hydrothermal synthesis, pyrolysis, and sol–gel process, all of which require high annealing temperatures to promote the crystallization of metal oxides; so, they are not suitable for the preparation of flexible devices. In order to fabricate OSCs on flexible substrates, it is necessary to develop new electron transport layers with high electron mobility, high transparency, and low operating temperature.

Functional graphene composites, such as TiO_2-rGO or ZnO-rGO, can serve as electron transport layers that can meet these requirements.[12] In the aqueous phase, GO is mixed with TiO_2 or ZnO, and then the TiO_2-rGO or ZnO-rGO composites can be prepared by heating at 180°C for reduction, followed by centrifugation and drying. In the preparation process of OSCs, the dispersion of these two composites can be directly spin coated on the acceptor material, and after heating and drying at 80°C, the TiO_2-rGO or ZnO-rGO composite for the electron transport layer can be obtained (Fig. 6.10(a)). Studies have shown that TiO_2 or ZnO is bonded covalently to the produced rGO during the thermal reduction of GO. In TiO_2-rGO or ZnO-rGO composites, a wide band gap of ~3.2 eV and a valence band of ~7.24 eV contribute to forming an efficient hole-blocking layer while the relatively well-matched energy levels of the TiO_2 and ZnO conduction bands match the work function of graphene, forming an efficient energy transfer layer for fast electron extraction [Fig. 6.10(b)]. The comparative experimental results show that the

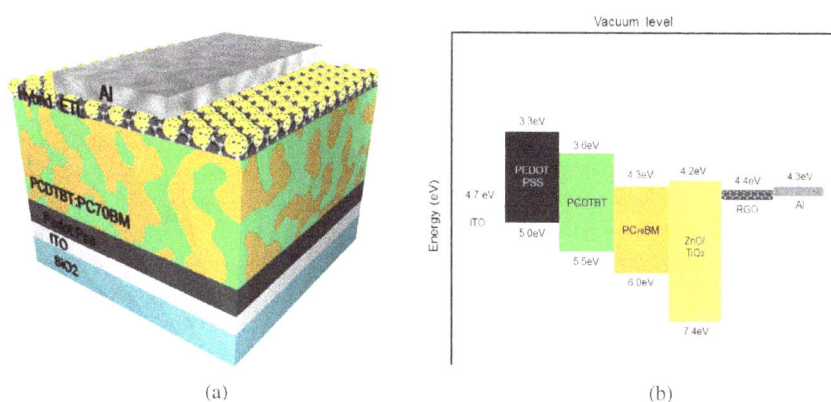

(a)   (b)

Figure 6.10.   OSCs using TiO_2-rGO or ZnO-rGO composite as electron transport layer[12]: (a) device structure and (b) band diagram for the materials used in OPV devices.

addition of rGO can effectively reduce the resistance of TiO_2 or ZnO materials for electron transport, thereby enhancing the performance of the devices.

6.1.2 *Dye-sensitized solar cells*

Dye-sensitized solar cells (DSSCs) are photoelectric conversion devices with high efficiency, simple structure, and low cost. The general configuration of DSSCs is shown in Fig. 6.11. They consist of current collectors (ITO or FTO conductive glass), photoanodes (semiconducting metal oxide, such as TiO_2), photo-sensitizers loaded on the photoanode (dye molecules), electrolytes with redox mediator (such as I^-/I^{3-}), and counter electrodes with catalytic activity (cathode, Pt, or Au).[13] When sunlight falls on a DSSC device, the present dye molecules on the surface of the TiO_2 layer (which behaves like an electron transport layer) absorb the incident photons and consequently excite the electrons. The excited electrons from the excited state of the dye are rapidly injected into the transparent current collector through nanoporous anode. Through the external circuit, electrons reach the counter electrode. The regeneration of the ground state of the dye takes place when S^+ oxidizes the liquid electrolyte of iodide, I^- to form triiodide, I^{3-} and neutralizes the state of sensitizer. I^- diffuses to the counter electrode (cathode) with catalytic ability, receives electrons from the external circuit from the cathode, and is reduced to I^-. Again, the oxidized mediator (I^{3-}) diffuses toward the counter electrode

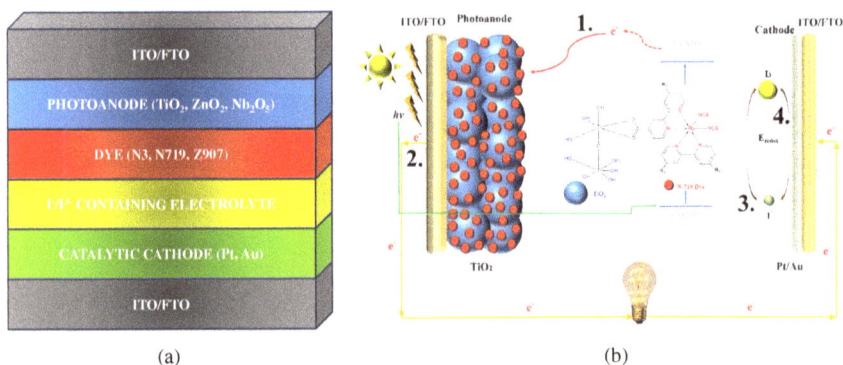

(a)	(b)

Figure 6.11. Schematic diagram of DSSC structure and schematic of the charge transfer reaction in DSSCs.[13]

with catalytic activity, receives electrons from the external circuit, and finally reduces to I⁻, thus completing the cycle.

Functionalized graphene materials can be utilized to optimize the components of the device, such as current collectors, anodes, dye molecules, electrolytes, and cathodes. Functionalized graphene materials can be incorporated to enhance the light absorption of DSSCs and reduce energy loss in the process of transporting electrons, leading to improved photoelectric conversion efficiency of DSSCs. In order to improve the collection efficiency of the light, one can start by improving the light transmittance of the photoanode and enhancing the light absorption of the dye molecules. Electron transport is governed by the redox mediators in DSSCs: The redox mediator in the electrolyte regenerates the oxidized dye molecules, and the oxidized form of the mediators is then regenerated at the cathode. Therefore, in order to improve the electron transport efficiency and reduce transmission loss in DSSCs, the device must meet the following requirements: (i) fast diffusion of the redox mediator between the working and counter electrodes; (ii) the regeneration process should be fast compared to the process of oxidation of the dye, resulting in a steady flow of electrons; and (iii) the reduction rate of redox mediators at the anode should be slow, while the reduction rate at the cathode should be fast. Otherwise, redox mediators will consume the electrons on the anode, resulting in a lack of electrons flowing to the external circuit, after which the electrons cannot be rapidly extracted from the cathode. In this regard, the device cannot efficiently receive electrons from the external circuit, resulting in the breakdown of the device.

6.1.2.1 *Photoanodes*

The application of functionalized graphene materials as photoanodes in DSSCs improves the electron transport from photoexcited dye molecules to the external circuit. Currently, TiO_2 is the most commonly used photoanode material; as an *n*-type semiconductor, it can transport the photoexcited electrons of the dye to the external circuit. A major challenge limiting the photoelectric conversion efficiency of DSSCs is the competition between charge transport of photogenerated electrons across the TiO_2 nanoparticle networks and charge recombination within the device. The electron–hole recombination is considered to significantly reduce the efficiency of the device. Therefore, methods such as modifying the morphology of TiO_2, doping to tune the band gap of TiO_2, and hybridization with carbon materials are commonly used to suppress the recombination of electrons and holes. Incorporating graphene into the TiO_2 layer to form an efficient route for charge extraction can significantly suppress recombination and improve charge transport. Nanocrystalline TiO_2 can anchor on the graphene flake compactly through physisorption and electrostatic interaction to form highly conductive bridges between adjacent TiO_2 particles. The interconnected graphene network acts as an electron transfer medium to rapidly shuttle electrons out of the TiO_2 photoanode, thus suppressing the recombination of photogenerated electron–hole pairs (excitons).[14] However, a strong van der Waals force exists on the surface of larger graphene flakes, which will cause them to aggregate in the composite, resulting in a decrease of TiO_2-graphene contact, thereby affecting their effectiveness as additives. By controlling the size and morphology of graphene nanomaterials, by cutting them into nanoribbons, one can address this problem. While retaining the excellent photoelectric properties of graphene, GNRs can be well dispersed in TiO_2 nanoparticles and are not easy to aggregate. Small amounts of GNRs are incorporated into TiO_2 mesoporous films to give TiO_2-GNR hybrid photoanodes. At the optimum GNR concentration (0.005 wt.%), the DSSCs yield the highest conversion efficiency of 7.18%, which is 20% higher than that of the control device made of a bare TiO_2 photoanode.[15] This is due to the fact that the hybridized photoanode can efficiently transfer the photoexcited electrons to the external circuit with the assistance of GNRs.

6.1.2.2 *Catalytic counter electrodes (cathodes)*

As mentioned earlier, DSSCs use a redox process of redox mediators to achieve charge transportation. A platinum counter electrode (cathode) catalyzes the reduction of I^{3-} to I^- after electron injection, thereby recycling the redox mediator. Frustratingly, the reduction of I^{3-} can also occur at the surface of TiO_2 nanoparticles, and the competing redox reaction reduces the overall device efficiency. Improving the electrocatalytic efficiency of the counter electrode by inhibiting competing redox reactions is a crucial step toward efficient DSSCs — lowering the rate at which the redox mediators are reduced at the anode or accelerating the reduction rate at the cathode. Therefore, the optimization or replacement of the commonly used Pt counter electrodes is an effective way to solve the problem. Functionalized graphene materials, such as heteroatom-doped and/or defect-rich graphene materials, have been demonstrated to exhibit excellent electrical conductivity and catalytic ability, which can efficiently reduce redox mediators. N-doped graphene nanosheets can be used as metal-free electrocatalysts for efficient reduction of $Co(bpy)_3^{3+}$ to replace the Pt counter electrode in DSSCs.[16] It is worth noting that the charge transport resistance of this N-doped graphene electrode (1.73 Ω cm^2) is much lower than that of the Pt electrode (3.15 Ω/cm^2), and its electrochemical stability is also better than that of the Pt electrode. The DSSCs with the optimized N-doped graphene counter electrodes have a higher FF of 74.2% and cell efficiency of 9.05%, whereas those of the DSSCs using Pt-CE are only 70.6% and 8.43%, respectively. N-doped-carbon materials can produce local strains in a hexagonal carbon network, thus leading to structural deformations; the additional lone pair of electrons of nitrogen atoms can bring negative charges with respect to the delocalized π-system of the sp^2-hybridized carbon framework, which can enhance the electron transfer ability and electrocatalytic activities. In addition, the defects of graphene can also serve as active sites for catalysis, but a large number of defects and a great amount of doping will reduce the conductivity of graphene. Therefore, it is necessary to optimize the structure of graphene materials to achieve a balance between high electrical conductivity and high catalytic activity.

6.1.2.3 *Graphene-based dyes*

Dyes are core components of DSSCs, responsible for the maximum absorption of the incident light. Light-absorbing dyes used for DSSCs should possess (i) a band gap capable of harvesting a majority of the solar spectrum, (ii) a conduction band energy that facilitates rapid charge injection to the anode, (iii) long-term photostability, and (iv) good adherence to the surface of the semiconducting anode. To date, ruthenium complexes are the most successful dyes. Although their energy conversion efficiency has already exceeded 10%, in order to further improve device performance, it is necessary to develop new dye molecules. As mentioned earlier, pristine graphene is a zero-bandgap material, but with chemical functionalization or doping, its semiconducting properties (*p*- or *n*-type) can be induced, making graphene a promising photoabsorber material for solar cells. As shown in Fig. 6.12, rGO modified with triphenylamine-based dyes (TPAs) can be used as a photosensitizer for DSSCs.[17] In this functionalized rGO, the TPA acts as the light-absorbing excitation unit to generate excitons, and then a fast quenching of the dye-excited state to

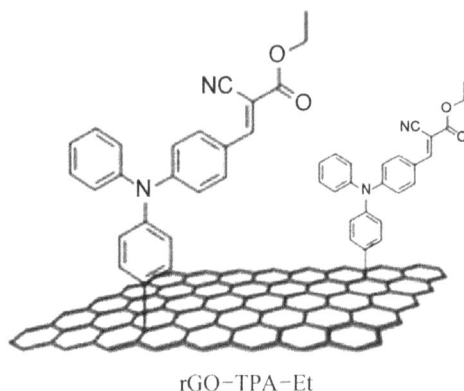

rGO-TPA-Et

Figure 6.12. Structure of TPA-modified rGO as a photosensitizer for DSSCs.[17]

the ground state is observed when bound covalently to rGO. The electrons generated by photoexcitation from the dye can be rapidly injected into the TiO$_2$ conduction band, resulting in efficient separation of electron–hole pairs.

6.1.2.4 *Electrolyte additive*

As one of the crucial components in DSSCs, the redox mediators in the electrolyte are in charge of providing electrons to the oxidized dye molecules and reducing them to the ground state, thus completing the regeneration of the dye molecules. Currently, the most commonly used electrolyte is in a liquid phase, but it has problems that are associated with volatile solvents and leakage, which destroy the long-term operational stability of DSSCs. Most importantly, the low diffusion of the charge carriers of the electrolyte hinders the fast reduction of dye molecules and the reduction of redox mediators on the cathode surface. Therefore, the electrolyte must be optimized to improve the performance of the device. Due to its extremely high electron mobility [200,000 cm^2/(V·s)], graphene can be added to electrolytes to improve the performance of DSSCs. Adding graphene nanofibers made from nanosheets to the electrolyte containing redox mediator Co(III)/(II) can greatly improve the charge transfer of the electrolyte and the electrocatalytic ability of redox mediators.[18] When graphene nanosheets are added to liquid electrolytes (redox mediator I$^-$/ I^{3-}), the resistance of the electrolyte is significantly reduced and the efficiency of DSSCs can reach 9.26%, which is 20% higher than that of the device with the non-graphene electrolyte solution. However, it should be noted that the amount of graphene should be controlled. An excessive amount of graphene will result in the aggregation of graphene in the electrolyte, which will reduce the conductivity of the electrolyte solution, thus degrading the performance of the device.

6.1.3 *Perovskite solar cells*

Perovskite solar cells (PSCs) have developed rapidly in recent years due to their large light absorption coefficient, high charge carrier mobility, long carrier diffusion length, and high efficiency. At present, the photoelectric conversion efficiency of PSCs has increased to an impressive value of 25.7% using novel fluorene-based hole transport materials.[19] Perovskite refers to an organic–inorganic hybrid material with the structure of a perovskite crystal whose molecular formula is ABX_3 (A = organic cation, B = Pb, Cd, or Sn, X = I, Cl, or Br). The structure of the PSC device includes a transparent electrode (usually conductive glass such as FTO), hole transport layer, perovskite material, electron transport layer, and metal electrode (Fig. 6.13). The working principle of PSCs is similar to that of OSCs, which also have active materials with perovskite structures that absorb solar irradiation (light) to excite electrons, resulting

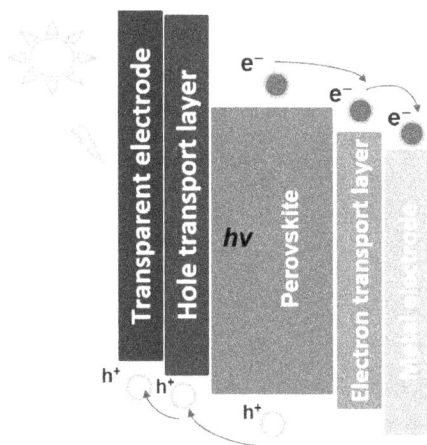

Figure 6.13.   Schematic of structure and working principle of PSCs.

in electron–hole pairs. The electrons are transferred to the metal electrode through the electron transport layer, and the holes are transferred to the transparent electrode through the hole transport layer. The perovskite crystal material possesses a high light absorption coefficient, a low manufacturing cost, and simple processing, which are beneficial in solar cells, but the material's poor stability, serious degradation at high temperatures, and potential metal contamination restrict its further development. To overcome these existing problems of PSCs and further improve their performance, it is necessary to optimize the device structure and materials. Similar to other types of photovoltaic devices, it is obvious that functionalized graphene materials can find important applications in PSCs and improve their performance; for example, graphene can be used in the electron transport layer, hole transport layer, and protective layer.

6.1.3.1 *Electron transport layers*

The electron transport layer is one of the important components of PSCs. The electron transport layer is responsible for extracting electrons from the perovskite layer and transporting them to the collecting electrode, while blocking hole transport, thereby suppressing charge carrier recombination and, hence, improving device efficiency. The electron transport layer material should (i) have a matching work function with the LUMO energy level of the perovskite crystal, (ii) exhibit a high electron mobility, (iii) effectively block hole transport to the metal electrode, and (iv) present excellent chemical compatibility with perovskite crystals and sound interfacial compatibility with metal electrodes. Due to its exceptional electron conductivity, graphene functional materials can be used to improve the performance of the electron transport layer. The electron transport layer synthesized by the addition of rGO in [6,6]-Phenyl-C_{61}-butyric acid methyl ester (PCBM) could significantly increase the conductivity of PSCs and improve the electron extraction from the perovskite to the metal electrodes, leading to higher short circuit current density (J_{sc}) and FF values [Fig. 6.14(a)].[20] At the same time, rGO stabilizes the PCBM/perovskite interface with the degradation rate significantly suppressed, making the rGO-doped devices highly stable under continuous solar illumination in ambient conditions. Functionalized graphene materials, such as graphene quantum dots (GQDs), can also be used to improve the performance of electron transport layer materials. SnO_2 is a commonly used electron transport layer material; however, the numerous trap states

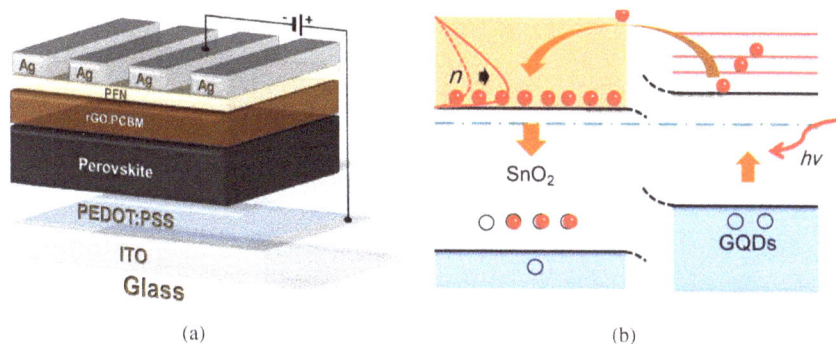

(a)   (b)

Figure 6.14. Example of functionalized graphene as an electron transport layer in PSCs: (a) schematic device architecture of the fabricated PSCs using rGO as electron transport layer material[20] and (b) schematic of the hot electron transfer from GQDs to SnO$_2$ under illumination.[21]

in low-temperature solution-processed SnO$_2$ will reduce the PSCs' performance and result in serious hysteresis. The electronic properties of SiO$_2$ are significantly improved by adding a small amount of GQDs.[21] When the device works, the photogenerated electrons in GQDs can transfer to the conduction band of SnO$_2$. The transferred electrons from the GQDs will effectively fill the electron traps and improve the conductivity of SnO$_2$, which is beneficial for improving the electron extraction efficiency and reducing the recombination at the electron transport layer/perovskite interface [Fig. 6.14(b)]. Therefore, the PSC device fabricated with SnO$_2$ and GQDs could reach the highest steady-state conversion efficiency of 20.23% with very little hysteresis.

6.1.3.2 *Hole transport layers*

The hole transport layer is another important component for PSCs. The hole transport layer is a *p*-type material, which plays a crucial role in the selective extraction of photogenerated holes from the perovskite layer and transportation to the collecting electrode, and also provides good ohmic contact between the perovskite active layer and the collecting electrode. Hence, an effective hole transport layer material should have (i) high hole mobility, (ii) a work function that matches the HOMO of the perovskite materials, (iii) good solubility and film-forming properties, (iv) good transparency in the visible region, and (v) low cost. The *p*-type organic

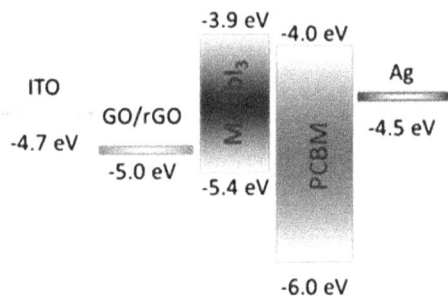

Figure 6.15.   Energy level structure of PSCs using rGO/GO as hole transport layer.[22]

hole conductor, PEDOT:PSS, has been most commonly used as the hole transport layer in state-of-the-art PSCs. However, PEDOT:PSS usually limits the stability and efficiency of PSCs due to its electrical inhomogeneity, great acidity, and hygroscopic nature. GO and rGO are used as hole transport layer materials for PSCs because of their matching work function and good compatibility with ITO (Fig. 6.15).[22] Studies have shown that GO exhibits extremely high hole-extraction performance in PSCs, but devices based on GO hole transport layer display lower efficiencies than those based on rGO. If GO is reduced to rGO, the hole extraction ability is suppressed, but the rGO-based devices show energy conversion efficiency superior to that of the GO devices (16.0% vs 13.8%); when the hole injection from perovskite to GO occurs rapidly, hole propagation from GO to the indium-doped tin oxide (ITO) substrate becomes a bottleneck to overcome, which leads to a rapid charge recombination that decreases the performance of the GO device relative to the rGO device. The hole extraction rate coefficient of rGO is lower than that of GO but higher than that of PEDOT:PSS. In this regard, a suitable hole extraction rate can realize efficient hole extraction without causing holes to accumulate in the interface, thus greatly improving the performance of the device — the conversion efficiency is of the order rGO (16%) > PEDOT:PSS (14.8%) > GO (13.8%).

6.1.3.3 *Transparent conductive electrodes*

In PSCs, the transparent conductive electrode should have a high optical transmittance in the visible region for more light to enter the cell and high electrical conductivity for efficient transportation of photogenerated

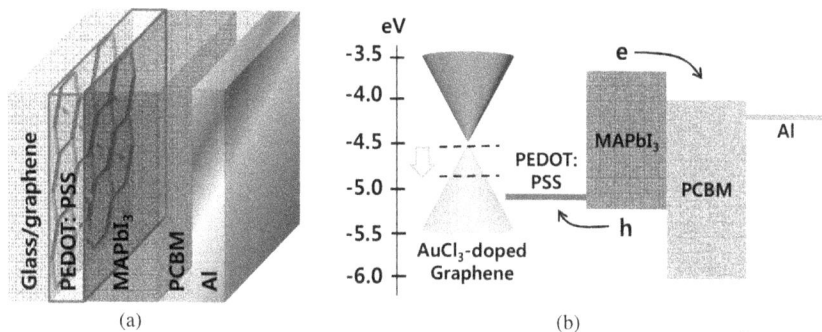

Figure 6.16. Examples of PSCs using functionalized graphene as transparent electrodes[23]: (a) schematic device structure of *p-i-n*-type PSCs using $AuCl_3$-G as a transparent electrode and (b) energy band diagram of the PSCs.

current to the external circuit. In this regard, the high electrical conductivity and high optical transmittance in the visible region of indium tin oxide (ITO) have led to its wide use as a transparent conductive electrode in *p-i-n*-type PSCs. However, the ITO transparent conducting electrode is known to be degraded by the acidic PEDOT:PSS hole-transporting layer, thereby deteriorating the long-term stability of PSCs. Graphene has been regarded as one of the most promising candidates adopted as transparent conducting components in PSCs due to its high transmittance in visible-near infrared regions, excellent flexibility, good chemical stability, and easy adjustment of Fermi level by doping. The Fermi level of graphene coincides with the Dirac point; hence, graphene has a zero band gap. However, the Fermi level can be shifted to the top of its valence band by graphene doping, and when it is lower than the Dirac point, it exhibits a *p*-type semiconductor nature. After doping, the work function of graphene can well match the HOMO energy level of the hole transport layer material. As Fig. 6.16(a) shows, $AuCl_3$-doped graphene ($AuCl_3$-G) can be used as a transparent electrode for *p-i-n*-type PSCs.[23] The doping of $AuCl_3$ reduces the transmittance of graphene; however, it effectively adjusts the work function of graphene [Fig. 6.16(b), from ~70 Ω/sq to ~890 Ω/sq], and the graphene electrode shows excellent hole mobility and greatly improved sheet resistance compared to the pristine graphene. As a result, the conversion efficiency of $AuCl_3$-G-based PSCs is 17.9%.

EFGnPs-F

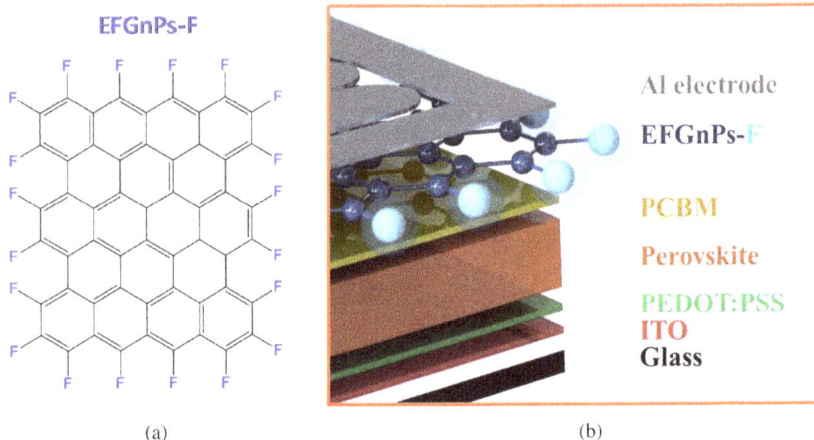

(a)

(b)

Figure 6.17. Example of PCSs using functionalized graphene as a protective layer[24]: (a) schematic structure of edge-selectively fluorine functionalized graphene nanoplatelets (EFGnPs-F) and (b) schematic device diagram of PSCs using EFGnPs-F as a protective layer.

6.1.3.4 *Protective layers*

It is well known that perovskite thin films can be easily hydrolyzed and decomposed from dark brown to yellowish thin films under a humid air environment. The hydrophobic nature of functionalized graphene makes it an ideal protective layer to improve the lifetime and stability of PSCs. As shown in Fig. 6.17(a), the mechanically exfoliated graphene nanosheets are fluorinated to obtain edge-selectively fluorine functionalized graphene nanoplatelets (EFGnPs-F).[24] The fluorine substitutions at the edge make EFGnPs-F yield super-hydrophobic properties. EFGnPs-F are coated on the electron transport layer PCBM to give the protective layer to the device, as shown in Fig. 6.17(b). With EFGnPs-F, after 30 days' exposure to an ambient humidity of ~50% at room temperature, PSCs are able to maintain 82% of their initial conversion efficiency.

6.1.4 *Photocatalytic applications*

Photocatalysis aims to harvest (part of) the energy of sunlight and use it to drive chemical reactions through a coupled process and, subsequently, utilize solar energy to promote catalytic conversion. Photocatalytic

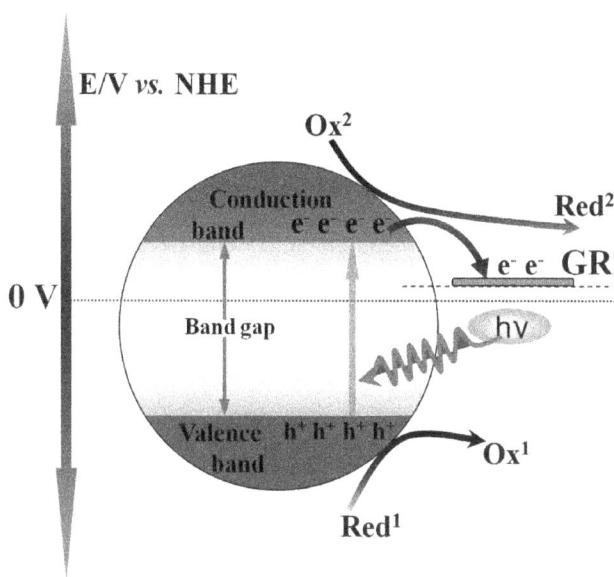

Figure 6.18.   Schematic diagram of the photocatalytic process over the composites of the graphene semiconductor.[25]

technology has been proven to be an effective way to solve environmental pollution and energy crises. Generally, photocatalytic reaction processes primarily involve three main steps (Fig. 6.18)[25]: (i) electron–hole pair generation via the adsorption of photons by a semiconducting material, (ii) charge carrier separation and transportation to the surface of photocatalysts (photoelectrodes), and (iii) reduction/oxidation reaction occurring on the surface of photocatalysts (photoelectrodes). The photoexcited electrons with a highly active reducing ability provide a driving force for the reaction (also known as overpotential). The photocatalytic performance of semiconductors can be enhanced by the following methods: (i) adjustment of the band gap or extension of the limiting wavelength of absorption through photosensitizers, (ii) inhibition of the carrier recombination, and (iii) increment of photoabsorption by increasing the surface active sites of materials, which is expected to further promote the reactions.

Single-layered graphene is a zero-gap semimetal that exhibits exceptionally high carrier mobility, a homogeneous 2D planar structure, and high transparency, making it an ideal material for loading/hybridizing semiconductors and receiving/transmitting electrons generated by photoexcitation, thus significantly improving the catalytic performance of semiconductor materials. As shown in Fig. 6.18, in a graphene/semiconductor photocatalyst, graphene plays an important role in accepting, storing, and subsequently shuttling photogenerated electrons, thereby promoting the separation and transfer of charge carriers (electron–hole pairs) for improved artificial photoredox reactions.[25] Notably, the overall photocatalytic performance of graphene–semiconductor composites is not the only issue of the electronic conductivity contribution of graphene to improving the charge carrier separation and transfer. Other factors, e.g., the surface area, mass transfer kinetics and local ensemble environment, could have a synergistic influence on the photocatalytic redox processes.

In order to improve the catalytic performance of graphene/semiconductor composites, one can start from the following aspects: (i) Improve the quality of graphene and construct a graphene-based 3D macrostructure to optimize the electronic conductivity of graphene (3D architectures have a large accessible surface area, aggregation resistance, interconnected conductive framework, fast mass kinetics, and a special microenvironment); (ii) strengthen the interfacial contact between graphene and semiconductors, and optimize the atomic charge carrier transfer pathway across the interface; and (iii) conduct systems materials engineering of graphene-semiconductor composite photocatalysts, that is, the integration of individual components into a complete and functioning system while simultaneously optimizing the resulting interfaces to design an efficient artificial photosynthetic materials system. Currently, the most studied and most valuable photocatalytic reactions are the photocatalytic conversion of water and carbon dioxide.

6.1.4.1 *Photolysis of water*

Photolysis of water is the splitting of water molecules in the presence of photons into H_2 and O_2. Under photoexcitation, the semiconductor provides electrons for the reduction of H^+ to H_2 and provides holes for the oxidation of H_2O to O_2; thus, hydrogen and oxygen can be generated while water is split. The standard Gibbs free energy for this reaction is +237.2 kJ/mol, suggesting that the potential overall thermodynamic barrier for a water-splitting reaction is 1.23 V. The ideal band gap of the semiconductors should be around 2.0 eV for the effective utilization of solar energy. Meanwhile, the bottom of the conduction band (CB) must be located at a more negative potential than the reduction potential (H^+/H_2), while the top of the valence band (VB) must be positioned more positively than the

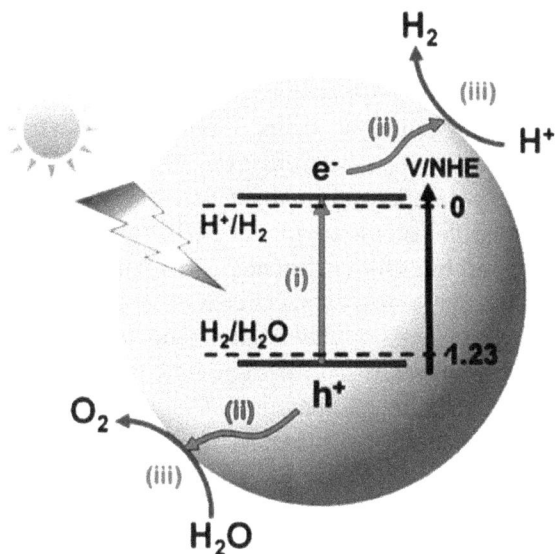

Figure 6.19. The principle and main process of photocatalytic oxygen generation from water.[26]

oxidation potential (H_2O/O_2). As shown in Fig. 6.19, the bottom of the conduction band must be more negative than the reduction potential of H^+ to H_2 (0 V vs. normal hydrogen electrode (NHE)), while the top of the valence band must be more positive than the oxidation potential of H_2O to O_2 (1.23 V vs. NHE).[26] Only in this way can water molecules receive electrons generated by photoexcitation, be reduced to H_2, and then transfer electrons to holes generated by photoexcitation to be oxidized to O_2. Commonly used semiconductor materials include TiO_2, P25, CdS, SiC, C_3N_4, $BiVO_4$, TaOH, ZnCdS, MoS_2, Ni, and $ZnIn_2S_4$. Unfortunately, few single semiconductors satisfy both requirements of energy level and band gap. Therefore, the commonly used photolysis of water to produce hydrogen is actually a half-reaction. The electrons generated by photoexcitation are captured by water molecules to reduce H^+ to H_2, while the holes generated by photoexcitation are used to oxidize other substances, such as alcohols, lactic acid, and S^{2-}/SO^{2-}.

In the photocatalytic water-splitting system, graphene materials can be used as electron acceptors and carriers, photocatalysts, and photosensitizers. Graphene-based materials are often used to accept and transfer electrons generated by the photoexcitation of semiconductor materials. In semiconductors, light absorption creates energetic electrons and electron vacancies or holes — actual resultants from the photolysis of water. However, if electron and hole transport is not rapid enough after effective separation, then recombination occurs. Graphene exhibits high electron mobility, which can transport electrons over their 2D structures and effectively promote the separation of photogenerated electron–hole pairs while inhibiting electron–hole recombination. The effectiveness of graphene as an electron acceptor and transporter depends on its intrinsic electrical properties and the effective large-area contact between graphene and semiconductor materials. Therefore, the quality of graphene needs to be improved to obtain better electron mobility. At the same time, graphene should be well hybridized with semiconductor materials. Compared with graphene prepared by the CVD method or exfoliation method, rGO is more frequently used to fabricate composite photocatalysts with semiconductor materials. This is because rGO is easy to prepare on a large scale, and it is easy to use the solution method to compound with semiconductor particles. On the other hand, covalent chemical bonding formed at the interfaces between the semiconductor and graphene with oxygen-containing groups could also narrow the band gap by introducing new energy levels within its electronic structure, thus leading to an enhanced photocatalytic activity.

Doping heteroatoms or inducing defects is another way for graphene to improve its catalytic performance. Doping governs the electrical properties of graphene, and the doped graphene turns to be either *n*-type or *p*-type semiconductor, therefore enhancing its photocatalytic efficiency. After the introduction of heteroatoms, the original lattice symmetry of graphene is broken, and its π^* anti-bonding orbital and π bonding orbital degenerate to open a band gap. This makes graphene a semiconductor material with light-absorbing ability and photocatalytic ability. The introduced heteroatoms can also serve as catalytic active sites of graphene functional materials. In addition, the defects of graphene materials can also serve as catalytic active sites. Therefore, adjusting the structure of graphene is also crucial to enhance its catalytic performance.

6.1.4.2 *Photocatalytic CO_2 reduction*

Photocatalytic reduction of CO_2 is another important form of converting solar energy into chemical energy. Carbon dioxide is earth's most important greenhouse gas that drives global climate change. The conversion of

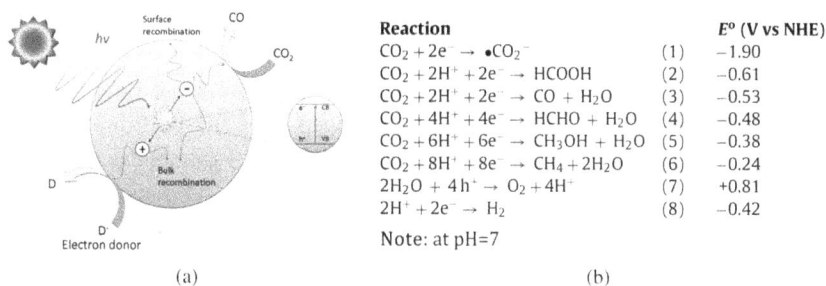

Reaction		E^0 (V vs NHE)
$CO_2 + 2e^- \rightarrow \bullet CO_2^-$	(1)	-1.90
$CO_2 + 2H^+ + 2e^- \rightarrow HCOOH$	(2)	-0.61
$CO_2 + 2H^+ + 2e^- \rightarrow CO + H_2O$	(3)	-0.53
$CO_2 + 4H^+ + 4e^- \rightarrow HCHO + H_2O$	(4)	-0.48
$CO_2 + 6H^+ + 6e^- \rightarrow CH_3OH + H_2O$	(5)	-0.38
$CO_2 + 8H^+ + 8e^- \rightarrow CH_4 + 2H_2O$	(6)	-0.24
$2H_2O + 4h^+ \rightarrow O_2 + 4H^+$	(7)	$+0.81$
$2H^+ + 2e^- \rightarrow H_2$	(8)	-0.42

Note: at pH=7

(a)   (b)

Figure 6.20. Photocatalytic CO_2 reduction[27]: (a) schematic diagram of photoexcitation and electron transfer process and (b) pathways for the generation of solar fuels and the related potentials.

CO_2 into fuels or valuable chemicals is able to utilize CO_2 and mitigate the effect of global warming. Photocatalysis is a promising approach for the conversion of CO_2. Similar to photocatalytic water splitting, photocatalytic CO_2 reduction utilizes photoexcited electrons from semiconductors to react with CO_2 and converts them to other substances, leaving behind holes that are then consumed by other reducing substances (Fig. 6.20).[27] The commonly used photocatalysts for CO_2 reduction are TiO_2, ZrO_2, Ga_2O_3, Ta_2O_5, $SrTiO_3$, $CaFe_2O_4$, $NaNbO_3$, $ZnGa_2O_4$, Zn_2GeO_4, and $BaLa_4Ti_4O_{15}$. The conversion of CO_2 into different substances requires different redox potentials, which are determined by different reaction paths, and the number of electrons required for each reaction path is also different. Being able to discharge multiple electrons in one shot with protons is important to improve reaction efficiency. Thus, generating sufficient electron–hole pairs, separating charges efficiently, and providing active catalytic sides are the chief factors for CO_2 photoreduction.

Hybridizing graphene with the abovementioned semiconductor materials provides us with promising opportunities to design artificial photosynthesis systems with high performance. In these composite materials, excellent electron conductivity, unique 2D morphology, and high transparency of graphene make it an ideal platform to assemble semiconductor components and accept/transport photogenerated charge carriers, thereby improving the efficiency of the photocatalytic processes. Upon illumination of light, the photoexcited electrons are absorbed by the semiconductor and rapidly flow into the graphene material, leaving behind holes in the semiconductor to achieve effective separation of electron–hole pairs, thereby increasing the concentration of effective carriers. For example, the intrinsic UV-active charge generation (photocurrent) of pure TiO_2 is enhanced by a factor of 10 by incorporating rGO.[28] In addition, as electrons injected into graphene migrate on the surface of graphene, they will react with CO_2 molecules adsorbed on graphene. The 2D planar structure of graphene provides a larger reaction site for CO_2 molecules, thereby improving the conversion efficiency of CO_2. The incorporation of graphene can also tune the nature of catalytic materials to achieve selective photoreduction of CO_2. For example, rGO-coated gold nanoparticles (rGO-AuNPs) show excellent selectivity for HCOOH (>90%).[29]

In addition to the abovementioned catalytic reactions, graphene composites have also been widely used in the degradation of organic pollutants in water and selective photocatalytic chemical transformations, such as photocatalytic selective oxidation of alcohols into corresponding aldehydes using graphene–TiO_2 nanocomposites.

6.1.5 *Summary*

In order to obtain graphene-based photosynthesis systems with excellent performance, firstly, it is necessary to choose an appropriate type of functionalized graphene material and precisely control the material structure, which requires the development of a more efficient preparation method for functionalized graphene materials. Secondly, to optimize the morphology of graphene/semiconductor composite, one should aim at the preparation methods of composite materials, the interface modification of graphene and functional materials, and the structure and morphology of composite materials. Thirdly, one should optimize the device structure to further enhance photocatalytic performance. Finally, from the viewpoint of the system's material engineering, designing and fabricating graphene-based photocatalytic system, in which all the components operates well, plays a crucial role in achieving an efficient reaction system.

6.2 Energy Storage

The most important form of commercial energy is electricity, which is central to many parts of life in modern societies. With the development of society, electrical energy storage is required to support the increasing demand for energy infrastructure, transportation, and consumer electronics. In recent years, the demand for portable electronic devices has skyrocketed, with smartphones, laptops, and electric vehicles becoming an integral part of our daily lives. As a result, the need for electrochemical energy storage devices with high energy density as well as high power density has become more pressing than ever. At present, the most important and most promising energy storage devices include supercapacitors (SCs), lithium-ion batteries (LIBs), and other rechargeable batteries, such as lithium-sulfur batteries (Li-S batteries), sodium-ion batteries (Na-ion batteries), and lithium-oxygen batteries (Li-O$_2$ batteries). The whole performance of these energy storage devices relies on several similar components including two electrodes containing electrochemically active materials, electrolytes, and separators. Although their working principles are different, developing electrode materials and electrode systems with excellent electrochemical activity is an effective way to improve a device's performance. Ideal electrode materials should offer large effective contact areas for electrolyte/electrode interfaces, short electron/ion transfer paths, ease of functionalization, and excellent electrolyte wettability. Graphene materials have been widely used as electrode materials for SCs, LIB, and other secondary batteries due to their high specific surface area, high electrical conductivity, and high flexibility. In addition, graphene's morphology and structure make it ideal for energy storage use, for example, its large accessible surface area, a large number of exposed active sites, and fast reaction kinetics. Meanwhile, a variety of functionalized graphene materials can satisfy the demands of electrode materials utilized in high energy density and high power density devices, for instance, porous 3D graphene, defect-free graphene, heteroatom-doped graphene, metal or metal oxide-doped graphene, and other graphene composites.

6.2.1 *Supercapacitors*

Supercapacitors have the characteristics of high power density, fast charge/discharge rates, and long life; they are a kind of electrochemical energy storage device with great potential. There are two main classes of supercapacitors based on the charge storage mechanism: electrical double-layer capacitors (EDLCs) and pseudocapacitors. In these two kinds of SCs, functionalized graphene materials are mainly used to fabricate the electrodes of the devices to boost the energy density, power density, rate performance, and cyclability of the devices. In recent years, on the basis of these two basic types of SCs, hybrid capacitors have been developed, such as asymmetric capacitors, that combine the performance benefits of EDLCs and pseudocapacitors. The basic working principles of these capacitors are based on those of EDLCs and pseudocapacitors. In the following, we mainly focus on EDLCs and pseudocapacitors.

6.2.1.1 *Electric double-layer capacitors*

EDLCs store charge electrostatically using reversible adsorption of ions of the electrolyte onto active materials that are electrochemically stable and have high accessible specific surface area (SSA), such as carbon-based active materials. The EDLC device consists of the following main parts: two electrodes, an electrolyte, a diaphragm, and (or) a current collector (Fig. 6.21).[30] When a voltage is applied between the two electrodes of the ELDC, the two electrodes are polarized, and the positive/negative ions of the electrolyte are adsorbed through electrostatic interaction, and an electric double layer (EDL) is formed at the electrode/electrolyte interface. In EDLCs, the electrode's active materials are electrochemically stable and do not undergo any Faradaic processes. Thanks to this working principle, the EDLCs has some distinct advantages such as higher power density induced by a fast charging/discharging rate (in seconds) and a long cycle life (4100 000 cycles) when compared to batteries and fuel cells. However, the energy density in EDLCs is much lower than that in

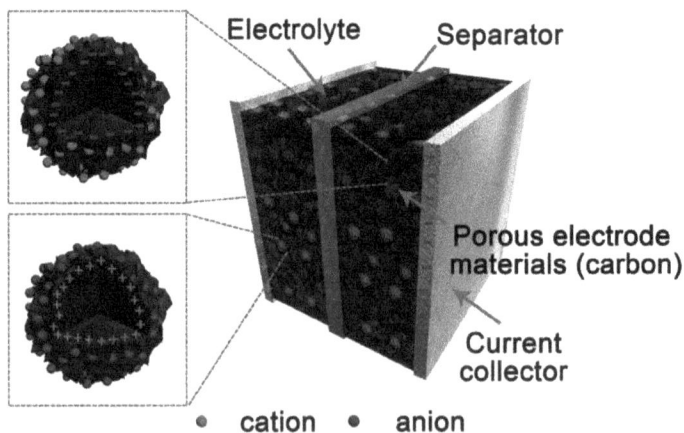

Figure 6.21. Structure of EDLCs.[30]

batteries. Therefore, EDLCs are considered a complement for batteries especially in high-rate applications such as electric vehicles or hybrid electric vehicles.

The key to reaching high capacitance by charging the double layer lies in using electrodes with large specific surface areas and high electrical conductivity. In this regard, graphene and graphene-based materials are attractive due to their unique mechanical and electrical properties as well as exceptionally high surface area (2630 m²/g). The intrinsic capacitance of single-layer graphene is reported to be ~21 mF/cm² and, theoretically, SCs based on graphene materials could achieve an EDL capacitance of ~550 F/g if all the surface area can be used.[109] Therefore, graphene has been widely used to make electrodes of EDLCs.

The structures and compositions of graphene materials can be varied considerably by using different fabricating and processing methods, such as the number of layers, size, functional groups, wrinkles, defects (holes or heteroatom doping), and shape, leading to a different specific surface area, light transmittance, and conductivity. For example, mechanical exfoliation can produce very high-quality graphene with a fairly low concentration of defects and the basic structure of graphite; thus, the resistance at room temperature is low (<300 Ω/sq). Reduction of GO can produce rGO with more defects and few oxygen-containing groups, leading to poor electrical conductivity (>300 Ω/sq). However, it is difficult to obtain perfect-quality graphene due to the limitations of current fabrication methods and synthesis procedures. Therefore, methods such as controlling the specific surface area, manipulating the internal pore size, enhancing the electrical conductivity, and adjusting the macrostructure (such as 1D nanowires, 2D films, and 3D sponge structures) are adopted to improve the performance of graphene materials in ELDCs.

Another reason for the low specific capacity of graphene in practical applications is the aggregation and restacking of graphene due to the π-π interaction between sheets. The aggregation of graphene reduces its accessible surface area and generates contact resistance between layers, which reduces the conductivity of the material, resulting in a decrease in the performance of the capacitor.

Graphene-based EDLCs exhibit the inherent characteristics of EDLCs, that is, high power density but low energy density. Currently, many scientific publications have focused on increasing the energy density without compromising other characteristic advantages of the supercapacitor such as the power capability. It is generally believed that binder-free graphene electrodes with a large thickness, high surface area and volume ratio, and excellent mechanical properties can meet the requirements of high power density as well as high energy density at the same time. For example, EDLCs using direct-laser-reduced rGO film as the electrode material present high energy density while maintaining high power density.[31]

GO films are made by drop casting GO dispersion onto a flexible substrate (a DVD media disc).[31] Under the laser irradiation of a commercial CD/DVD drive, the GO film changes from a golden brown color to black as it is reduced to rGO [referred to as laser-scribed graphene (LSG)]. The laser causes the simultaneous reduction and exfoliation of GO sheets and produces an open network of rGO. This network structure not only prevents the agglomeration of graphene sheets but also creates open holes, which helps facilitate the electrolyte's accessibility to the electrode surfaces. In addition, the LSG films show an excellent conductivity of 1738 S/m as opposed to 10 to 100 S/m for activated carbons, the state-of-the-art material used in commercial devices. Additionally, LSG shows excellent mechanical flexibility with only ~1% change in the electrical resistance of the film after 1000 bending cycles. Thus, LSG can be directly used as EC electrodes without the need for any additional binders or conductive additives. Meanwhile, it can also be used as the current collector in the EC. The supercapacitor with LSG as the electrode and current collector (LSG-EC) can exhibit energy densities of up to 1.36 mWh/cm^3, a value that is approximately two times higher than that of the commercial supercapacitor AC-EC (2.75 V/44 mF). Additionally, LSG-ECs can deliver a power density of ~20 W/cm^3, which is 20 times higher than that of the AC-EC and three orders of magnitude higher than that of the 500-μAh thin-film lithium battery.

6.2.1.2 *Pseudocapacitors*

A pseudocapacitor also includes electrodes, diaphragms, and electrolytes, but its working principle is different from that of EDLCs (Fig. 6.22).[30] In addition to the EDL, the energy storage mechanism also includes the fast and reversible faradaic process (redox reaction) between the electrode active material and the electrolyte. In principle, compared to EDLCs, pseudocapacitors exhibit higher capacitance and energy density due to the coexistence of electrostatic and electrochemical storage occurring at the surface and also near-surface areas of the electrodes, but relatively low

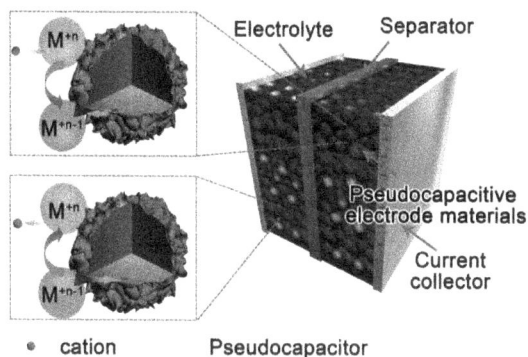

Figure 6.22. Device structure of a pseudocapacitor.[30]

power density due to low electrical conductivity of materials and slow electron transfer kinetics. The commonly used electrode active materials for pseudocapacitors include conductive polymers [polyaniline (PANI), polypyrrole (PPy), poly (3,4-ethyelene dioxythiophene) (PEDOT), poly-thiophene (PTh), and poly (p-phenylene vinylene) (PPV)] and metal oxides (RuO_2, MnO_2, Fe_3O_4, Co_3O_4, V_2O_5, CeO_2, NiO, and ZnO). Despite having the advantages of higher specific capacitance and energy density, pseudocapacitive materials generally suffer from low stability and short lifetimes because of their degradation and aging by swelling and shrinking during redox reactions.

In order to overcome these drawbacks and improve the performance of pseudocapacitive capacitors, graphene is often hybridized with these pseudocapacitive materials to make graphene composites. This is because graphene is exceptionally attractive due to its high electrical conductivity, large specific surface area, low cost, high mechanical strength and flexi-bility, and tunable porous microstructures as well as easy fabrication of the electrodes. The type of graphene commonly used in the preparation of composites includes graphene prepared by the CVD method and exfolia-tion method, GO, and rGO. Among these types of graphene, rGO is more widely used to prepare composite electrode materials for SCs. While rGO shows good conductivity, it also has a small number of oxygen-containing groups, such as hydroxyl, epoxide, carbonyl, and carboxyl. rGO presents excellent hydrophilicity due to the existence of the oxygen-containing groups, which not only makes it easier to fabricate composites with metal oxides and several conducting polymers but also allows easier electrolyte wetting. In addition, rGO can be functionalized by incorporating functional groups through various oxygen-containing groups. These functionalized groups can serve as redox centers to promote the pseudocapacitance of graphene composites. In addition, graphene doped with heteroatoms (such as N, O, and S) is also commonly used to prepare graphene composites because the introduction of heteroatoms provides additional redox centers.

The unique 2D planar structure of graphene can be used as a platform to support metal oxides or wrap conducting polymers. There are several fabrication methods to prepare graphene/metal oxide composites, such as electrodeposition, hydrothermal, co-precipitation, and spray pyrolysis. Graphene has applications for fabricating composites with polymers by *in situ* polymerization, self-assembly, and layer-by-layer assembly. The introduction of graphene materials not only improves the stability and conductivity of pseudocapacitive materials but also improves the specific capacitance, energy density, and power density of pseudocapacitive materials by contributing additional redox sites and optimizing the structure of composite materials. In addition, the structure of graphene composites — dimensions, sizes, orientations, and pore structures — can also greatly affect the performance of pseudocapacitors. The porous structure can increase the specific surface area of the composite material, provide more conductive channels for electrolyte ions, increase the number of active centers accessible to the material, and increase the loading capacity of redox species. All of these are beneficial to improve the performance of pseudocapacitors. For example, 3D graphene foam is prepared by the CVD method using 3D nickel foam as a template, which can be used as the electrode material of pseudocapacitors after loading NiO.[32] In this 3D composite, NiO can serve as a redox material and largely prevents the aggregation of graphene during electrochemical charge–discharge cycling. In addition, the porous 3D structure offers free channels to allow effective interaction between the electrolyte and the active materials. Here, reliable electrical contact between the graphene sheets and the current collectors allows rapid charge transfer from active materials to the current collector, while the highly porous 3D graphene foams with large surface areas facilitate the easy accessibility of electrolyte ions to NiO active materials. Due to the aforementioned advantages, pseudocapacitors using this 3D graphene composite as electrode material demonstrate good rate capability and superior stability.

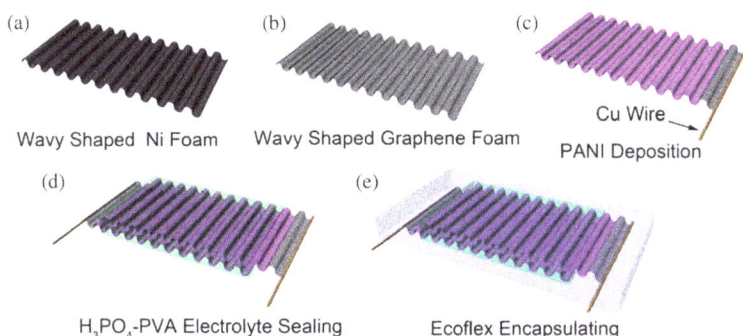

Figure 6.23. Schematic description of stretchable all-solid-state supercapacitor fabrication.[33]

In addition, the emerging field of wearable and bio-implantable electronics has necessitated the development of flexible and stretchable pseudocapacitors that inevitably require materials and electrodes to be bendable, rollable, foldable, and/or stretchable. As shown in Fig. 6.23, a kind of stretchable all-solid-state pseudocapacitor configuration with two slightly separated wavy shaped electrode and gel electrolyte is fabricated, and such supercapacitor demonstrates a high potential as an energy storage device for stretchable and wearable electronics.[33] Ni foam manually made into a wavy shape is used as a template, on which graphene is grown by the CVD method. Finally, a porous graphene sheet with a wavy shape is obtained after removing the Ni from the HCl solution. As a redox-active material, PANI is deposited onto the graphene sheet to fabricate the pseudocapacitive electrode. The H_3PO_4/PVA gel is used as a solid electrolyte to separate the two pseudocapacitive electrodes. A final flexible all-solid-state pseudocapacitor is obtained after encapsulation with elastic rubber. The supercapacitor exhibits a maximum specific capacitance of 261 F/g. Electrochemical cycle testing with the supercapacitor showed 89% capacitance retention over 1000 charge–discharge cycles at a current density of 1 mA/cm². The bending and stretching tests showed that the supercapacitor maintained high mechanical strength and high capacitance simultaneously, even under a strain of 30%.

6.2.1.3 *Micro-supercapacitors*

Micro-supercapacitors are in a planar form in contrast to the conventional sandwich structure: The interdigitated positive and negative planar electrodes are fabricated on the substrate, and the electrolyte fills the tiny gaps between the electrodes. This structural feature makes it possible to manufacture ultrathin, flexible, and microscale devices on the substrate. This satisfies the current demand for portable devices for ultrathin, flexible on-chip energy storage. In addition, due to the short distance between adjacent electrodes, the electrolyte path length between the two electrodes is also extremely shortened, so that anions and cations can transfer between the two electrodes at an extremely high speed, endowing the device with extremely high power density and rate performance. As mentioned earlier, electrode materials prepared with functionalized graphene materials have also been widely used in micro-supercapacitors due to their high specific surface area and high conductivity. The preparation methods of graphene for fabricating micro-supercapacitors include inkjet printing, photolithography, printing, and photo-engraving. For example, micro-supercapacitors can be produced by direct writing on GO films using a consumer-grade DVD burner (Fig. 6.24).[34] The aqueous dispersion of GO synthesized using a modified Hummers' method is drop cast onto a DVD disc covered with polyethylene terephthalate (PET), and the GO film can be formed after drying. The GO-coated DVD disc is then inserted into the DVD optical drive for laser patterning designed using regular computer

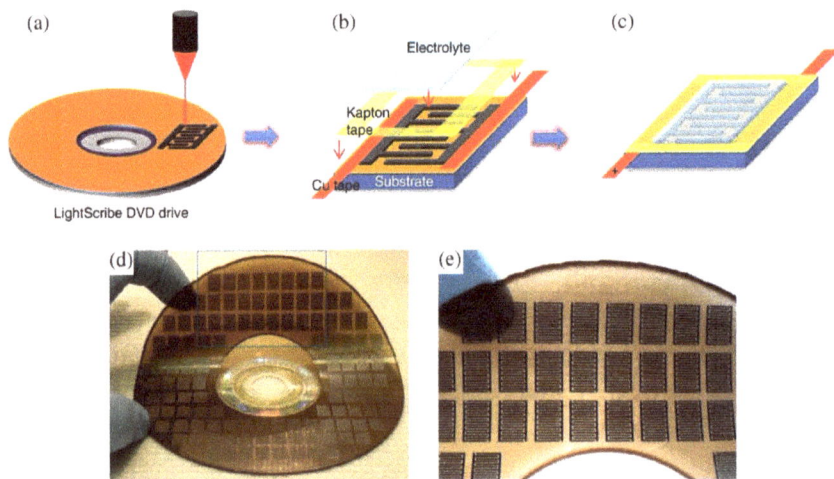

Figure 6.24. (a–c) Schematic diagram showing the fabrication process for a laser-scribed graphene micro-supercapacitor and (d, e) digital images of micro-devices.[34]

software. This laser irradiation process converts the GO to rGO, forming a micro-device, that is, the electrode of a micro-supercapacitor. Copper tape is applied to the electrode's edges to ensure good electrical contact between the devices and the electrochemical workstation. The interdigitated area is defined with polyimide (Kapton) tape to protect the contact pad from the electrolyte. Finally, the electrolyte is drop cast on the active interdigitated electrode area, and the flexible substrate loaded with microdevices is peeled off from the DVD disc to fabricate a large-scale flexible micro-supercapacitor. More than 100 micro-supercapacitors can be produced on a single disc in 30 min or less. Therefore, this method is suitable for large-scale preparation of micro-supercapacitors. In addition, these SCs exhibit superior performance with an ultrahigh power of ~200 W/cm^3. Microscale devices can be directly made on silicon wafers, demonstrating the potential for the next generation of complementary metal-oxidized semiconductor (CMOS) applications.

In micro-supercapacitors, the commonly used functionalized graphene materials are mainly based on graphene and rGO prepared by the CVD method. Compared with graphene, rGO is more suitable for the preparation of micro-supercapacitors by solution processing methods such as inkjet printing, printing, and photo-engraving, thus it is also most commonly used to prepare micro-supercapacitors. In addition, in order to improve the performance of electrode materials, other functionalized graphene materials have also been used as electrode active materials, such as graphene quantum dots, graphene/carbon nanotube composites, and graphene/metal oxide composites.

The structure of the device is also the focus of research, and electrode arrangement is crucial for the performance of superconductor devices. Currently, a variety of different device structures have been developed (Fig. 6.25).[35] The electrolyte of micro-supercapacitors includes a liquid state and a solid state. The major advantages when using solid-state

Figure 6.25. Micro-supercapacitors in different structures.[35]

electrolytes are the simple packaging and fabrication processes and lack of leakage, and it also can be made into an all-solid micro-supercapacitor, which is one of the objects of current research in this field. Substrates for micro-supercapacitors include polymer flexible substrates, glass, and silicon wafers. The optimization of all these device structures and electrode materials is done to improve the performance of the device, such as power density, energy density, cycling life, and frequency response.

6.2.2 *Lithium-Ion Battery*

Lithium-ion batteries are currently the predominant commercial form of rechargeable batteries and play a pivotal role in energy storage devices in various fields (such as power grids, electric vehicles, consumer electronics, and industrial defense). LIBs are electrochemical storage devices that are composed of negative (anode) and positive (cathode) electrodes, a porous separator (allowing Li ions to transport through), and an electrolyte (conducting Li ions during charging/discharging). As shown in Fig. 6.26, the positive and negative electrodes are attached to the

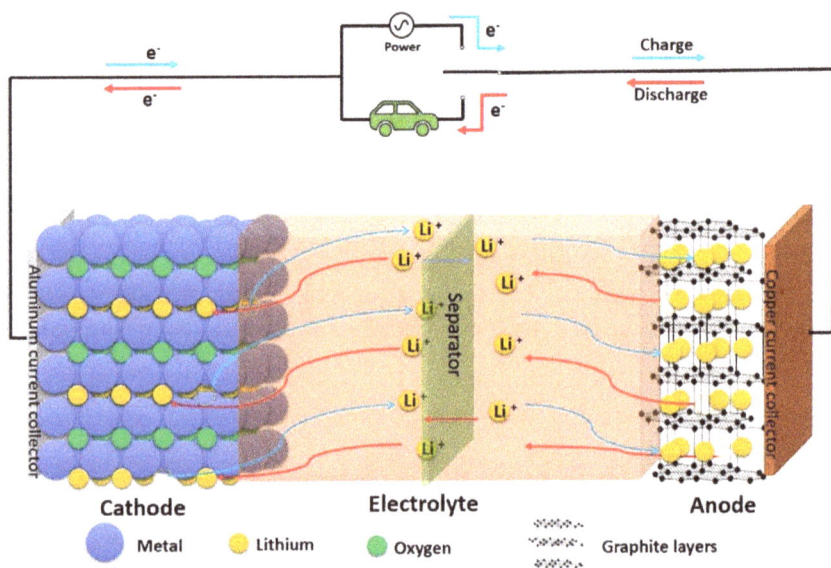

Figure 6.26. Schematic of the structure and working mechanism of a lithium-ion battery.[36]

corresponding current collectors and separated by a separator, and the electrolyte is filled between positive/negative electrodes.[36]

LIBs work through a "rocking-chair" mechanism to interconvert chemical and electrical energy, in which Li ions continuously shuttle between the Li-accepting anode and Li-releasing cathode. Take an LIB with a positive electrode of composite metal oxide $LiCoO_2$ and a negative electrode of graphite as an example. When charging, under the action of high positive potential, the trivalent cobalt ions (Co^{3+}) in the positive electrode material $LiCoO_2$ lose one electron and become tetravalent cobalt (Co^{4+}), and the electrons are transferred to the external circuit. Li^+ deintercalates from the $LiCoO_2$ lattice, enters the electrolyte, moves across the separator to the graphite anode, and intercalates in the anode. At the same time, the graphite anode accepts an electron from the external circuit and forms CLi_x with the embedded Li^+. When discharging, the graphite negative electrode releases an electron and transmits it to the external circuit. At the same time, Li^+ deintercalates from the anode material, enters the electrolyte, moves across the separator to the $LiCoO_2$ cathode, and intercalates into its lattice. At this time, the tetravalent cobalt (Co^{4+}) generated by the oxidation of trivalent cobalt (Co^{3+}) during charging is reduced to trivalent cobalt (Co^{3+}) by electrons from the external circuit. The overall charge/discharge reaction of the LIBs is shown in Eq. (8.1):

$$LiCoO_2 + C \underset{\text{discharge}}{\overset{\text{charge}}{\rightleftharpoons}} Li_{1-x}CoO_2 + CLi_x \qquad (8.1)$$

Due to this special working mechanism, LIBs have attracted great interest due to their high energy density, low self-discharge, low maintenance, and better safety compared to conventional batteries. However, LIBs still suffer from low power densities, long charging times, and poor cycle stability, due to which they are incapable of satisfying the ever-increasing demand for diverse applications in electronic devices and equipment. In order to improve the performance of LIBs, especially to further increase their energy density and power density, it is imperative to develop high-performance electrode materials. To simultaneously achieve such goals, it is necessary to develop new electrode materials with high electrical conductivity for fast electron transport, well-designed nanostructures for shortening diffusion distance of Li ions, and large accessible specific surface area for lithium storage. In this regard, graphene, due to its superior electrical conductivity, excellent mechanical flexibility, good chemical stability, and high surface area (2630 m^2/g), is expected to be a good candidate.

6.2.2.1 *Negative electrode (anode) material*

During the charging process of an LIB, the negative electrode material receives an electron from the external circuit and receives and stores the lithium ion transported from the positive electrode; when discharging, the negative electrode material releases an electron to the external circuit, and the stored lithium ions are released. Currently, graphite is the most used commercial anode material in LIBs. Graphite has a theoretical specific capacity of 372 mAh/g by forming LiC_6 upon Li intercalation between the stacked layers. It has been proposed that graphene can accommodate Li ions through an adsorption mechanism on both sides to form Li_2C_6 with a theoretical capacity of 744 mAh/g, which is twice that of graphite.

At present, among a variety of graphene materials (for example, CVD graphene, exfoliated graphene, and rGO), rGO is easy to prepare and functionalize on a large scale, and the distribution of oxygen-containing groups makes it easy to hybridize with other functional materials, thus rGO is the anode of choice for LIBs. As shown in Fig. 6.27, rGO is used as the negative electrode material in all-graphene LIBs (the positive and negative electrode materials are only made of graphene).[37] As mentioned earlier, rGO can be obtained by reduction of GO. The rGO exhibits a small amount of oxygen-containing functional groups combined with porous morphology. The pores in the rGO are formed by rapid gas

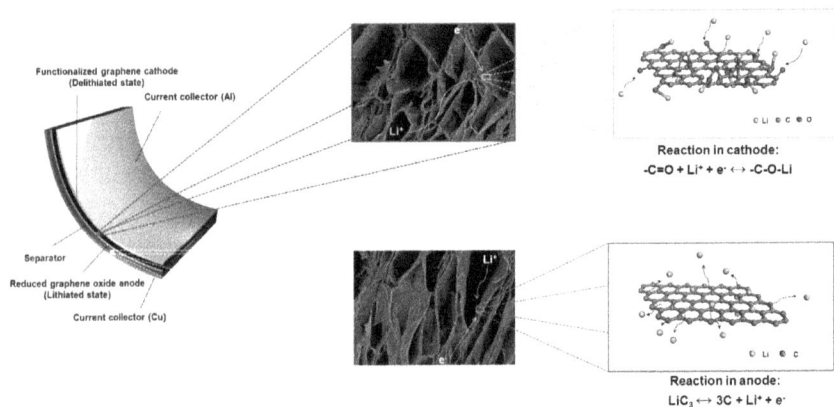

Figure 6.27. Schematic illustration of all-graphene battery and its electrochemical reaction.[37]

evolution under high vapor pressure and are interconnected and extended from the inside to the surface of the materials. The porous electrode structures are expected to be advantageous for electrochemical reactions because they enable greater penetration of the electrolyte into the interior of the electroactive materials. In all-graphene LIBs, the rGO anode is evaluated in the voltage range of 0.01 to 3.0 V at a current density of 0.1 A/g, and a capacity of 540 mAh/g is reversibly obtained over 100 cycles. In this case, >72% of the theoretical capacity (744 mAh/g) is achieved with a Coulombic efficiency near 100% over 100 cycles. Because the rGO fabricated in this study is multilayer rather than single-layer graphene, it delivers a slightly lower capacity than the theoretical value. The multilayer structure reduces the accessible surface area for lithium ions, which lowers the utilization rate of the material.

Although rGO has been widely used as an anode material for LIBs, there are still some problems to be solved urgently for rGO-based anode materials. rGO-based anodes still suffer from high initial irreversible capacities (mostly >500 mAh/g) and low first-cycle Coulombic efficiencies (usually <60%). This is mainly attributed to the formation of a solid electrolyte interphase (so-called SEI, usually ~10–100 nm in thickness, arising from the irreversible decomposition of electrolyte to form a surface passivation layer around the electrode) as well as chemical reactions of Li ions with the residual oxygen-containing functional groups in rGO. The formed by-products adsorbed on the electrode surface can increase the energy barrier of the charge transfer reaction, resulting in capacity fading upon cycling particularly at high rates. During charging/discharging cycles, the residual oxygenated groups of rGO can also be progressively reduced. This leads to a restacking of graphene layers, thus lowering the Li storage capacity. In short, all side reactions consume Li and/or active materials, leading to a gradual decrease in storage capacity and thereby relatively low-capacity retention. In addition, large voltage hysteresis is often observed in rGO, which is possibly due to active defects, which react with Li at low voltages upon discharging, while the breakage of strong Li-defect bonds requires higher voltages during charging. In the case of rGO-based anode materials, no obvious voltage plateaus occur from the discharge curve, which implies that such batteries may fail to provide stable potential outputs. Finally, and more importantly, significant disordering and abundant defects inevitably compromise the electron transport and electrochemical stability of graphene sheets. This leads to low rate capability and low power density.

Therefore, it is highly advisable to tailor the structural parameters and active defects of graphene to an appropriate level to achieve consistent, high-performing LIBs. Based on these considerations, graphene sheets consisting of fewer layers (hence larger accessible surface area exposed), larger interlayer spacing, smaller sizes (hence more edges available), more defects, and more crumpled nanostructures (hence more nanocavities and nanovoids between sheets) could be beneficial in improving Li storage, cyclic performance, and rate capability (the ability to deliver capacity at high current rates) as evidenced by some theoretical and experimental studies.

In practical applications, graphene is often combined with other materials (e.g., carbon nanotubes, metal oxides, or semiconductor/metal materials such as silicon, germanium, and tin) to fabricate hybrid anodes, in which graphene is used as the active material of the composite or support material for other active materials.

Carbonaceous nanomaterials with high electrical conductivity, such as porous carbon, carbon nanofibers, and carbon nanotubes, are found to combine well with graphene for fabricating carbonaceous hybrid electrodes. The integration of carbonaceous nanomaterials with graphene enables promoted electron transport and ion diffusion and thereby reduces the charge transfer resistance; however, it is also highly conducive to enlarging interlayer spacing and preventing the restacking of graphene sheets during the electrode fabrication and cycling operation, thereby making full use of the interior space for Li-ion diffusion and storage. Finally, LIBs using such hybrid anodes exhibit higher specific capacities and better rate capabilities. In addition, porous carbon materials can serve as physical spacers, reducing the aggregation of graphene sheets; such porous nanostructures also have the ability to shorten the Li-ion transport path, accommodate large volume changes, and enlarge the electrode–electrolyte interface. This can lead to large capacities, high lithiation capability, and good cycling stability. The electrochemical performance can be further improved by using doped carbon materials as hybrid components, such as *N*-doped and *S*-doped porous carbon hybrids. *S*-doping further increases the electrochemical activity of porous carbon by increasing the number of nanopores and interlayer spacing.[38]

Metal oxides are currently important members of the anode family for LIBs, which have the advantages of high capacity, abundant reserves, low cost, and efficient fabrication. LIB anodes based on metal oxides are generally classified into three main categories: insertion, conversion, and alloying types in terms of reaction mechanisms with Li.

1. *Insertion*: Insertion-type metal oxides typically include TiO_2, vanadium oxides (e.g., V_2O_5, V_6O_{13}, LiV_3O_8), lithium titanium oxides (e.g., $Li_4Ti_5O_{12}$, $LiTi_2O_4$, $LiCrTiO_4$, $SrLi_2Ti_6O_{14}$), and lithium phosphates (e.g., $LiTi_2(PO_4)_3$, $Li_3V_2(PO_4)_3$, $LiVPO_4F$). The fundamental electrochemical reactions for two typical insertion-type metal oxides are illustrated in Eqs. (8.2) and (8.3). Such Li-intercalating materials usually undergo a so-called "topotactic reaction" process which involves the insertion (intercalation) and extraction (deintercalation) of Li ions into and from the host lattice with minor modification of the crystal structure. Insertion-type anodes thus exhibit long lifetimes, low irreversible capacity loss, and small volume variation. They are currently limited to low inherent electrical conductivities, low specific capacities, high intercalation potentials, and interparticle aggregation.

$$M_xO_v + nLi^+ + ne^- \leftrightarrow Li_nM_xO_v \quad (8.2)$$

$$Li_mM_xO_v + nLi^+ + ne^- \leftrightarrow Li_{m+n}M_xO_v \quad (8.3)$$

2. *Conversion*: Most metal oxides belong to conversion-type materials and include binary compounds (M_xO_y, M: Mn, Fe, Co, Ni, Cu, Mo, Ru, etc.), ternary (binary metal) oxides ($A_xB_yO_z$, A or B: Mn, Fe, Co, Ni or Cu, A≠B), and complex oxides. In principle, conversion-type anodes undergo a reversible "conversion (redox)/displacement" electrochemical reaction, which involves the formation and decomposition of Li_2O along with the reduction to and oxidation of metallic NPs, as expressed in Eq. (8.4). Due to multi-electron reactions, conversion-type anodes can deliver large capacities and high energy density. Unfortunately, they are limited by many drawbacks such as poor conductivity, high redox potential, large potential hysteresis between oxidation and reduction, large volume changes, severe interparticle aggregation, unstable SEI, low cycling stability, and short lifetimes.

$$M_xO_v + 2yLi^+ + 2ye^- \leftrightarrow xM + yLi_2O \quad (8.4)$$

3. *Alloying*: Alloying-type metal oxides include binary tin oxides (such as SnO and SnO_2), ternary tin oxides $M_xSn_nO_y$ (M: Mg, Mn, Fe, Co,

Zn, Ca, Sr, Ba, Y, Nd, etc.), and mixed oxides ZnM_2O_4 (M: Fe, Co, and Mn). During the charging process, these metal oxides involve electrochemical reduction to the respective metals by Li ions, followed by the formation of Li alloys. These oxides usually participate in conversion and alloying reactions, as shown in Eqs. (8.5) and (8.6) using SnO_2 as a typical example. However, alloy anodes in LIBs have been limited by poor cyclic capability and high capacity fading due to the huge volume variation during alloying/de-alloying processes.

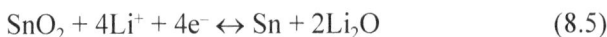

$$SnO_2 + 4Li^+ + 4e^- \leftrightarrow Sn + 2Li_2O \qquad (8.5)$$

$$Sn + 4.4Li^+ + 4.4e^- \leftrightarrow Li_{4.4}Sn \qquad (8.6)$$

Over the past several years, to alleviate the problems mentioned earlier, metal oxide/graphene composites have been developed in an intensive effort to improve the overall performance of metal oxide anodes in LIBs. A wide variety of synthetic methods have been investigated: (i) *ex situ* hybridization of pre-synthesized metal oxides and graphene or GO, followed by reduction; (ii) co-reduction of metal-containing precursors and GO sheets by chemical, hydrothermal, solvothermal, microwave, and/or thermal annealing; (iii) *in situ* growth of metal oxides onto graphene or GO, followed by reduction; and (iv) post-deposition of nanostructured metal oxides onto graphene using electrochemical and thermal evaporation methods. When fabricating graphene-based composites, metal oxides occur in various nanostructured forms such as nanotubes, nanowires, nanosheets, core-shell nanostructures, and hollow nanoparticles.

Direct mechanical mixing of graphene sheets with pre-synthesized metal oxides is the simplest method for producing composite electrodes in which diverse nanostructures typically coexist. In this system, graphene serves as an electrically conductive additive that helps construct an effective electron transfer network throughout the overall electrode. However, nanostructured metal oxides and graphene sheets are prone to aggregate into large ones, thereby showing limited improvement in the electrochemical performance. For a layered structure, metal oxides are alternated with graphene sheets to form a self-assembled layer-like composite, which can be further processed into a macroscopic paper or film by vacuum filtration or solution casting. The most commonly used method is based on electrostatic self-assembly between charged metal oxides or metal precursors and functionalized graphene/GO sheets.

In addition, sandwich-like graphene/metal oxide composite anode materials can be fabricated using graphene as a template. In this composite, metal oxides are immobilized between graphene sheets. Unique textural features of layer- and sandwich-structured composite anodes allows them to be free of additives and even metal current collectors. This allows high storage capacities to be obtained per mass and volume. Such hybrid anodes are mechanically flexible and have promising applications in portable, wearable, and implantable devices. In addition, graphene-wrapped metal oxides have been mainly fabricated by *in-situ* intercalation of metal-containing precursors into graphene or GO sheets followed by chemical and/or thermal treatment. Wrapping metal oxides with flexible graphene sheets can effectively prevent the detachment and agglomeration of pulverized metal oxides and accommodate the large stress and strain changes of the electrode during cycling. This architecture facilates electrical connectivity of the entire anode, and provides shortened path for rapid Li+ ions and electrons diffusion, thus effectively improving the sepecific capacity and rate performance of the anode materials. For example, a 3D graphene/Fe_2O_3 aerogel with porous Fe_2O_3 nanoframeworks well encapsulated within graphene is designed and fabricated, which could be dietly used as flexible anode for LIBs, as shown in Fig. 6.28.[39] This aerogel is fabricated by an excessive metal ions induced self-assembly and spacialy confined Ostwald ripening strategy: that is, small crystals or sol particles dissolve and redeposit onto larger crystals or sol particles. The hierarchical structure offers a highly interpenetrated porous conductive network and intimate contact between graphene and porous Fe_2O_3 as well as abundant stress buffer nanospace for effective charge transport and robust structural stability during electrochemical processes. The obtained free-standing 3D graphene/Fe_2O_3 aerogel is directly used as an anode for LIBs and shows an ultrahigh capacity of 1129 mAh/g at 0.2 A/g after 130 cycles and outstanding cycling stability with a capacity retention of 98% after 1200 cycles at 5 A/g.

Graphene can also be hybridized with metals (Sn) or semi-metals (Si and Ge) to fabricate anode materials for LIBs. Metals (Sn) or semi-metals (Si and Ge) can electrochemically react with lithium in the electrolyte (containing lithium ions) to form alloys, which have the ability to store lithium (Eq. 8.7). Typically, the theoretical specific capacities of fully lithiated silicon ($Li_{4.4}Si$), germanium ($Li_{4.4}Ge$), and tin ($Li_{4.4}Sn$) are calculated to be as high as 4200, 1624, and 994 mAh/g, respectively. Therefore, they can be used as anode materials for high-performance LIBs.

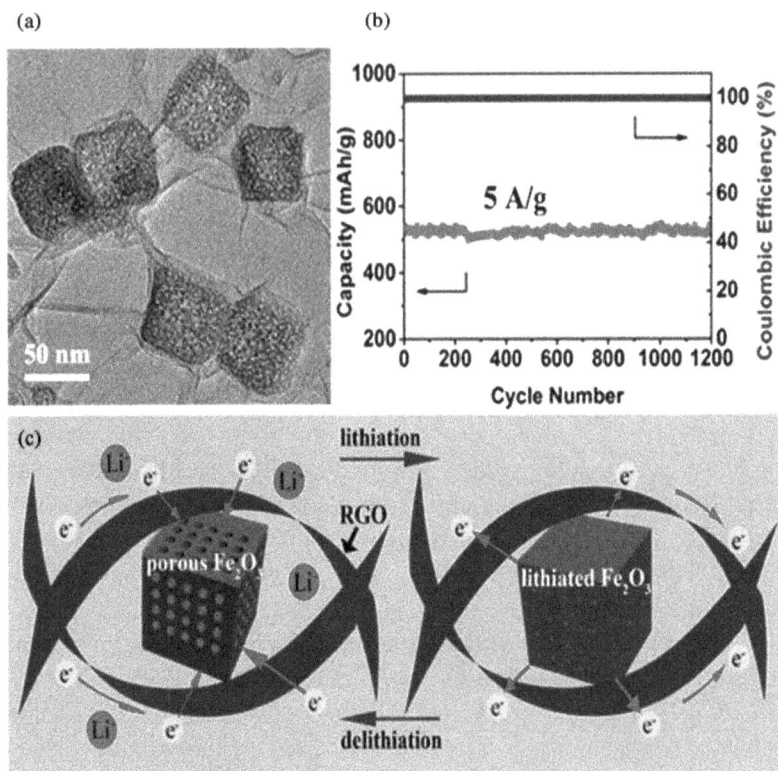

Figure 6.28.   Porous Fe_2O_3 nanoframeworks well encapsulated within 3D graphene are used as anode materials for LIBs[39]: (a) morphology of the composite; (b) cycle performance of the composite anode material; and (c) schematic of delithiation/lithiation reactions of the composite anode material.

$$M + nLi^+ + ne^- \leftrightarrow Li_nM \qquad (8.7)$$

However, the huge volume variations of these metals (Sn) or semi-metals (Si and Ge) upon lithiation/delithiation can produce ultrahigh internal stress that induces dramatic pulverization of electroactive materials. In this case, electrode cracking and fracture as well as an unstable SEI layer therefore bring about low-capacity retention, poor cyclability, and reversibility. Composites made of graphene and these materials with high Li storage capacity can suppress the volume expansion and pulverization of these materials, and the large specific surface area of graphene enables more active sites and electrolyte immersion, thus improving the cycling stability, specific capacity, and rate performance of the composite anodes. For example, protecting silicon films with single-layer graphene can improve the performance of silicon anode materials.[40] As shown in Fig. 6.29, silicon film (~200 nm or ~2 μm) is deposited on the carbon nanotube microfilm (CNM) current collector by DC sputtering, and then the single-layer graphene is grown by a CVD process on the silicon film to fabricate a 2D graphene/silicon/CNM (Gr-Si-CNM) composite film. In this composite film, the underlying CNM substrate is highly flexible and allows the Si film to expand and contract during lithiation/delithiation cycles, and the graphene capping layer also provides an interconnecting elastic web that toughens the Si film and reduces its tendency to pulverize and delaminate. In addition, the graphene capping can also stabilize the SEI film. Therefore, these Gr-Si-CNM composite films exhibit long cycle life (>1000 charge/discharge steps) with an average specific capacity of ~806 mAh/g. The volumetric capacity average after 1000 cycles of charge/discharge is ~2821 mAh/cm^3, which is 2 to 5 times higher than what is reported in the literature for Si nanoparticle-based electrodes.

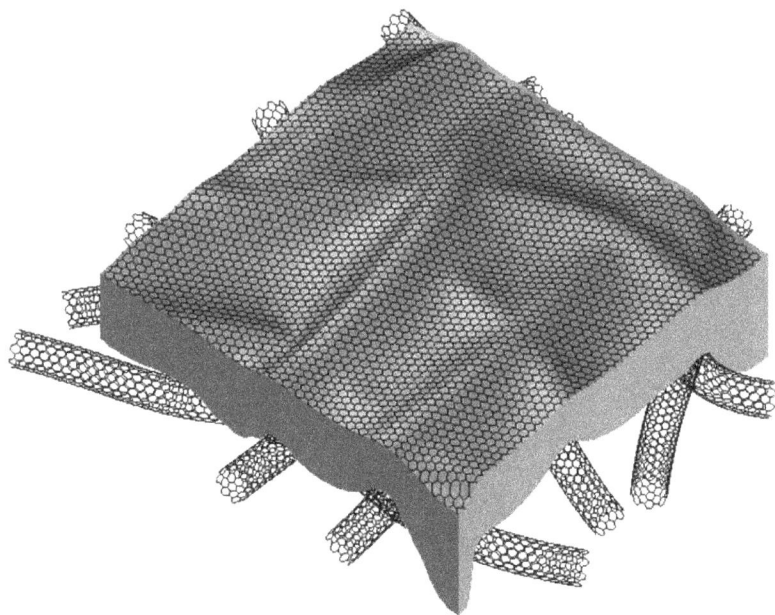

Figure 6.29.   Schematic representation of Gr-Si-CNM composite film.[40]

6.2.2.2 *Positive electrode (cathode) materials*

The cathode of LIBs is typically a transition metal compound that serves as the host for reversible Li-ion intercalation/deintercalation during charge/discharge processes and is composed of metal oxides, such as VO_2, $Li_3V_2(PO_4)_3$, $LiMnPO_4$, Li_2FeSiO_4, $LiNi_{0.5}Mn_{1.5}O_4$, and $LiMn_2O_4$. Among them, ternary cathode materials (referring to ternary metal oxides) including nickel-cobalt-aluminum oxide (NCA, such as $LiNi_{0.8}Co_{0.15}Al_{0.05}O_2$) and nickel-cobalt-manganese oxide (NCM, such as $LiNi_{0.6}Co_{0.2}Mn_{0.2}O_2$ or $LiNi_{0.8}Co_{0.1}Mn_{0.1}O_2$) have been widely reported to exhibit higher specific capacity, and these materials show a promising potential for applications in LIBs. The important characteristics of the cathode include the voltage at which it exchanges lithium, the amount of reversible Li-ion intercalation, the stability of the intrinsic material, and the transportation of electron conduction and Li-ion diffusion. The first two factors determine the energy density (usually high voltage leading to high energy density) and the latter factors limit the lifetime and rate performance of LIBs. However, practical applications of these cathode materials are limited by several issues such as low electrical conductivity, slow Li-ion diffusion, interparticle aggregation, large volume expansion, and irreversible phase transitions.

In order to improve the performance of metal oxide cathode materials, metal oxides are often mixed with other materials (such as graphene) to form composite materials. As mentioned previously, pristine graphene cannot be used directly to increase the storage capacity of Li ions. However, graphene can be used as an additive of metal oxide cathode materials to improve the performance of cathode materials; for example, it does provide a large surface to anchor and separate metal oxides, high electrical conductivity, and mechanical support to the cathode.

Recently, graphene-based composite cathode materials have been prepared by co-precipitation, hydrothermal, solvothermal, spray-drying, thermal annealing, sol-gel, and photothermal methods. Similar to anodes of metal oxide/graphene composites, these cathode composites can also form wrapped, anchored, encapsulated, layered, mixed, and sandwich-like architectures. For example, graphene nanosheets can be used as an additive to improve the performance of oxide cathode materials. As shown in Fig. 6.30, a novel processing strategy is developed utilizing an ethyl cellulose-stabilized dispersion of primary nanostructured lithium manganese oxide (nano-LMO) nanoparticles and graphene nanoflakes to realize a nano-LMO/graphene composite cathode with substantially improved packing density and active material loading.[41] In the composites, graphene nanoflakes offer advantages as a conductive additive including improved electrical conductivity, mechanical resilience, and high aspect ratio. As shown in Fig. 6.30(b), graphene is coated on the nano-LMO particle surfaces. During the charging/discharging, a thin and stable SEI layer is formed on graphene, which stabilizes the LMO/electrolyte interface, eliminates undesirable side reactions of the electrolyte/electrode, and suppresses Mn dissolution, thus ultimately affording excellent cycling stability. In addition, the enhanced charge transfer resulting from nano-LMO and the highly conductive graphene network further yield excellent rate capability (~75% capacity retention at 20 C rate) and unprecedented electrochemical performance at low temperatures with nearly full capacity retention at −20°C.

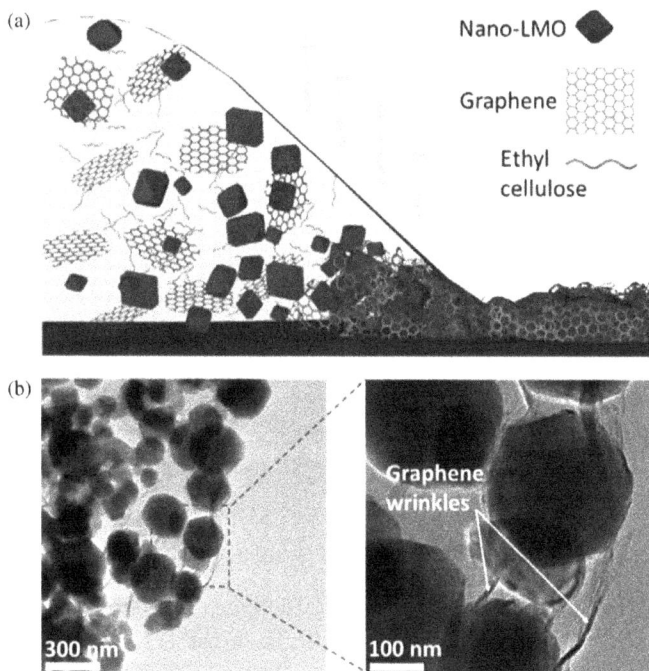

Figure 6.30. Schematic fabrication and characterization of nanostructured composite cathodes of Ethyl cellulose-stabilized graphene nanoflakes and nano-LMO[41]: (a) schematic illustration of fabrication and (b) morphology of the composites.

Functionalized graphene materials can also be used as cathode active materials for LIBs. Studies have shown that oxygen-containing groups on graphene, such as hydroxyl, carboxyl, and carbonyl, can serve as active sites for lithiation. Therefore, the existence of residual oxygen-containing groups on rGO prepared by the reduction of GO allows lithium storage activity, and rGO can be used as a cathode active material for LIBs. The all-graphene LIBs as described in the anode material section use partially reduced GO as the cathode active material.[37] Porous partially reduced GO can be obtained in a one-step annealing process at low temperature (120°C) in a controlled atmosphere. Within this voltage range between 1.5 and 4.5 V at a current rate of 0.1 A/g, the cathode initially delivers approximately 150 mAh/g at an average discharge voltage of 2.5 V. The capacity increases slightly to ~200 mAh/g over 100 cycles, likely because of the gradual activation of functional groups within the electrode.

Analysis indicates that the C=O functional group acts as a redox center for the storage of Li ions — reacting with Li ions to form Li–O–C. Note that the graphite oxide did not show satisfactory electrochemical activity as a cathode despite the presence of functional groups, which is attributable to its low electrical conductivity. It is expected that partial reduction of the graphite oxide increases electrical conductivity while maintaining an appropriate amount of C=O redox centers to be used as a cathode material, as Fig. 6.31 shows.[37] This outstanding cyclability is attributed to C=O/C–O faradaic surface reactions, which resemble pseudocapacitors that do not accompany a significant lattice expansion or contraction. Therefore, this cathode active material exhibits a high specific capacity upon repeated cycles with a Coulombic efficiency of ~93%. Due to remarkably fast Li storage capability, the rate capability of the graphene cathode is excellent, and a capacity of greater than 100 mAh/g is delivered at a current density of 3000 mA/g.

6.2.2.3 *Other metal-ion secondary batteries*

LIBs have dominated the field of electrochemical energy storage for the last 20 years. It still remains one of the most active research fields. However, there are difficult problems still surrounding LIBs, such as high cost, unsustainable lithium resources, and safety issues. Rechargeable batteries based on alternative metal elements (Na, K, Mg, Ca, Zn, Al, etc.) can provide relatively high power density and high energy density. The

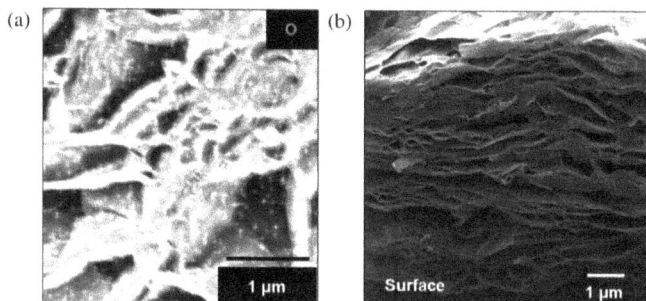

Figure 6.31. Cathode active material synthesized by partially reduced GO[37]: (a) FE-SEM image of the functionalized graphene cathode, merging with O atoms acquired by EDS mapping analysis and (b) FE-SEM images of the functionalized graphene cathode.

structure and working mechanism of these metal-ion secondary batteries are similar to LIBs; cations serve as charge carriers and shuttle back and forth between the negative and the positive electrodes during charge–discharge cycles. At present, the research on these new metal-ion secondary batteries is still in its infancy, and its actual performance and practicability are not as good as those of mature LIBs. Current research is focused on improving the performance of these batteries and enhancing their practical deployment. Graphene functional materials can also be used as electrode materials for these novel metal-ion batteries to enhance their performance. In these new batteries, the role of graphene materials is similar to its role in LIBs.

6.2.3 *Other secondary batteries*

6.2.3.1 *Lithium-sulfur battery*

A lithium-sulfur battery (Li-S battery) is an electrochemical energy storage system with high energy density. As shown in Fig. 6.32, the basic structure of a Li-S battery is similar to that of an LIB, consisting of a positive electrode, a negative electrode, a separator, an electrolyte, and a current collector.[42] However, the materials and energy storage mechanisms of Li-S batteries are different from those of LIBs. Generally, a Li-S battery uses lithium as the negative electrode and sulfur as the positive electrode. During discharge, the lithium ions in the electrolyte migrate to the cathode

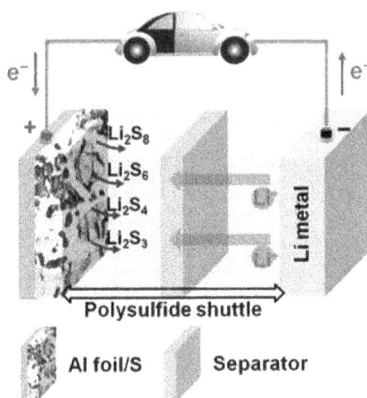

Figure 6.32. Schematic of the structure of a typical Li-S battery with electrode configuration.[42]

where the sulfur is reduced to lithium sulfide (Li_2S); actually, (the sulfur reduction reaction to lithium sulfide is much more complex and involves the formation of lithium polysulfides (Li_2S_x, $2 \leq x \leq 8$), while at the anodic surface, dissolution of the metallic lithium occurs, with the production of electrons and lithium ions. During charging, the sulfur in Li_2S loses electrons and is oxidized, while the lithium ions in the electrolyte get electrons from the negative electrode and are reduced to metallic lithium (Eq. (8.8)). Based on this principle, the electrochemical reduction of sulfur has a theoretical capacity of 1672 mAh/g, and Li-S batteries have emerged with a theoretical energy density of 2500 Wh/kg, which possess better performance than LIBs. In addition, the active material sulfur in Li-S batteries has the advantages of natural abundance, low cost, and low toxicity. Therefore, Li-S batteries have been considered to be the next-generation high-performance lithium batteries.

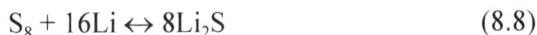

$$S_8 + 16Li \leftrightarrow 8Li_2S \qquad (8.8)$$

However, significant challenges remain that hinder the commercialization of this technology: (i) Elemental sulfur and solid sulfides (Li_2S and Li_2S_2) are both insulators for electron and ion transport, resulting in enlarged internal resistance and reduced rate capability; (ii) severe volumetric change (~80%) of S during charge and discharge process will gradually decrease the mechanical integrity and stability of the electrode over cycling; (iii) polysulfide anions formed as reaction intermediates in the charge–discharge process are highly soluble, and they are easily reduced to solid precipitates while migating through the seperator to Li anode during cycling, resulting in the loss of active mass (sulfur). At present, the strategy of encapsulating sulfur with carbon materials is considered to be one of the effective methods to address these issues. The encapsulation of sulfur with carbon materials is supposed to enhance the electrical conductivity of sulfur, trap soluble Li_2Sn intermediates, and accommodate volume variation of electrodes during cycling. Among these carbonaceous materials, the application of graphene in Li-S batteries is very promising due to its unique 2D structure, high conductivity, and superior mechanical flexibility, and it is widely used for fabricating sulfur cathodes, separators, and current collectors for Li-S batteries.

One way to deposit sulfur onto graphene is by taking advantage of the strong interaction between sulfur and carbon and directly impregnating aggregated graphene sheets with melted sulfur. Graphene-based sulfur cathodes exhibit high electrical conductivity and inhibit the shuttle effect of polysulfides between the cathode and anode. Since pristine graphene is non-polar, its poor compatibility with the polar sulfide Li_2S/Li_2S_2 (discharge product) will result in the separation of graphene and Li_2S during cycling, causing rapid degradation of capacity. Therefore, graphene is often functionalized — heteroatom doping, hydroxylation, and amination — so that it has polar sites or groups, such as nitrogen atoms, hydroxyl groups, or amino groups. After functionalization, the interactions between graphene materials and sulfur/polysulfides are enhanced, effectively inhibiting the separation of graphene materials and sulfur active materials and improving the cycle performance of the battery.

In nitrogen-doped graphene, the spin density and charge distribution of carbon atoms will be influenced by the neighbor nitrogen dopants, resulting in the formation of active centers for the chemisorption of sulfur and enhancement of the adsorption ability of both sulfur and polysulfides. Nitrogen doping can promote chemical bonding between carbon scaffold and sulfur chains upon heat treatment during the sulfur loading process, which enables uniform distribution of sulfur in the carbon host initially as well as after redeposition, improving cycling performance. Nitrogen doping can also greatly enhance the adsorption of soluble lithium polysulfide intermediates, which can effectively retard the diffusion of the lithium polysulfides and trap them within the cathode, thus improving both cycling stability and Coulombic efficiency. Importantly, in contrast to widely used non-conductive adsorbents, nitrogen-doped carbon is highly conductive and thus allows for direct redox and utilization of adsorbed material, rather than requiring desorption and diffusion of polysulfides to an electrochemically active surface. For example, highly crumpled nitrogen-doped graphene sheets are used to improve the performance of sulfur cathodes.[43] As shown in Fig. 6.33, highly crumpled nitrogen-doped graphene sheets are prepared by using GO as the precursor and cyanamide as the nitrogen doping source in the presence of a pore-forming agent via a simple thermally induced expansion strategy. This nitrogen-doped graphene sheet shows an ultrahigh pore volume (5.4 cm^3/g) and specific surface area (1158 m^2/g) with the pore size ranging from 2–50 nm. As expected, the nitrogen-doped graphene sheet sulfur cathode shows a high capacity of 1227 mAh/g and a long cycle life (75% capacity retention after 300 cycles) at a high sulfur content of 80 wt.%. Thanks to the high capacity and high sulfur loading, an areal capacity of 5 mAh/cm^2 is achieved on the electrode level at such a high sulfur content.

Graphene with a microporous or mesoporous structure shows a larger specific surface area, better mechanical strength, and faster mass transfer and charge transport capabilities, which can further improve the conductivity of the sulfur cathode, better accommodate the volume expansion of sulfur, and inhibit the diffusion of polysulfides. In addition, due to its superior conductivity, flexibility, and compatibility with sulfur, graphene can be directly fabricated into electrodes with sulfur without the need for metal current collectors and polymer binders. In the case of Li-S batteries based on such electrodes, the removal of metal current collectors and

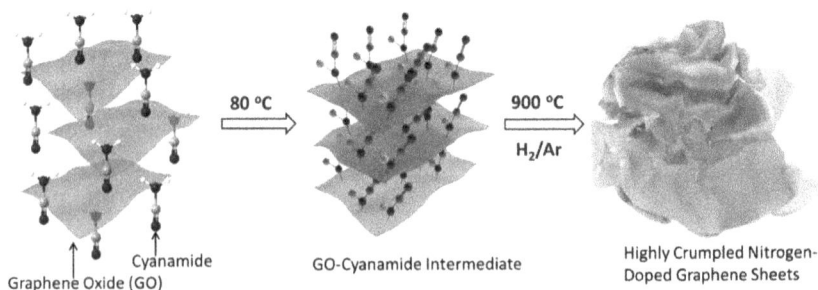

Figure 6.33. Schematic illustration of the synthesis of highly crumpled nitrogen-doped graphene sheets.[43]

binders contributes to the improved practical energy density because of the lighter weight.

Different kinds of functionalized graphene can be used to fabricate an all-graphene structure for the sulfur cathode (Fig. 6.34).[44] In this sulfur cathode, highly porous graphene (HPG) is used as a support for the active material. HPG is a kind of rGO, which has a high specific surface area (771 m²/g), a high pore volume of 3.51 cm³/g, and a broad pore size distribution ranging from 1 to 60 nm. The large pore volume of HPG enables the accommodation of a high amount of sulfur, achieving a high sulfur content of 80 wt.% in the S/HPG composite.

The partially oxygenated graphene (POG) not only has good electrical conductivity (square resistance of 25 Ω/sq) but also contains a moderate number of oxygen-containing functional groups, which can be used as a polysulfide adsorption layer in the sulfur cathode. In addition, the areal density of the POG interlayer is only 0.2 mg/cm (the thickness is about 3 µm), much lighter than the sulfur loading in the cathode (about 5 mg/cm), ensuring a high energy density.

The highly conductive graphene (HCG) prepared by an intercalation-exfoliation technique shows a high conductivity with a low square resistance of 3 Ω/sq, and it is used as a current collector in the sulfur cathode. The HCG current collector enables improved adhesion to the sulfur due to its surface roughness and thus lowers the impedance and polarization of the Li-S battery. On the other hand, it contributes to the improved practical energy density of the cell because of its light weight (12 µm in thickness, 1 mg/cm² in areal density). Benefiting from the all-graphene structural design, the cathode exhibits a high initial discharge capacity of 1500 mAh/g at 0.34 A/g and retains a reversible discharge capacity of 841 mAh/g after 400 cycles.

In Li-S batteries, commercial polymer separators, such as polypropylene (PP), have large pore sizes, which cannot prevent the diffusion of

Figure 6.34. Schematic of the all-graphene structural design of the sulfur cathode.[44]

polysulfides from the sulfur cathode to the lithium anode, resulting in a decrease in battery performance. The insertion of a functionalized graphene material interlayer can both significantly decrease the resistance of the sulfur cathode and enhance active material utilization by trapping the soluble lithium polysulfides within the carbon interlayer. The functionalized graphene layer exhibits an appropriate pore structure, which will not influence the diffusion of electrolyte ions to the sulfur active material, and can be used as an interlayer between the sulfur cathode and the separator. In addition, the graphene interlayer also provides faster electron transport sites to enhance the rate performance of the battery. As mentioned earlier, the wrinkled nitrogen-doped graphene (Fig. 6.33) has a strong adsorption capacity for polysulfides and can be used as an interlayer between the sulfur cathode and the separator. A nitrogen-doped graphene interlayer is prepared by coating nitrogen-doped graphene sheets on a separator (Celgrad 2325) using PVDF as the binder (Fig. 6.35). The thickness of the nitrogen-doped graphene sheet layer is controlled at 10 μm, and a better interlayer structure can be obtained when the mass loading is 0.4 mg/cm². At this time, the introduced interlayer can not only effectively trap soluble polysulfides to enhance the performance of the battery but also avoid the problematic weight increment of the battery due to the introduction of the interlayer.

The use of electrochemically active graphene functional materials as interlayers can not only effectively trap polysulfides but also provide extra capacity to compromise the capacity loss brought about by the

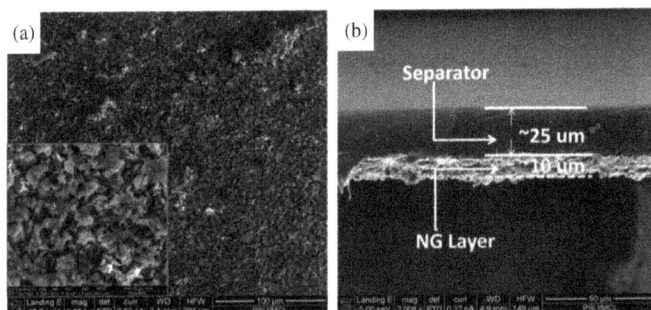

Figure 6.35. SEM images of the nitrogen-doped graphene sheet-coated separator[43]: (a) low- and high-magnification (inset) SEM images (top view) of the graphene sheet-coated separator and (b) cross-sectional SEM images of the graphene sheet-coated separator.

introduction of the interlayer. As shown in Fig. 6.36, the *N*, *S*-co-doped graphene sponge can not only be used as an electroactive interlayer between the sulfur cathode and separator but can also provide additional capacity.[45] The *N*, *S*-co-doped graphene sponge is prepared by reducing GO through an *in situ* oxidation/reduction process using thiocarbohydrazide as a reducing agent. The carbon surface that is enriched in oxygen, nitrogen and sulfur by doping, and the *N*, *S*-co-doping can significantly enhance the binding of lithium polysulphides as compared to the undoped or single *N*/*S*-doping graphene, thus leading to better cycle performance. The *N*, *S*-co-doped graphene sponge is a kind of rGO, which has high electrical conductivity and is able to improve the rate performance of the battery. In addition, the *N*, *S*-co-doped graphene sponge provides additional capacity (about 30% of the total capacity) for the battery due to doping or encapsulation of sulfur. Due to the aforementioned benefits, when this *N*, *S*-co-doped graphene sponge is used as an electrochemical interlayer for Li-S batteries, the reversible capacity of the battery can reach 2193 mAh/g at 0.2°C, and an impressive value of 829.4 mAh/g is still obtained at higher rates of 6°C.

6.2.3.2 *Lithium-air (oxygen) batteries*

Lithium-air (oxygen) batteries are electrochemical energy storage devices that provide a greatly increased theoretical specific energy, holding significant potential and prospects. They use lithium metal as the negative

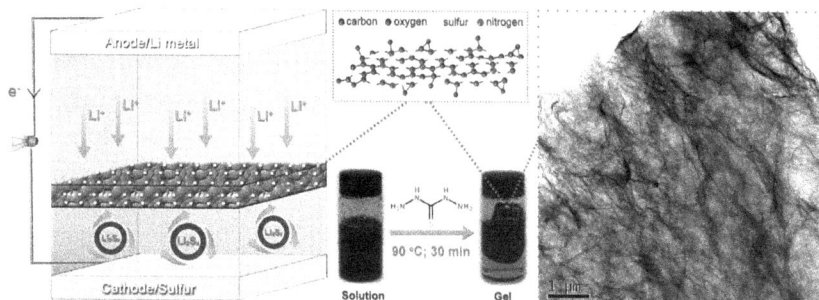

Figure 6.36. Schematic representation of the synthesis process, microstructure, and assembled Li-S batteries with interlayer of the N,S-co-doped graphene and TEM images of the sample.[45]

electrode (anode) material and oxygen in the air as the positive electrode (cathode) reactant. Theoretically, due to the inexhaustible oxygen in the air, the capacity of lithium-air batteries is limited only by the metal lithium electrodes. The theoretical specific capacity of Li-air batteries is extremely high (~5200 Wh/kg including the oxygen mass or ~11,400 Wh/kg excluding oxygen mass). A typical Li-air battery is mainly composed of four components: a lithium metal anode, a separator, an electrolyte, and an anode with a porous structure. Electrolytes of Li-air batteries can be divided into four types: non-aqueous liquid, solid-state, aqueous, and hybrid electrolytes. The chemistry at the cathode differs depending on the electrolyte. Non-aqueous liquid Li-air batteries have been widely studied in the past decades because they can provide extremely high energy density with a simple battery structure.

The working principle of a non-aqueous Li-air battery is featured in Fig. 6.37. During discharge, the cathode undergoes an oxygen reduction reaction (ORR), and Li ions move from the anode to the cathode to combine reduced O_2 molecules, forming Li_2O_2 and depositing it on the cathode surface. During charging, the cathode undergoes oxygen evolution

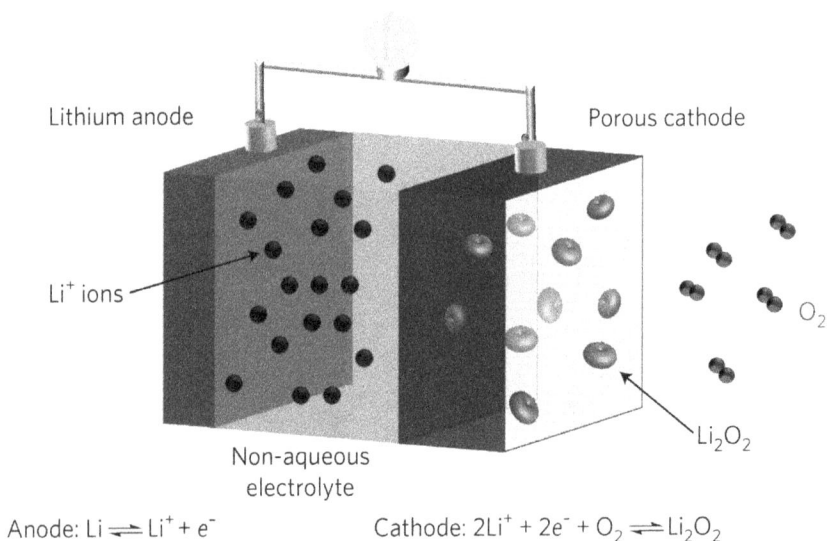

Anode: $Li \rightleftharpoons Li^+ + e^-$ Cathode: $2Li^+ + 2e^- + O_2 \rightleftharpoons Li_2O_2$

Figure 6.37. Schematic of a typical non-aqueous Li-air battery and the electrochemical reactions on positive and negative electrodes.[46]

reaction (OER), and this process proceeds in reverse with Li ions moving back and O_2 being released. Based on this working principle, the theoretical specific capacity of non-aqueous Li-air batteries is about 3500 Wh/kg.

However, in practice, due to the limit of the material system, the measured energy density of Li-air batteries is still significantly lower than the theoretical value. During the charge/discharge process, the activity of OER/ORR at the cathode is not high and the overpotential during the reaction is too high, resulting in low energy efficiency. A high overpotential can cause the decomposition of the electrolyte, thus decreasing the life of the battery. Therefore, to improve the performance of Li-air batteries, it is necessary to increase the OER/ORR activity of cathode materials and reduce the overpotential.

In addition, the composition and porosity of cathode materials are also crucial to improve the performance of Li-air batteries. Porous cathode of the Li-air batteries provides space for the electrochemical reaction and produced Li_2O_2 in the discharging process. Therefore, conductive and highly stable porous gas diffusion cathode is expected to improve the performance of Li-air batteries.

A suitable pore size distribution is also required. Studies have shown that mesopores (2~50 nm) are conducive to enhancing the ORRs, and pores that are too large or too small are not suitable; if the pores are too small, they are readily blocked by Li_2O_2 formation, leading to a sluggish or terminated diffusion of O_2 and large-sized pores might be flooded with the electrolyte, preventing the oxygen from diffusing into them. It can be concluded that high-performance cathode materials are required to exhibit high specific surface area, good conductivity, and excellent OER/ORR activity. Therefore, due to their high chemical stability, superior electrical conductivity, and extremely large specific surface area, graphene materials are ideal candidates for cathode materials in Li-air batteries. At present, graphene and its derivatives have been extensively investigated as electrodes for Li-air batteries.

6.2.3.3 *Support material for cathodes*

In Li-air batteries, graphene and noble metals/metal oxide with high catalytic activity are often hybridized as cathode materials. In most cases, graphene is employed as a support to achieve a good dispersion of metal/metal oxides and improve charge transfer during the charge/discharge process. Graphene with a large high specific surface area and electrical conductivity can effectively improve catalytic activity. Graphene has an open framework structure, so it can very readily load catalytically active materials by *in situ* growth technology. In this process, graphene and the active materials are bonded covalently or the generated oxygen atoms serve as bridges between them, which can greatly improve electron transport during charging and discharging.

Functionalized graphene, such as rGO, can be used to support Pt, Pd, and Ru nanoparticles to enhance the OER ability of these metals during charging. rGO has many oxygen-containing functional groups and defects so that it can hybridize with these noble metals via strong bonding. After incorporating graphene, the OER overpotentials of different noble metals decrease greatly during charging, which is most significant in the case of the Ru-rGO composite (from 4.3 V to 3.5 V); Ru nanoparticles in the composite facilitate thin film-like or nanoparticulate Li_2O_2 formation during ORR.[47] During charging, this type of Li_2O_2 decomposes at lower overpotentials. Therefore, the Ru-rGO composite cathode exhibits enhanced cycling stability in Li-air batteries.

The 3D functionalized graphene materials not only show high conductivity and excellent flexibility but also have a specific surface area and more exposed active sites. Therefore, such graphene materials as support materials can significantly improve the conductivity of metals or metal oxides, thereby improving the performance of cathode materials. For example, 3D NiO-graphene foam (NiO-GF) and Ni-graphene foam (Ni-GF) are used as the cathodes of Li-air batteries and show an extremely high specific capacity performance of 25,986 mAh/g and 22,035 mAh/g, respectively (current density is 0.1 A/g).[48] These composite materials are synthesized by GO, $Ni(NO_3)_2 \cdot 6H_2O$, and $CO(NH_2)_2$ via the hydrothermal method and heat treatment. In these composites, NiO or Ni nanoparticles are uniformly distributed on the 3D network structure of graphene foam (Fig. 6.38). As a 3D reticular material, GF can improve the efficiency of electron transfer and provide a higher surface area for NiO and Ni to grow on, which offers more active sites for catalysts. Meanwhile, NiO and Ni are capable of effectively reducing the energy barrier of the reaction from Li and O_2 to Li_2O_2 and improving the efficiency of this reaction. Therefore, using the 3D reticular material, which is made by combining rGO and NiO or Ni, as the cathode material of the Li-air battery is an effective strategy for future energy storage systems.

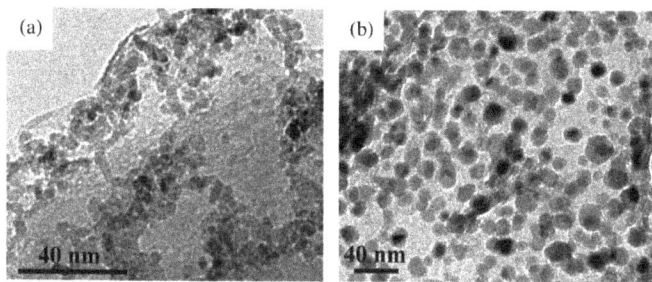

Figure 6.38. TEM images of (a) NiO-GF and (b) NI-GF.[48]

6.2.3.4 *Cathode active material*

Graphene prepared by the chemical method contains many edges and defect sites, which can serve as catalytically active sites for ORR and OER. The fine-structure modification of graphene edges and defects can be controlled by the preparation process. Graphene materials with rich edge and defect structures can be used as metal-free ORR/OER catalysts, so they are often used as cathode materials for Li-air batteries.

In addition, chemically doped graphene can effectively adjust the charge density and surface chemical properties of graphene, resulting in a significant increase in its electrochemical activity. For example, N-doped graphene is active in catalyzing ORR. Nitrogen atoms are highly electronegative, which makes the nearby carbon atoms the active sites for Li/O_2 adsorption, which is beneficial to the nucleation of Li_2O_2 on the graphene surface. Graphene containing oxygen functional groups (hydroxyl, epoxide, carbonyl, and carboxyl) also has ORR and OER catalytic activity. Hierarchical porous graphene can be used as a metal-free cathode active material for Li-air batteries. For example, graphene nanosheets rich in defects and oxygen functional groups (hydroxyl, epoxide, and carboxyl) can be obtained by thermal expansion and simultaneous reduction [Fig. 6.39(a)].[49]

Using functionalized graphene and binder materials (DuPont Teflon PTFE-TE3859 fluoropolymer resin aqueous dispersion, 60 wt.% solids) as raw materials, hierarchically porous functionalized graphene as the electrode is fabricated using a microemulsion solution. As shown in Fig. 6.39(b) and 6.39(c), in this electrode, the functionalized graphene

Figure 6.39. Functionalized graphene as an example of positive electrode active material for lithium-air (oxygen) batteries[49]: (a) schematic structure of functionalized graphene and schematic diagram of catalytic reaction (carbon atoms are gray, oxygen atoms are red, and hydrogen atoms are white); (b) TEM image; and (c) high-resolution TEM image of functionalized graphene.

aggregates into loosely packed "broken egg" structures, leaving large interconnected tunnels that continue through the entire electrode depth. During discharge, the robust large tunnels can function as "highways" to supply oxygen to the interior parts of the air electrode, while the small pores on the walls are the "exits" which provide triphase (solid–liquid–gas) regions required for oxygen reduction. Density functional theory (DFT) results show strong interactions between the redox-reaction product Li_2O_2 and defective sites of graphene (Fig. 6.39(a)), indicating that the Li_2O_2 particles most likely prefer to nucleate and grow around the defective sites; however, in the vicinity of those defective sites, the aggregation of Li_2O_2 clusters is energetically unfavorable. Therefore, the deposited Li_2O_2 would form isolated nanosized "islands" on the functionalized graphene's surface. The limited size or thickness of the reaction products with preferred growth points may also improve the rechargeability of Li-air batteries because it prevents the continuous increase of electrode impedance and provides better access for a catalyst during the charging process. Thanks to its hierarchically porous structure and lattice effect, this graphene cathode delivers an extremely high capacitance (15,000 mAh/g) during discharge.

6.2.4 *Summary*

In summary, functionalized graphene materials have been widely used in various electrochemical energy storage devices, such as SCs, LIBs, Li-S batteries, and lithium-air batteries. In these devices, functionalized graphene materials are mainly used to fabricate electrodes. The purpose of introducing functionalized graphene materials is to further improve the energy density, power density, and cycle stability of energy storage devices. In order to achieve these goals, a variety of means and strategies have been developed, such as developing graphene functionalization methods, controlling the morphology and structure of graphene functional materials and graphene composite materials, and creating a device structure design based on functionalized graphene materials. From this discussion and analysis, it can be concluded that high-performance graphene electrode materials need to have properties such as high specific surface area, high conductivity, high storage density, suitable defect/functional groups, and optimized composite structure and morphology.

Although functionalized graphene materials have made some progress in the field of energy storage, their application in this field is still in its infancy. In order to achieve greater breakthroughs and progress, further efforts are needed including the following: (i) develop *in situ* characterization and detection technologies and further study the mechanism of charge transfer and storage in functionalized graphene materials, (ii) employ functionalization or hybridization to prevent the aggregation and restacking of graphene materials during use, and (iii) design and optimize the morphology and microstructure of graphene composite materials from the system level. In addition, the structure of graphene-based devices and the compatibility of components in devices with graphene materials must also be considered.

6.3 Biomedicine

Functionalized graphene materials show great application potential in the field of biomedicine due to their diverse properties. Functionalized graphene materials have a very high specific surface area and a variety of functionalized groups, so they can efficiently adsorb or connect a variety of molecules and functionalized polymers for drug or gene delivery and medical treatment. Functionalized graphene also has a variety of acoustic, optical, electrical, magnetic, and semiconducting properties, which can be used for bioimaging and to fabricate different types of biosensors. Some types of functionalized graphene also have NIR absorbance due to which they can be used as photothermal agents for *in vivo* cancer treatment.

Due to the different preparation methods and structural compositions, graphene can be divided into graphene oxide (GO), reduced graphene oxide (rGO), and graphene prepared by the CVD method. These graphene materials with different structures and groups are applicable in different fields of biomedicine.

GO has abundant hydrophilic groups (carboxyl, epoxide, and hydroxyl) at the edge of the sheet and defect sites, and it also contains sp^2- and sp^3-hybridized hydrophobic carbon frameworks; thus, this special structure makes GO exhibit an amphipathic nature. Abundant hydrophilic groups make it highly dispersible in water or polar solvents. The hydrophobic carbon framework is able to complex hydrophobic or aromatic molecules through hydrophobic interaction or π-π interaction. A variety of oxygen-containing groups (carboxyl, epoxide, and hydroxyl) on GO not only enable it to complex cations (such as metal cations) and hydrogen bond donors/acceptors through electrostatic interaction and hydrogen bonding but also provide rich chemical reaction sites to attach other molecules through covalent bonds. In addition, defect sites produced during the preparation of GO can be used to adsorb proteins and DNA/RNA.

rGO can be prepared from GO by thermal, electrochemical, and chemical reduction. Compared with GO, it has a relatively complete aromatic skeleton; compared with graphene prepared by the CVD method, rGO retains some oxygen-containing groups (carboxyl, epoxide, and hydroxyl). Therefore, rGO has relatively balanced physical and chemical properties among the three types of graphene materials — there are chemical reaction sites on its surface, and it shows great solubility in solvents while exhibiting excellent photoelectric and mechanical properties.

Graphene prepared by the CVD method has high conductivity due to the relatively intact conjugated aromatic skeleton, fewer defects, and lack of oxygen-containing groups. In addition, its aromatic framework is rich in π electrons, so it is often used to make biosensors. However, graphene prepared by the CVD method cannot be dispersed in water, so it is not suitable for nanomedicine or use as a nanocarrier.

Functionalized graphene materials with diverse structures and rich functionalities can also be prepared by functionalizing these three kinds of graphene. These functionalized materials not only have the advantages of graphene itself but also present superior biocompatibility and rich biological functionality.

If graphene-based materials are expected to be used in the biomedical field, their biocompatibility and cytotoxicity have to be taken into account. Recent research has shown that graphene and graphene composites are potentially toxic to the human body. This is mainly manifested in two aspects: cytotoxicity and genotoxicity. Studies have shown that graphene-based materials can damage cell walls: (i) They cause physical damage to cell walls or extract phospholipids from cell walls. (ii) Graphene composite materials can interact with DNA/RNA or destroy DNA/RNA, thereby affecting physiological activities of normal human. The toxicity of graphene composites depends on the type (GO, rGO, and G), the nanosheet size, functional groups, injection methods, dosage, and concentration of graphene. By tuning these factors, the toxicity risk of graphene composites in biomedical applications is controllable. However, in order to further understand the working principle of graphene and its composites in biomedical applications and their impact on the human body, in-depth and long-term studies on their biocompatibility and potential toxicity are necessary.

6.3.1 *Drug delivery*

Due to their high specific surface area, controllable biocompatibility, and rich chemical modification methods, graphene-based materials have been used as carriers for drug delivery. Commonly used graphene carriers are based on GO and rGO. This is because GO and rGO are easy to prepare in large quantities; the oxygen-containing groups presented at the edge or surface of the sheets not only exhibits excellent solubility in water but also provides abundant active sites for further functionalization. Structurally, the main difference between graphene and GO is the presence of oxygenated groups on the surface of the latter, which results in better solubility in aqueous media, easier handling, and a richer surface chemistry. The hydrophobic sp^2 and sp^3 zones of GO make it possible to adsorb drug molecules through hydrophobic interactions or π-π interactions.

Functionalization of GO or rGO by covalent or non-covalent methods can greatly improve their aqueous dispersibility and biocompatibility, enhance their cell membrane permeability, and reduce cytotoxicity. The covalent functionalization method is brought about by reactions between the oxygen-containing groups on GO and biocompatible polymers (such as polyethylene glycol (PEG), chitosan, and polyethylenimine); non-covalent functionalization is achieved through hydrophobic interaction and π-π interaction between functional molecules and graphene-based materials.

Functionalized graphene materials can be used as delivery platforms for drug molecules and genes, transporting them from *in vitro* to *in vivo*. After the drug molecules are transported into the cell by the graphene functional material, the drug molecules are released due to the change in the environmental conditions of the material or external stimuli (such as

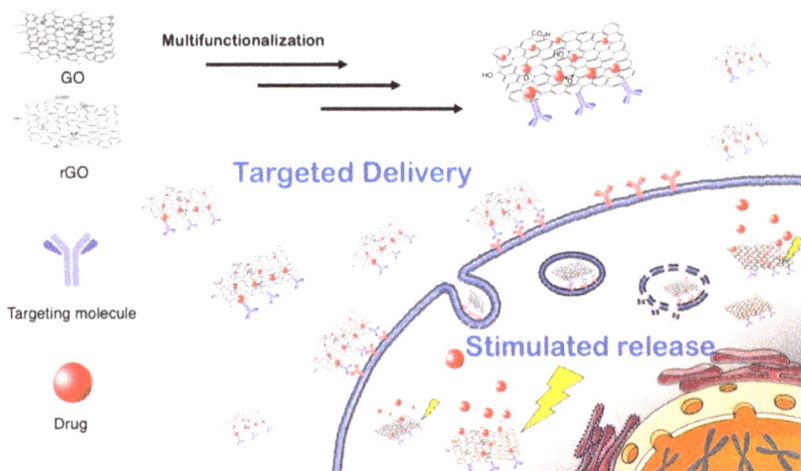

Figure 6.40.   Representative illustration of the applications of GO and rGO in drug delivery.[50]

pH variation or light). Active targeting of GO constructs favors the accumulation of the nanomaterial in the desired tissues, enhancing its therapeutic results while decreasing the side effects. Figure 6.40 shows a complete scheme of graphene functional material as a carrier for drug delivery[50]: (i) functionalization of GO or rGO, (ii) loading of targeting molecules (make the carrier localize to the target cells, which is not necessary), (iii) loading of drug molecules, (iv) the composite loaded with drug molecules transported via endocytosis into the cells, and (v) release of drug molecules under external stimuli.

For example, GO modified with dopamine (DA) can be used for the delivery of the anticancer drug methotrexate (MTX) into the DA-receptor positive human breast adenocarcinoma cell line.[51] DA is the targeting molecule of the positive human breast adenocarcinoma cell line, which is attached via covalent bond to GO, and the produced graphene composite can deliver MTX into DA receptor positive human breast adenocarcinoma cell line (Fig. 6.41). Then, the electrostatic interaction and hydrophobic interaction between the drug molecule MTX and the GO composite are used to load the MTX molecule onto the DA/GO composite. The approximate pKa value of MTX is 4.7 and when the environmental pH is lower

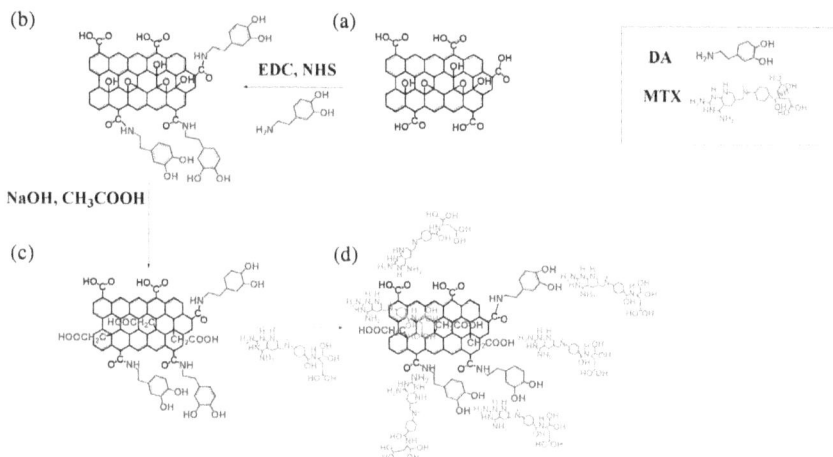

Figure 6.41. Schematic of synthesis process of DA-modified GO (DA-nGO) and the loading of MTX on DA-nGO.[51]

than 4.7, $-NH_2$ of MTX forms $-NH_3^+$ since the pKa distribution of GO is higher than 3.35; the carboxyl groups on GO sheets are negatively charged ($-COO^-$). Therefore, there is a strong electrostatic interaction between $-NH_3^+$ of MTX and $-COO^-$ of GO. Moreover, there is a hydrophobic interaction between the basal plane of nGO and the aromatic parts of MTX. Driven by these two kinds of interactions, MTX molecules are absorbed into the DA-modified GO composite. When the compounds loaded with drug molecules move to the target cells (positive human breast adenocarcinoma cell line), the environment becomes neutral pH. The $-NH_3^+$ groups of MTX are altered to $-NH_2$, and therefore the electrostatic interactions between $-NH_3^+$ groups of MTX and $-COO^-$ of GO are decreased, which accelerates the drug release at neutral pH conditions. So far, the delivery and release of the anticancer drug MTX have been completed.

In addition to drug delivery, graphene functional materials can also be used to deliver genes to diseased cells for gene therapy. Gene therapy is a medical technology that aims to produce a therapeutic effect through the transfer of genetic material into cells. Functionalized graphene materials transfer genes in a manner similar to drug delivery. The driving force of gene loading on functional materials is mainly the electrostatic force between the anionic phosphate backbone of the gene and the cationic surface of the complex. However, it is worth noting that the gene may be decomposed by enzymes during the transfer process, and the gene itself cannot efficiently penetrate the cell membrane. Therefore, in order to improve the cell transfection efficiency, it is necessary to inhibit the decomposition of the gene during the transfer process and improve its membrane permeability. Studies have shown that GO complexes can improve the permeability of genes (such as siRNA or plasmid DNA) to penetrate cells and can effectively protect genes from their breakdown.

For example, poly-L-lysine (PLL) and tetrapeptide RGDS (Arg-Gly-Asp-Ser) modified GO (GO-PLL-RGDS) is synthesized and used to deliver small interfering RNA (siRNA) induced with vascular endothelial growth factor (VEGF), VEGF-siRNA for cancer therapy (Fig. 6.42).[52] PLL is a water-soluble cationic polymer with excellent biocompatibility, which can improve the water solubility and biocompatibility of the nano-complex. Tetrapeptide RGDS can actively target tumor via the interaction of RGD with the integrin $\alpha_v\beta_3$ overexpressed on the cytomembrane of cancer cells, thus enhancing the tumor targeting efficiency of gene carrier GO-PLL-RGDS. Studies have shown that 10 μg of GO-PLL-RGDS complex could load 1 μg of VEGF-siRNA, and the release is slow and sustained. The expression of VEGF-siRNA protein is down-regulated by 51.71%. Moreover, GO-PLL-RGSD complex exhibits low cytotoxicity in the MTT assay.

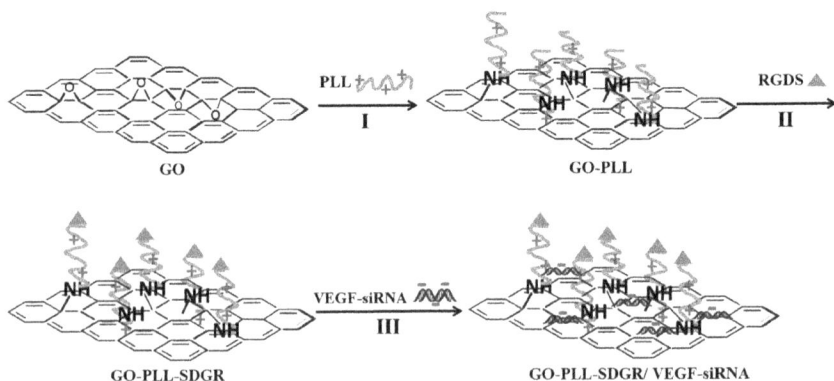

Figure 6.42. Schematic diagram for GO-PLL-SDGR and GO-PLL-SDGR/VEGF-siRNA preparation.[52]

6.3.2 *Biosensors*

In the field of biomedicine, rapid and sensitive detection of biomolecules and biological tissues is of great significance for clinical diagnosis and in healthcare. In living cells or the human body, real-time imaging of specific cells or tissues and real-time visual monitoring of molecular activities and biological processes in living cells are important means to study the physiological activities of cells and tissues and have a profound impact on biomedicine and biological science. In order to achieve the abovementioned goals, the development of efficient and sensitive biosensors and bioimaging technologies is important. The basic operating mechanism of these technologies is to convert the specific recognition of the target (molecule, cell, or biological tissue) into measurable signals such as light, electricity, and magnetism.

Among biosensing and bioimaging materials (organic/inorganic nanoparticles, carbon nanotubes, and metal oxide nanosheets), graphene-based materials have the following unique advantages and great application potential: (i) Graphene has exceptionally high carrier mobility, high carrier density, and high specific surface area, making it suitable for detection with high signal-to-noise ratio. (ii) Almost every atom in 2D planar structure of graphene is exposed to the environment, making it extremely responsive to local chemical or electrical disturbances, thus showing extremely high sensitivity to analytes. (iii) The 2D planar structure of graphene also provides a larger detection area and a homogeneous surface for efficient and unique functionalization. (iv) Some sp^3-hybridized carbons are embedded in the sp^2 framework in functionalized graphene (such as GO), which provides a place for the recombination of electron–hole pairs, so that the functionalized graphene exhibits photoluminescent properties and can be used as a fluorescent label for bioimaging. (v) The fluorescence wavelength range of graphene materials is extremely wide (from UV to NIR). (vi) The sp^2-hybridized carbon in graphene materials can also quench nearby fluorescent substances like other graphite materials, and the quenching efficiency is much higher than that of commonly used organic quenchers. (vii) The dual roles of GO as both a fluorophore and a quencher allow GO to be used not only as an energy donor but also as an energy acceptor for fluorescence resonance energy transfer (FRET) sensors and bioimaging. (viii) Graphene shows prominent features in Raman spectra, with characteristic vibrations ranging from 1000 to 3000 cm^{-1}, manifested as D peaks, G peaks, and 2D peaks; thus, graphene can also be used for Raman imaging. In addition, graphene also has high biocompatibility and is suitable for use in an *in vivo* biosensing system with low toxicity.

6.3.2.1 *Biosensors*

A complete graphene-based biosensor system usually includes three parts: target, receptor, and transducer. Among them, the target is an analyst (such as small molecules, proteins, cells, and nucleotides); the receptor is the recognition element, usually a bioactive molecule with specific interaction with the target (such as enzymes, antibodies, or nucleotides); and the transducer is a graphene functional material responsible for converting chemical information from the recognition event into a measurable signal (such as light, electricity, or magnetism). However, the receptor is omitted in diverse biosensors since the direct interaction of the analyte with the graphene material produces measurable changes in its properties. Therefore, graphene material can be used as both the receptor and transducer.

6.3.2.2 *Intracellular detection*

Graphene biosensors can monitor the target in the living cells, and convert the recognized chemical signals into measurable signals, such as optical or magnetic signals. As shown in Fig. 6.43, a dye-labeled peptide nucleic acid (PNA)/GO composite (PNA-GO) biosensor is able to detect miRNA in living cells.[53] In this biosensor, miRNA is the target; dye-labeled PNA serves as both a receptor and a fluorescent indicator; and GO serves as both a carrier and a fluorescent quencher. The fluorescence of the PNA is quenched by GO when the dye-labeled PNA fluorescent and GO are hybridized to fabricate a biosensor. When the biosensor enters the cell, the PNA will recognize the specific target miRNA and form a complex, resulting in the detachment of PNA from GO. Since the PNA is detached from GO, the quenching effect of GO disappears, and the fluorescence of the dye-labeled PNA is recovered; that is, the complex of the dye-labeled PNA and target miRNA emits fluorescence. There are many considerations in the composition of this biosensor; PNA is a non-natural nucleic acid analog in which the backbones are held together by uncharged amide bonds rather than the negatively charged-phosphodiester bonds as in DNA. PNA as a probe for miRNA sensing offers many advantages including high sequence specificity, high loading capacity on the GO surface

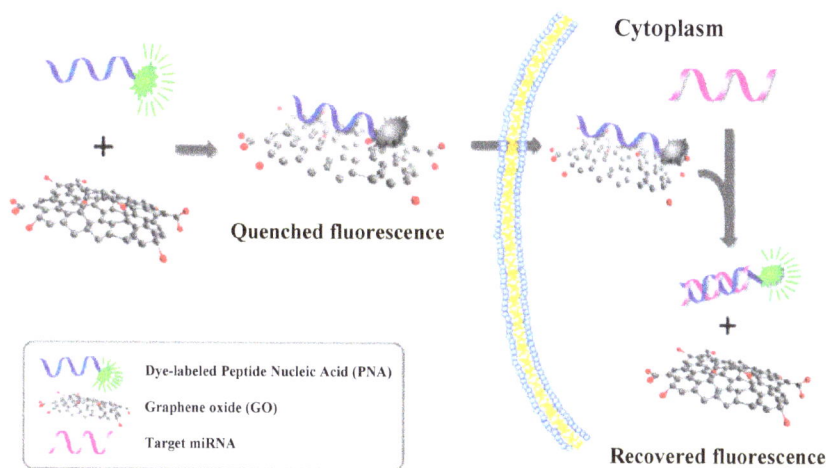

Figure 6.43. Dye-labeled PNA/GO complex biosensor for miRNA detection in living cells.[53]

compared to DNA and resistance against nuclease-mediated degradation. The thermal stability of the duplex of PNA and DNA or RNA is better than that of their counterparts DNA:DNA and RNA:RNA, and it exhibits high target specificity for target RNA. This PNA also shows good stability and will not be degraded by enzymes during transportation and recognition. In addition, the PNA-GO biosensor itself shows extremely efficient intracellular delivery and does not require other transfection agents without the assistance of an additional transfection reagent. Thanks to these advantages, the PNA-GO sensor allows the detection of specific target miRNAs with the detection limit as low as ~1 pM in a living cell.

6.3.2.3 *In vitro detection*

Graphene-based biosensors dedicated to *in vitro* diagnosis are usually graphene-based electronics. Among them, graphene-based field effect transistors (FETs) are most commonly used. As shown in Fig. 6.44, a GFET consists of a semiconducting channel between two metal electrodes, the drain and source electrodes, through which the current is injected and collected. Graphene materials are often used as channel materials, and they could be excellent building blocks in the fabrication of FET biosensors. Firstly, graphene has highly sensitive electronic properties,

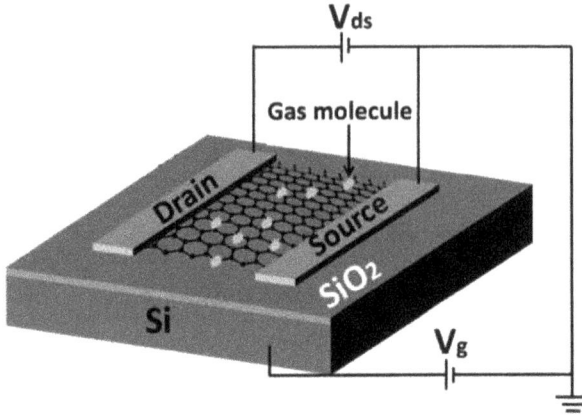

Figure 6.44.   Schematic illustration of the structure of the graphene-based FET (GFET).[54]

which can improve the sensitivity of the created FET biosensors. Secondly, graphene has a 2D structure with a high surface area, which can improve the capture of biomolecules and enhance the signal-to-noise ratio of detection. Thirdly, graphene is easy to modify, and therefore a lot of biomolecules like DNA, protein, peptides, and antibodies can be bound onto graphene through both non-covalent and covalent interactions.

The working principle of GFETs is the conductance change of graphene-based channels, which causes a change in the current between source and drain electrodes. Factors affecting the conductance of the graphene-based channel mainly include changes in device gate voltage, doping of graphene materials, carrier scattering, and changes in local dielectric properties. When the graphene-based channel is in contact with the target, the electrical properties of the graphene-based channel will change in the following ways, thereby affecting the change of the electrical signal of the device.

1. *Change in gate potential*: When the measured substance is charged, an electrical double layer is formed by the redistribution of ions on the graphene surface and an electrical double-layer capacitance is produced, thereby affecting the conductivity of the graphene material.
2. *Effect of doping*: Many molecules, especially those with aromatic rings, can be adsorbed on graphene surfaces and have a strong

interaction with graphene. Direct charge transfer occurs between them, thereby changing the electronic structure of graphene; that is, doping changes the electrical properties of graphene.

3. *Changes in local dielectric properties*: The complex of graphene with the target will change the local dielectric constant and local ionic strength of graphene, which will cause a change in graphene carrier intensity.

4. *Scattering of carriers*: Adsorbed substances can cause carrier (electron or hole) scattering, leading to a reduction of graphene carrier mobility and changes in its conductivity. In addition, detection of the target will also cause changes in the environmental pH or graphene lattice. All of these effects will affect the electrical properties of graphene materials and be detected by devices.

As shown in Fig. 6.45, GFETs are used to detect DNA based on target recycling and self-assembly signal amplification.[55] Graphene prepared by

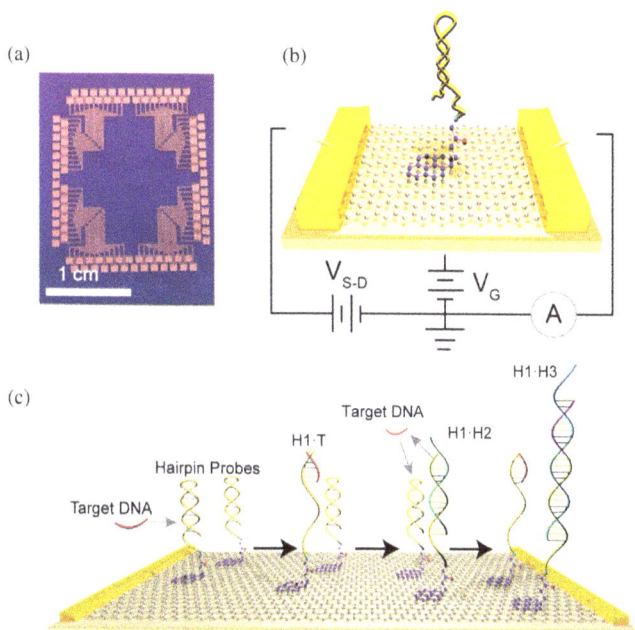

Figure 6.45.   Images and schematic illustration of GFETs for detection of DNA[55]: (a) optical image of a GFET; (b) device structure; and (c) detection mechanism.

the CVD method is used as a channel material to support *1*-pyrenebutyric acid *N*-hydroxysuccinimide ester (PBASE) through π-π interaction. Aminated hairpin DNA is grafted to PBASE as a probe using esterification reaction [Fig. 6.45(b)]. Hairpin DNA (H1) is metastable at 95°C and can be opened by target DNA (T) via self-assembly and base pairing to form a complex H1·T. When the device is working, the H1-loaded device is exposed to the target DNA (T) and three helper DNAs (H2, H3, and H4). The target DNA (T) triggers the opening of the hairpin DNA (H1) to form a complex H1·T. The protruding segment of H1·T is bound to the toehold of hairpin H2 to initiate the strand displacement reaction to form the complex H1·H2 and release DNA (T). The dissociated T is recycled to trigger additional self-assembly cycles as described earlier, leaving a stable H1·H2 duplex. The complex H1·H2 can trigger a hybridization chain reaction (HCR) of H3 or H4 to form a more complicated complex H1·H2·H3·H4. In this way, the presence of the target DNA (T) can be circularly used to trigger HCR, leading to long-nicked double-stranded polymers for the amplification of DNA products that can be detected by GFET through chemical gating. Thanks to this HCR, compared to the traditional single-stranded probe GFET device, the detection sensitivity of this GFET device is improved by a factor of 20,000 or more.

In addition to FETs, sensors for *in vitro* detection also include electrochemical biosensors that use recognition-induced electrochemical information as response signals, such as cyclic voltammetry biosensors, differential pulse voltammetry sensors, square-wave voltammetry sensors, and electrochemical impedance spectroscopy sensors. There are also mechanical sensors that use the mechanical changes generated by recognition as detection signals. In short, no matter which sensor is used, the specific recognition of the target by the functionalized graphene material is converted into measurable signals such as light, electricity, and force.

6.3.2.4 *Wearable diagnostic devices*

There has been considerable development in wearable, flexible, and stretchable electronic devices that monitor physiological signals and motion activities in the past decade. Continuous monitoring of human blood pressure, pulse, and limb movement is of great significance in disease diagnosis, treatment, and postoperative rehabilitation. Pressure sensors are among the most important wearable devices. Graphene materials have great application potential in the field of wearable pressure sensors

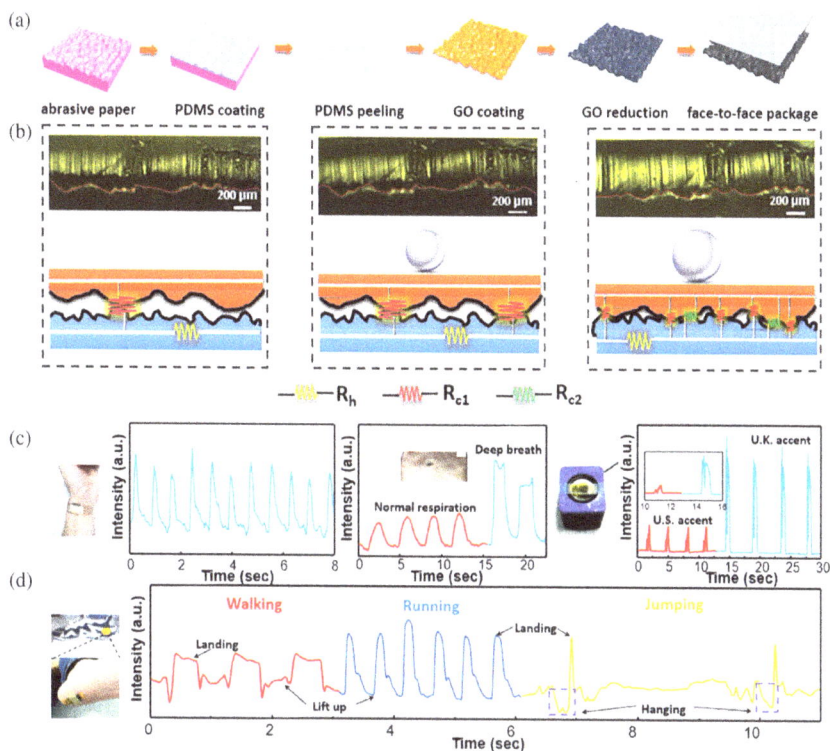

Figure 6.46. Graphene-based pressure sensor[56]: (a) fabrication process; (b) photographs and schematic illustrations of circuit models; and (c–d) applications of the graphene pressure sensor for various physiological signals' detection.

due to their flexibility, good mechanical properties, and light weight. Inspired by the epidermis tissue structure in human skin, a graphene pressure sensor with random distribution of spinosum exhibits pressure detection with high sensitivity ($25.1\,\text{kPa}^{-1}$) in a wide linearity range (0–$2.6\,\text{kPa}$).[56] The reticular layer of the dermis with a spinosum surface consists of two types of touch/pressure receptors, which are extremely sensitive to low-intensity external stimuli. Inspired by this microstructure, a highly sensitive graphene pressure sensor can be prepared. As shown in Fig. 6.46(a), abrasive paper is used as a template to prepare polydimethylsiloxane (PDMS) film, showing the spinosum surface with a random height distribution; GO is then coated on the surface of the PDMS film and a high temperature is applied to reduce the GO. Finally, a face-to-face package is used to prepare pressure sensors with different roughness surfaces. When different pressures are applied to the device, the distance between the double-layer films will change, making the contact points between the two layers of materials different and then changing the resistance between the two layers of materials. Different pressures lead to the change of resistance, and the detection of pressure can be achieved as the mechanical signal is converted into an electrical signal (Fig. 6.46(b)). Owing to the excellent high sensitivity, the pressure sensor exhibits tremendous utility in the detection of human physiological signals, voice, and motion activities (Figs. 6.46(c) and 6.46(d)).

6.3.3 *Bioimaging and phototherapy*

6.3.3.1 *Bioimaging*

Numerous physiological activities in the human body, such as biomass adsorption and diffusion, cell proliferation, apoptosis, and metabolism, are closely related to the biological and physiological state of the organism. Therefore, visual observation and differentiation of cells from different tissues and in different states are of great significance in clinical diagnosis. Bioimaging materials can bind to cancer cells. The metabolic activity of cancer cells is significantly higher than that of normal cells, which can enrich the imaging material in cancer cells. Compared with other cells in the organism, the enrichment of the imaging material highlights the cancer cells. Based on this principle, bioimaging of cancer cells can be realized. Graphene-based functional materials have large specific surface area and are easy to functionalize, so they are easily modified by small dye molecules, polymers, nanoparticles, drug molecules, and biomolecules for cell bioimaging applications.

6.3.3.2 *Fluorescent labeling*

When illuminated, the concentration of fluorescent graphene-based materials in tumor cells (the targeting effect of graphene materials or the enrichment effect of tumor cells) is higher than that in normal cells, which can emit stronger fluorescence and be highlighted. Graphene materials modified with PEG and NIR fluorescent dye (Cy7) can be enriched in tumor cells in mice for fluorescence labeling. Amine terminated six-arm branched PEG is conjugated to GO sheets

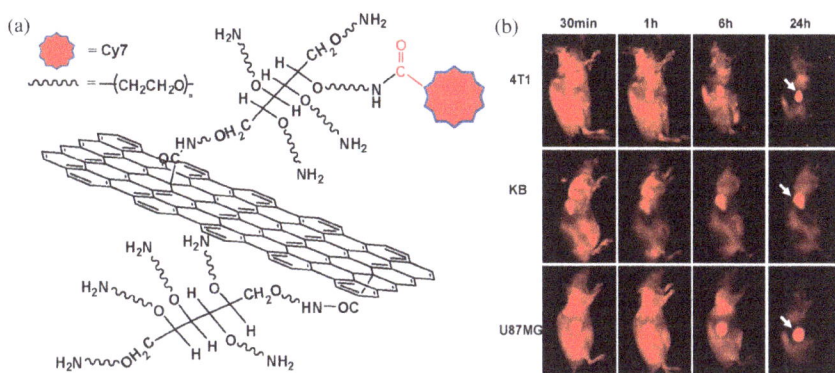

Figure 6.47. (a) A scheme of GO with PEG functionlization and labeled by NIR fluorescent dye Cy7, and (b) in vivo behaviores of NGS-PEG-Cy7.[57]

via amide formation, and the Cy7 dye, a commonly used NIR fluorescent dye, is then covalently conjugated to the PEGylated GO via the formation of an amide bond, thus giving NGS-PEF-Cy7 [Fig. 6.47(a)].[57] In this composite, GO is used as an NIR light-absorbing material and carrier. GO functionalized with PEG exhibits high solubility and stability in physiological solutions. The cyanine fluorescent dye Cy7 is used for fluorescence imaging. When the graphene composite material is injected into the veins of mice, the composite material is enriched in the mice tumor cells due to the highly active metabolism of the tumor cells and enhanced permeability and retention effect of the material to the tumor cells. When irradiated with a 704-nm laser, the graphene composite shows a higher intensity fluorescence (740~790 nm) due to its high concentration in tumor cells, which is highlighted in Fig. 6.47(b).

6.3.3.3 *Surface-enhanced Raman imaging*

Fluorescence imaging is the most commonly used bioimaging method, but its high excitation energy, photobleaching effect, and broad excitation/absorption peaks limit its application. On the contrary, Raman spectroscopy based on molecular vibrational and rotational scattered light does not require higher-energy incident light. When it is used in bioimaging, it can work only with low-energy incident light. Compared with other materials, graphene has special Raman scattering signals: D peak, G peak, and 2G

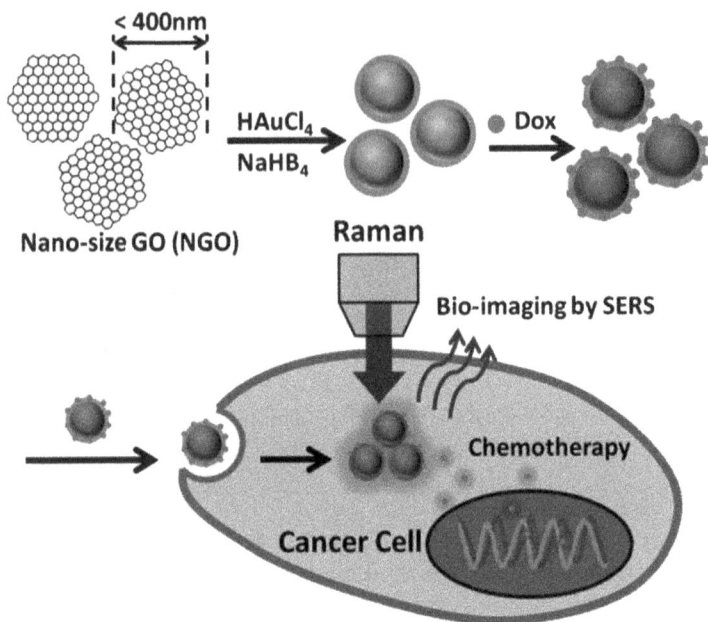

Figure 6.48. Illustrative mechanism of SERS-based bioimaging and anticancer drug delivery by using Au@NGO.[58]

peak; the intensity of Raman scattering can be significantly enhanced by hybridizing with metal nanoparticles. Therefore, graphene-based composites have been used for the highly sensitive detection of biomolecules and bioimaging. A gold nanoparticle/graphene composite (Au@NGO) is fabricated by wrapping GO prepared by Hummers' method on gold nanoparticles, which can be used for surface-enhanced Raman scattering (SERS) imaging of cancer cells (HeLa) (Fig. 6.48).[58] In this composite, GO with characteristic Raman absorption is used for imaging; gold nanoparticles with surface plasmon resonance are used to enhance the Raman signal of GO. The Au@NGO nanoparticles enter HeLa cells through endocytosis, which are mainly distributed in the cytoplasm and some of them are found in the membrane compartments. HeLa cells containing Au@NGO nanoparticles exhibit stronger Raman scattering, and the signal intensity is much higher than that of cells containing only nano-size GO (NGO). In addition, since the Au@NGO nanoparticles can efficiently enter cancer cells, they can also serve as a carrier to deliver and release doxorubicin (Dox) into cancer cells.

6.3.3.4 *Phototherapy*

Because of strong absorption in the NIR region, graphene-based materials are also often used in photothermal therapy (PTT) and photodynamic therapy (PDT).

PTT is a therapy method that converts light into heat energy to kill the cancer cells. First, a material with high photothermal conversion efficiency is injected into the human body, which is enriched in tumor cells, and then a light source (usually NIR light) is used to irradiate the tumor site. The material absorbs light energy and converts it into heat, which kills cancer cells. The NGS-PEG mentioned earlier is a material that can be used in photothermal therapy.[57] Mice with 4T1 tumor cells were intravenously injected with NGS-PEG (a dose of 20 mg/kg), followed by irradiation with an 808-nm laser at the power density of 2 W/cm^2 for 24 h. The surface temperature of tumors on NGS-injected mice reached ~50°C after laser irradiation, in contrast to the ~2°C surface temperature rise for irradiated tumors on uninjected mice. All irradiated tumors on mice injected with NGS disappeared 1 day after laser irradiation, leaving the original tumor site with black scars, which disappeared about 1 week after treatment. No tumor regrowth was noted in this treated group over the course of 40 days, demonstrating the excellent efficacy of NGS-PEG-based *in vivo* PTT. It is worth mentioning that this composite shows no noticeable toxicity and no damaging effect on the tissues of mice.

PDT is based on the dynamic interaction between a photosensitizer, light with a specific wavelength, and molecular oxygen, promoting the selective destruction of the target tumor tissue. The excited photosensitizer can transfer energy to the surrounding oxygen and produce highly active singlet oxygen. The singlet oxygen reacts with the molecules in the adjacent cancer cells to produce cytotoxicity, resulting in cell damage and even death. Composites fabricated by upconversion nanoparticles (UCNPs), PEGylated nanographene oxide (NGO), and phthalocyanine (ZnPc) have been used for photothermal killing of human cervical cancer cells.[59] The preparation process is shown in Fig. 6.49. Rare earth metal oxide nanoparticles with upconversion luminescence are covalently grated onto PEG-modified GO through covalent bonds, and then phthalocyanine is loaded to fabricate a functional composite UCNPs-NGO/ZnPc. UCNPs with fluorescent properties enable functional composites to be used for *in vivo* fluorescence imaging. NGO can absorb NIR light and emit heat so that functional composites can be used for PTT. ZnPc as a

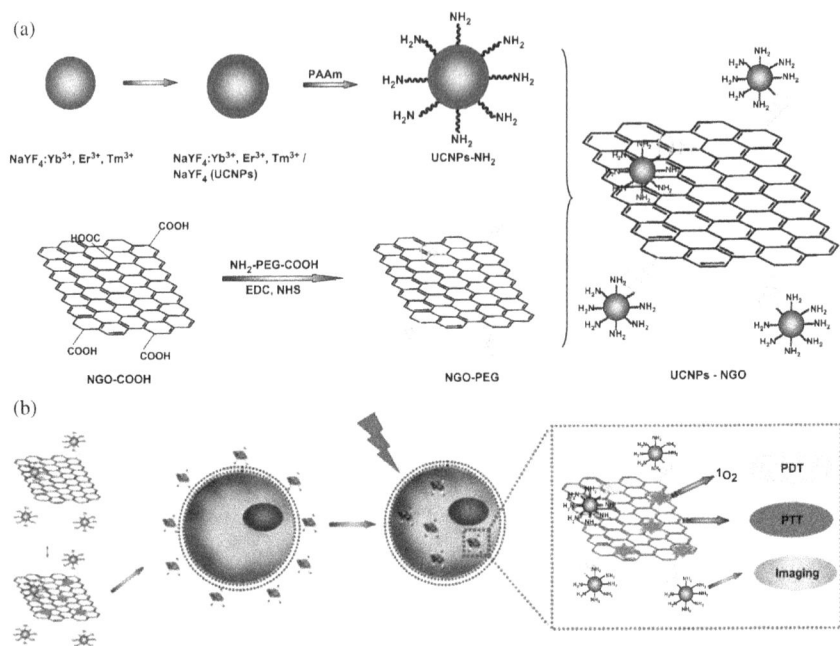

Figure 6.49. (a) Schematic illustration of the synthetic procedure for UCNPs-NGO: Numbers of core-shell structured UCNPs being covalently grafted with NGO via bifunctional polyethylene glycol; (b) Schematic illustration of UCNPs-NGO/ZnPc as a multifunctional theranostic nanoplatform for cancer treatment. NGO/ZnPc as a multifunctional theranostic nanoplatform for cancer treatment.

photosensitizer allows functional composites to be used for PDT. The viability of cancer cells injected with UCNPs-NGO/ZnPc composites is significantly decreased after irradiation using a 630-nm laser at a power density of 50 mW/cm² for 24 h. This is due to the production of singlet oxygen when the composites are irradiated. Highly active singlet oxygen reacts with cancer cells and kills them.

6.3.4 *Summary*

Though functionalized graphene materials have made much progress in the field of biomedicine, this research is still in its infancy. In order to promote the development of the application of graphene-based composite materials in the field of biomedicine, explore greater progress, and implement different practical applications, more in-depth research is needed.

First of all, it is necessary to further investigate the biotoxicity, cytotoxicity, and biocompatibility of functionalized graphene materials. Secondly, the *in vivo* targeting mechanism of graphene composites needs to be further clarified, and more target materials with better performance are required. Meanwhile, the mechanism of the interaction between functionalized graphene materials and biomolecules and cells should be further clarified. Finally, controllable and facile methods for large-scale preparation of graphene composite materials should be developed, and low-cost, high-sensitivity, and fast-response graphene-based portable sensors should be explored.

6.4 Environmental

As mentioned earlier, GO has also been applied as an absorbent for the removal of organic/inorganic pollutants in wastewater through *n-n* interaction, hydrogen bonding, hydrophobic interaction, and electrostatic interaction. There are many kinds of pollutants in industrial, agricultural, and domestic wastewater, which are mainly divided into three categories: chemical, physical, and biological pollutants (Fig. 6.50). Therefore, simply utilizing GO to deal with so many kinds of pollutants is impossible;

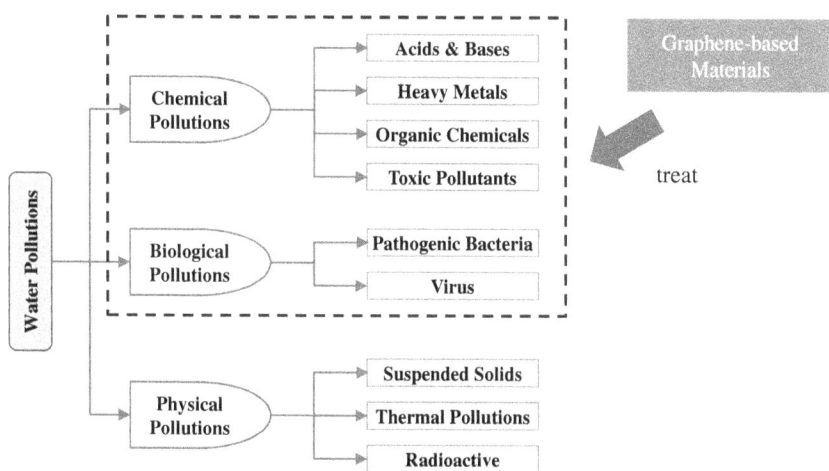

Figure 6.50. Types of water pollution and pollutants that are mainly treated by graphene-based materials.

thus, it is necessary to develop functionalized graphene materials with better capabilities. Functionalized graphene materials are mainly used to deal with chemical and biological pollutants of wastewater, especially organic and heavy metal pollution. The main organic pollutants in wastewater include organic pesticides, polychlorinated biphenyls, halogenated aromatic hydrocarbons, ethers, monocyclic aromatic compounds, phenols, polycyclic aromatic hydrocarbons, and nitrous acid; heavy metal pollutants mainly include cadmium, mercury, lead, arsenic, chromium, copper, zinc, thallium, nickel, and beryllium.

In addition to the abovementioned interactions, the mechanism of functionalized graphene materials for wastewater treatment also includes magnetic attraction and chelation (Fig. 6.51).[60] The main process with which graphene materials capture pollutants under a magnetic field is as follows: A composite made of graphene and magnetic nanoparticles absorbs pollutants through various interactions of graphene (π-π interaction, hydrogen bonding, hydrophobic interaction, and electrostatic

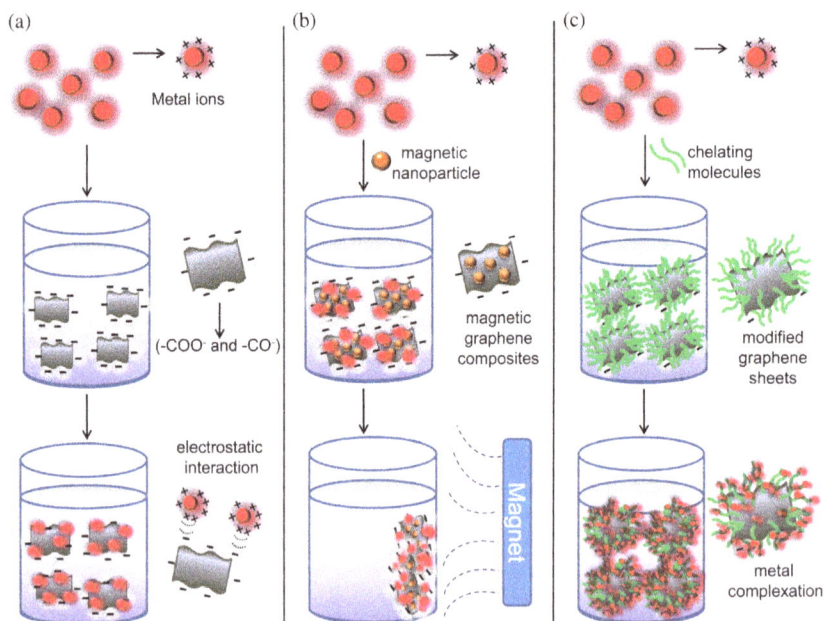

Figure 6.51.   Main strategies to apply graphene-based materials as adsorbents for the removal of metal ions from aqueous solutions[60]: (a) electrostatic interaction; (b) magnetic nanocomposites; and (c) conjugation with molecules.

interaction) and then a magnetic field is applied to remove the composites from aqueous solution. The process of chelation is as follows: Functionalized graphene materials with chelating molecules can complex metal ions in aqueous solution to form metal-ligand complexes, thus achieving the removal of metal ions from aqueous solution. In addition, the functionalized graphene materials with catalysts can catalyze the photodegradation of organic pollutants in wastewater and achieve water purification by decomposing organic pollutants. Oily pollutants can be removed through adsorption using graphene materials with a high specific surface area. Inorganic pollutants in water can be catalyzed and decomposed by functional graphene to produce compounds that are normally harmless to the environment.

6.4.1 *Removal of dyes*

Dye molecules in wastewater are generally organic molecules containing conjugated groups or anions and cations. As mentioned earlier, GO has a variety of active sites: anionic functional groups (—OH and —COOH), delocalized π-electron systems, hydrophobic regions, and epoxide groups. These active sites interact with different types of dye molecules in water through electrostatic interaction, π-π interaction, hydrophobic interaction, and hydrogen bonding interaction. In addition, most functionalized graphene materials have an extremely large specific surface area, so dye molecules can be adsorbed on graphene materials in large quantities. However, although carbon materials such as graphene can effectively adsorb dye molecules in water, most of the adsorbents are in the form of small particles, which are difficult to effectively remove from water. If they are left in the aqueous solutions, they may have an adverse impact on the environment. Therefore, it is necessary to effectively separate the graphene absorbents from water. Among various methods, magnetic separation is a fast and efficient way to separate nanoparticles from water. The composite material made of graphene and magnetic substances (such as Fe_3O_4) can not only adsorb dye molecules but also shows magnetic properties. After the composite adsorbs pollutants, it can be removed from the water by applying a magnetic field. For example, magnetically separable Fe_3O_4/rGO nanocomposite as an efficient photocatalyst, capable of degradation of several cationic and anionic dye molecules under direct sunlight irradiation is reported (Figure 6.52).[61]

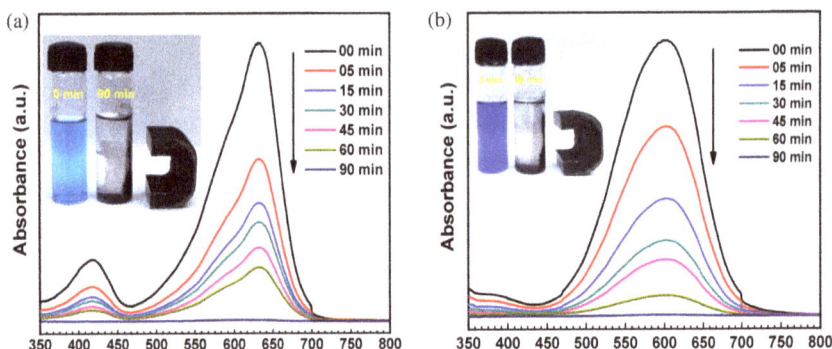

Figure 6.52. Removal of dye molecules by magnetic Fe_3O_4/rGO composites[61]: (a) methyl green and (b) methyl blue.

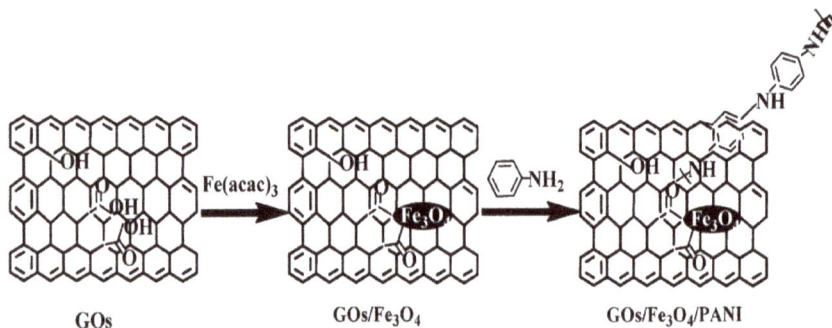

Figure 6.53. Synthesis route of GOs/Fe_3O_4/PANI composites.[62]

It should be noted that a composite of this composition is difficult to use continuously because the magnetic substance in the material is easily oxidized or decomposed during use, especially in an acidic environment. Therefore, conductive polymer materials with excellent chemical stability, such as polyaniline (PANI), are often used to wrap graphene and magnetic substances to fabricate multi-component composites to enhance the mechanical strength and adsorption properties of the composites, for example, GO/magnetite/polyaniline (GOs/Fe_3O_4/PANI) composites.[62] GO/ Fe_3O_4 synthesized by hydrothermal method is decorated with PANI by dilute polymerization to obtain GOs/Fe_3O_4/PANI composites (Fig. 6.53). The nitrogen-containing functional groups of PANI decorated on GOs/

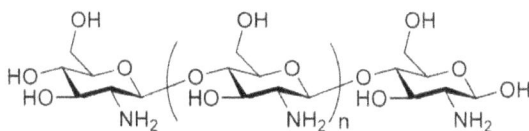

Figure 6.54.   Chemical structure of chitosan.[63]

Fe_3O_4/PANI provide numerous effective adsorption sites and thus increase the adsorption capacity. Moreover, PANI protects the Fe_3O_4 nanoparticles from dissolution and therefore improves the stability of GOs/Fe_3O_4/PANI in solution. This magnetic composite shows high adsorption capacity toward methyl orange (585.02 mg/g).

GO contains carboxyl groups that are negatively charged; thus, GO exhibits low affinity for anionic dyes due to the strong electrostatic repulsion between them, such as chitosan (CS), are incorporated with graphene to obtain a graphene composite with broad-spectrum adsorption capability.[63] CS is rich in two chelating groups, amino and hydroxyl (Fig. 6.54), which can provide composites with the ability to adsorb anions in water through electrostatic interaction. In addition, CS also shows excellent biocompatibility, biodegradability, and antibacterial properties.

GO/CS composite hydrogels can be fabricated by hybridizing CS and GO. There is a strong electrostatic interaction between CS and GO, so CS can be used as the cross-linking agent to connect GO. In these hydrogels, GO sheets remain less aggregated, showing a large surface area, and the interconnected pores allow adsorbate molecules to diffuse easily into the absorbent. The hydrogel can be used as column packing to fabricate a column, which is used to purify water by filtration, and the maximal adsorption capacities towards cationic methylene (MB) and anionic Eosin Y are both higher than 300 mg/g (Fig. 6.55). The adsorption capacities of MB on the hydrogel increase with GO content, whereas the adsorption capacities of Eosin increase with CS content. The results further confirm that the adsorption of cations takes place on GO, while that of anions takes place on CS. In addition, this composite hydrogel can effectively remove heavy metal ions such as Cu(II) and Pb(II) from water, the maximum adsorption capacities of which are 70 mg/g and 90 mg/g, respectively. This is also because both GO and CS are good absorbents for metal ions.

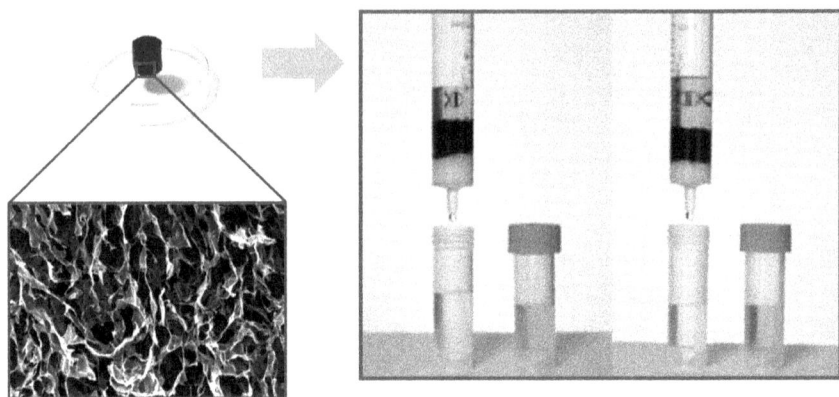

Figure 6.55. Removal of dyes (MB and Eosin Y) from water by filtration of GO/CS hydrogel.[63]

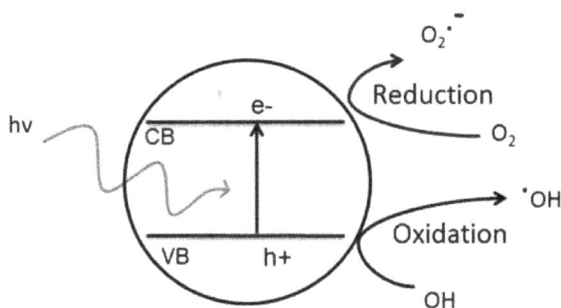

Figure 6.56. Schematic representation of the fundamental principle of semiconductor-based photocatalysts, where free electrons and holes are generated by photoexcitation.[64]

In addition to using the above adsorption methods to remove pollutants from water, graphene composites can also be used as photocatalysts to degrade organic molecules such as dyes. All the photocatalysts are basically semiconductors — there is an energy difference (band gap) between the catalyst's valence band (the energy band in which all valence electrons are located) and the conduction band (the energy band with higher energy than the valence band, and most energy levels are empty). When the photocatalyst absorbs the energy of light of a specific wavelength that matches its band gap, the electrons in the valence band of the photocatalyst will be excited to become free electrons in the conduction band, while the same number of holes are left in the valence band (Fig. 6.56). This creates negatively charged electrons and positively charged holes, which are called carriers. The free electrons and corresponding holes generated by the excitation are called electron–hole pairs. They either recombine or continue to move separately to the conduction and valence bands. The electron–hole pairs that are generated in this way migrate toward the surface where they can initiate redox reactions with adsorbates. Reactions with the positive holes are linked to oxidation and reactions with the electrons to reduction. In addition, hydroxyl radicals (OH^-) are produced when the photocatalyst is photoexcited in water. These OH^- radicals drive the oxidation of toxic pollutants in water to complete mineralization.

The limiting factor in semiconductor catalysis efficiency is the rapid recombination of photogenerated electrons and holes, which significantly affects practical applications. Therefore, the key to improving the catalysis efficiency is to enhance the separation and transport of catalyst carriers. Graphene-based materials are promising candidates for photocatalytic water remediation due to the high adsorption capacity of dyes, extended light absorption range, enhanced charge separation, and charge transport properties. In composite materials, graphene serves as a carrier for semiconductors, an adsorbent for pollutants, and an acceptor for charge carriers. For example, in a rGO/Ag/Fe-doped TiO_2 composite, rGO not only serves as a loading substrate for silver nanoparticles and iron-doped titania nanoparticles but also as the main adsorbent for pollutants.[65]

Under visible light irradiation, iron-doped titania nanoparticles absorb photoexcitation and generate free electrons and holes (Fig. 6.57). The photogenerated electrons in the conduction band are consumed by the surface-adsorbed oxygen molecules to produce superoxide anion (O_2^-) radicals, while the generated holes in the valence band are scavenged by surface hydroxyl groups to generate hydroxyl radicals ($\cdot OH$). Subsequently, these radicals can be used for the degradation of the pollutant. The Ag nanoparticles on the surface of the photocatalyst serve as acceptors for the free electrons, which can prevent the recombination of free electrons and holes generated by photoexcitation. In addition, the Fermi level of rGO is below the potential of the conduction band of TiO_2, and it is speculated that rGO may act as an electronic acceptor that could withdraw electrons from the valence band. In this research, the rGO/Ag/Fe-doped-TiO_2 reveals the highest MB solution degraded performance for which photocatalytic conversion efficiency reaches 95.33% in 150 min.

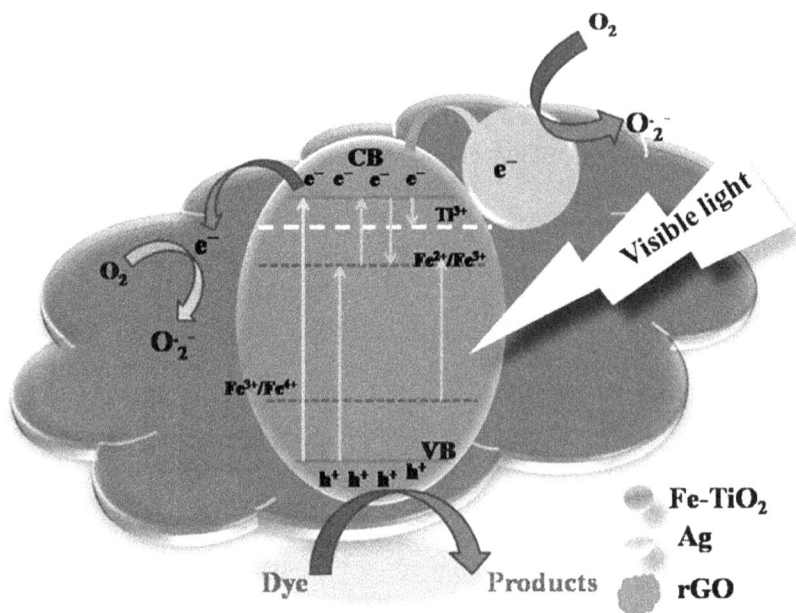

Figure 6.57. Schematic of the mechanism of charge separation for rGO/Ag/Fe-doped TiO$_2$[65]

6.4.2 *Removal of heavy metal ions*

Heavy industry, mining, and metallurgy will cause heavy metal pollution in water. Heavy metal ions, such as Cd^{2+}, Hg^{2+}, Pb^{2+}, As^{2+}, Mn^{2+}, Ag^+, Cu^{2+}, Co^{2+}, and Mn^{2+}, in wastewater seriously threaten human health and the ecological environment. Therefore, the treatment of water that is polluted by heavy metals is one of the major problems in the field of environmental protection. A commonly used method to deal with heavy metal pollution is to use adsorbents to absorb metal ions in wastewater. Carbonaceous materials, especially graphene materials, have the advantages of high specific surface area, corrosion resistance, excellent stability, rich in oxygen-containing groups, adjustable surface chemical properties, and easy large-scale production, making them the ideal adsorbents.

The most commonly used graphene-based adsorbents are GO-based materials. GO contains a large number of oxygen-containing groups ($-OH^-$, $-COO^-$ and $-O^-$). These oxygen-containing groups not only

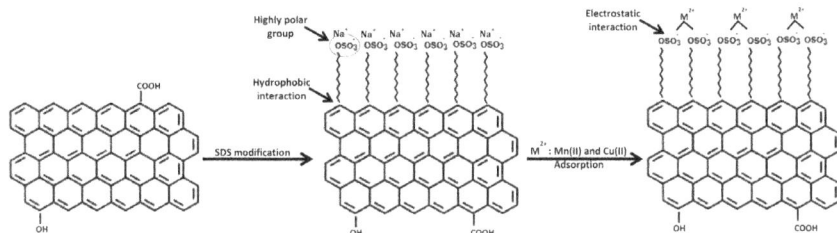

Figure 6.58. Schematic of adsorption of metal ions onto rGO/SDS self-assembled composites.[66]

endow GO with excellent hydrophilicity, making it dispersible in water, but also serve as active sites for hybridizing metal ions. Graphene-based materials adsorb metal ions and corresponding anions mainly through the following: (i) Electrostatic interaction takes place between anion oxygen-containing groups on graphene materials and metal cations. (ii) The sp^2 framework rich in delocalized π-electrons on graphene can donate electrons to metal cations, which can be regarded as Lewis bases, and metal cations can be regarded as Lewis acids. There is an acid–base interaction between them. (iii) Functional groups can be incorporated to form graphene composites, such as amino and hydroxyl groups, which interact with adsorbed cations. (iv) anion-π interaction takes place between anions such as phosphate (PO_4^-), perchlorate (ClO_4^-), halide ions (F^-, Cl^-, and Br^-) or ions with unpaired electrons and electron-deficient aromatic rings of graphene sheet. (v) Decomposition of inorganic anions which is catalyzed by graphene-based catalysts.

For example, when dissolved in water, rGO and sodium dodecylsulfate (SDS) tend to self-assemble into an amphiphilic composite (Fig. 6.58).[66] The driving force for assembly is the hydrophobic interaction between the alkyl moiety of SDS and the hydrophobic surface of rGO. Electrostatic interaction between bivalent metal ions and sulfate anion is in charge of the removal of heavy metal ions. In addition, the π-electron-rich surface of rGO can also adsorb metal cations through the cation-π interaction. The experimental data show that the maximum adsorption capacities for Cu^{2+} and Mn^{2+} on the rGO/SDS composite are 369.16 mg/g (pH = 5) and 223.67 mg/g (pH = 6), respectively.

Incorporating functional polymers capable of interacting with heavy metal ions and GO can produce graphene adsorption materials with better

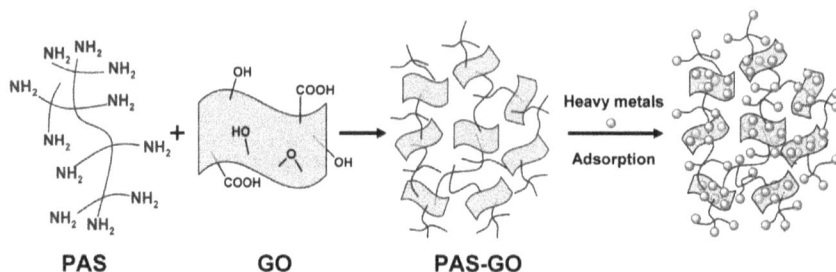

Figure 6.59.   Schematic illustration of the preparation of PAS-GO composite and adsorption process of heavy metal ions.[67]

adsorption capacity. Poly(3-aminopropyltriethoxysilane) (PAS) oligomers are cross-linked three dimensionally with GO nanosheets to obtain a high-performance adsorbent — PAS-GO composites (Fig. 6.59).[67] A three-dimensional cross-linking network is meant to prevent the aggregation of GO nanosheets, provide easy accessibility for foreign molecules, and introduce a large amount of amino functional groups for better adsorption of heavy metal ions. PAS-GO exhibited much higher adsorptivity toward Pb^{2+} with the maximum adsorption capacity of 312.5 mg/g at 303 K, and furthermore the maximum adsorption capacity increased with increasing temperature. It is worth mentioning that adsorption could be conducted in a wide pH range of 4.0–7.0. The adsorption mechanism includes the electrostatic interaction between anionic groups ($-COOH$ and $-OH$) on GO and metal cations, and coordination (also known as "chelation") between electron-rich groups on GO or amino groups on PAS and metal ions. The lone pair of electrons in nitrogen atoms of amine groups are shared with metal ions, leading to the formation of the coordination (or "chelate"), thereby adsorbing metal ions. Because of the difference in the chelation ability of different metal species, Pb^{2+} and Cu^{2+} are preferentially removed.

Magnetic materials (such as Fe_3O_4) can be incorporated with GO to fabricate magnetic graphene-based composites for the removal of heavy metal ions from wastewater. A GO-like 2D-carbon flake (CF) is synthesized using waste onion by thermal treatment, which is further decorated with crystalline Fe_3O_4 nanoparticles to obtain the Fe_3O_4@2D-CF composite (Fig. 6.60).[68] The adsorption capacity of the Fe_3O_4@2D-CF composite for As(III) is 57.47 mg/g. The adsorption mechanism mainly involves the substitution of Fe$-$OH groups and C$-$OH groups by arsenic species.

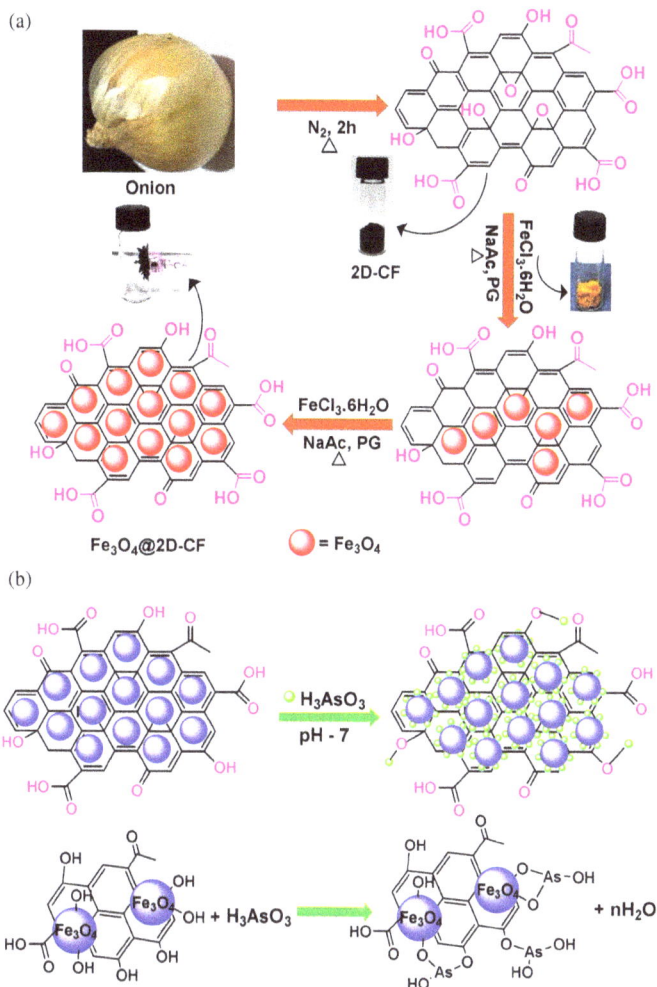

Figure 6.60.   Synthesis and magnetic behavior of Fe_3O_4@2D-CF composite.[68]

After arsenite is adsorbed by Fe_3O_4@2D-CF, it can be collected from the aqueous solution under an external magnetic field.

Magnetic chitosan and graphene oxide-ionic liquid (MCGO-IL) composites can effectively remove heavy metal chromium from water (Fig. 6.61).[69] The magnetic MCGO is prepared by mixing Fe_3O_4 wrapped in chitosan with GO, and then the composite is immersed in ionic liquid to obtain the MCGO-L composite. The ionic liquid not only increases the

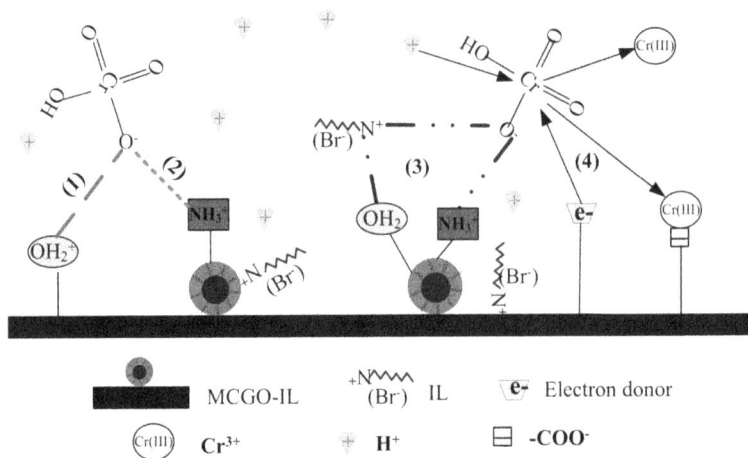

Figure 6.61. Proposed mechanism of Cr(VI) removal by MCGO-IL.[69]

water solubility of the composite material but also provides additional active adsorption sites for metal ions. The maximum adsorption capacity toward hydrogen chromate ($HCrO_4^-$) is 145.35 mg/g. The adsorption mechanism mainly involves electrostatic interaction and hydrogen bonding interaction between the composite and the metal ion. On the basis of the experimental results, a possible Cr(VI) removal mechanism is proposed, and it may consist of four steps: (i) the —OH protonated to formed —OH^{2+}, and then adsorption of Cr(VI) by electrostatic attraction; (ii) the formed of —NH_3^- to adsorbed Cr(VI) by electrostatic attraction; (iii) the cooperation between IL, the functional groups of the adsorbent, and the Cr(VI); (iv) Cr(VI) is reduced to Cr(III) with the assistance of π electrons on the carbocyclic six-membered ring of MCGO-IL, the release of Cr(III) species into solution by electrostatic repulsion between the protonated amine groups and the cation Cr(III), or the binding of Cr(III) species on MCGO-IL by electrostatic attraction between Cr(III) and negatively charged groups (—COO^-) of MCGO-IL.

Anions (such as NO_3^- and NO_2^-) often exist in water along with metal ions. Nitrate (NO_3^-) in water is non-toxic to the human body, but it can be reduced to NO_2^- in the environment or by bacteria in the stomach. NO_2^- induces a variety of diseases, such as blue baby syndrome and cancer, which are extremely harmful to the human body. Therefore, effective removal of NO_3^- and NO_2^- from water is an issue that must be highlighted in water treatment. Zero-valent iron nanoparticles (ZVINPs) are

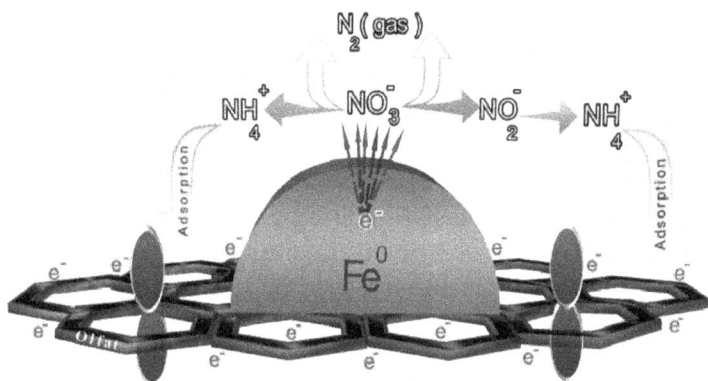

Figure 6.62.   Schematic representation of the possible mechanism of the catalytic reduction of nitrate ions by ZVINP/NG nanocomposites.[70]

supported on high surface area nanographenes (NGs) to prepare a ZVINP/ NG composite, which transforms NO_3^- into NH_4^+ and N_2; it is proven to be an ideal material for treating NO_3^- and NO_2^- in water (Fig. 6.62).[70] The results show that ZVINPs alone had low stability but greatly stabilized upon their support on the NG surface, which provided the ZVINPs with electrons and prevented their oxidation.

The transformation of NO_3^- into NH_4^+ and N_2 may be achieved through the following mechanisms: (i) Fe^0 reduces NO_3^- to NO_2^- (Eq. (8–9)) and then to NH_4^+ (Eq. (8–10)); (ii) NO_3^- is directly reduced to NH_4^+ (Eq. (8–11)); (iii) NO_3^- is directly reduced to NH_4^+ by Fe^0 (Eqs. (8–11) and (8–12)); (iv) NO_3^- is directly reduced to $N_2(g)$ by (Eq. (8–14)); and (v) NO_2^- is reduced to $N_2(g)$ by Fe^0 (Eq. (8–15)):

$$NO_3^- + Fe^0 + 2H^+ \rightarrow Fe_2^+ + NO_2^- + H_2O \qquad (8.9)$$

$$NO_2^- + 3Fe^0 + 8H^+ \rightarrow 3Fe_2^+ + NH_4^+ + 2H_2O \qquad (8.10)$$

$$NO_3^- + 4H_2 + 2H^+ \rightarrow NH_4^+ + 3H_2O \qquad (8.11)$$

$$4Fe^0 + NO_3^- + 7H_2O \rightarrow 4Fe(OH)_2 + NH_4^+ + 2OH^- \qquad (8.12)$$

$$3Fe^0 + NO_3^- + H_2O \rightarrow Fe_3O_4 + NH_4^+ + 2OH^- \qquad (8.13)$$

$$5Fe^0 + 2NO_3^- + 6H_2O \rightarrow 5Fe_2^+ + N_2(g) + 12OH^- \qquad (8.14)$$

$$3Fe^0 + 2NO_3^- + 8H^| \rightarrow 3Fe_2^+ + 2H_2O + N_2(g) \qquad (8.15)$$

There are many factors that affect the efficiency of graphene-based adsorbents in adsorbing metal ions in water, such as ionic strength in water, pH, and the number of oxygen-containing groups on GO. The influence of ionic strength on the adsorption efficiency in water is mainly determined by the competition between electrolytes (NaCl, KCl, and $NaClO_4$) in water and heavy metal ions on the surface of GO. Generally, the high concentration of electrolyte in water will reduce the adsorption efficiency of graphene-based materials for heavy metal ions, but there are exceptions; for example, the adsorption of Pb^{2+} by GO is not affected by $NaClO_4$. This may be due to the fact that the interaction between the adsorption sites on GO and Pb^{2+} is much higher than the interaction between the adsorption sites and the $NaClO_4$ electrolyte.

pH is the main factor affecting the adsorption capacity of GO-based materials. The adsorption behavior of GO depends on its point of zero charge (pH_{pzc}). When the pH of the solution is higher than the pH_{pzc} of GO ($pH > pH_{pzc}$), the hydroxyl and carboxyl groups on GO are anionized; that is, the surface of GO material is anionized. This enhances its electrostatic interaction with metal cations. In contrast, when the solution's pH is lower than the of GO ($pH < pH_{pzc}$), the hydroxyl and carboxyl groups on GO are protonated, and thus the cationized surface of GO lowers the adsorption capacity for metal cations. In addition, the pH of the solution also affects the adsorbed metal cations. When the pH is high, the metal ions may form hydroxides, such as $Me(OH)^+$, $Me(OH)_2$, or $Me(OH)_3^-$. The interaction between those metal ions and GO is weak depressing the adsorption capacity of GO to metal ions. Therefore, when GO materials are used to adsorb heavy metal ions, it is necessary to adjust the pH of the solution to anionize the surface of GO, and the metal ions exist in the form of Me^+. The pH is not an identical value, and different pHs of adsorption are required for different metal ions and different graphene materials.

6.4.3 *Removal of drugs*

The production process of drugs, domestic waste, and human metabolites can introduce drug molecules into water bodies, especially antibiotic drugs. Antibiotics are one of the greatest discoveries of the 20th century; they either kill the bacteria or slow down bacterial growth in the human body and are used to prevent and treat a number of diseases. However, abuse of antibiotics can result in compromised human immunity, certain

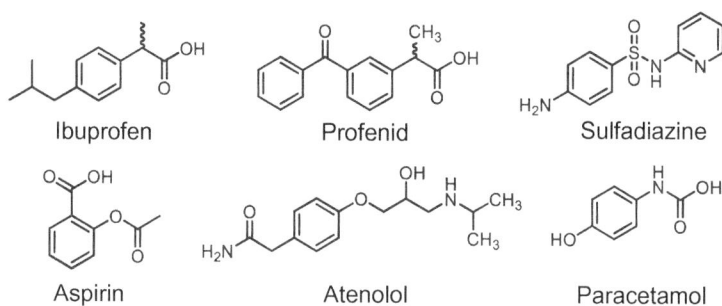

Figure 6.63. Chemical structures of some commonly used antibiotics.

poisonous side effects, and enhanced antimicrobial resistance. A high concentration of antibiotics in aquatic environment will have serious consequences on human health. Therefore, removal of drugs from wastewaters, especially antibiotic molecules, is inevitable in water treatment and water purification.

As shown in Fig. 6.63, commonly used antibiotics are mainly small organic molecules, most of which contain one or several functional groups such as amino, imino, hydroxyl, carbonyl, carboxyl, and aromatic rings; thus, they are easily protonated or anionized to be positively or negatively charged. Hence, the antibiotics easily interact with other molecules or materials through hydrogen bonding interaction, π-π interaction, electrostatic interaction, and hydrophobic interaction. Therefore, the treatment of antibiotics in aquatic environments is similar to that of organic dyes mentioned earlier, and antibiotics can be adsorbed or degraded by graphene materials through various types of interactions.

GO is encapsulated into environmentally benign sodium alginate (SA) to prepare a composite hydrogel (GO-SA-H) and an aerogel (GO-SA-A), which are used as adsorbents to remove ciprofloxacin (CIP, an antibiotic) from aqueous solutions.[71] The maximum adsorption capacities of GO-SA-H and GO-SA-A reach 86.12 and 55.55 mg/g, respectively. The driving force of adsorption is mainly electrostatic interaction between CIP and oxygen-containing groups (—OH and —COOH) on SA and GO, as well as π-π interaction between CIP and GO. The incorporation of GO improves the pore uniformity of the gels and decreases the pore sizes. In addition, the abundant oxygen-containing groups on GO and the conjugated structure containing π-electrons provide more adsorption sites for the composite gel. These factors are beneficial in improving the adsorption capacity of the gel.

The rGO/magnetite composites synthesized by an *in situ* reaction have proven to adsorb fluoroquinolone, CIP, and norfloxacin (NOR).[72] The maximum adsorption capacity for CIP and NOR is 18.22 mg/g and 22.2 mg/g, respectively. The dominant mechanisms in the adsorption are electrostatic interaction and π-π interaction between the antibiotics and the composites. Thermodynamic results show that this adsorption is a spontaneous exothermic process. The saturation magnetization of the composite in this work is 12 emu/g, which ensures fast magnetic separation during the adsorption experiments.

6.4.4 *Removal of oil*

Industrial wastewater containing oily substances and water-insoluble organic substances (such as toluene, benzene, methylene chloride, and chloroform) and accidental releases of crude oil have greatly endangered the ecological environment and public health. Among them, oil spills are the most dangerous. Crude oil is one of the most important energy sources used globally and the lifeblood of the global economy. Crude oil production, transportation, and processing occur all the time. This has resulted in frequent oil spill accidents during these processes, causing damage to the ecological environment, especially the aquatic environment, from which it is difficult to recover effectively. In order to solve these pollution problems, it is imperative to develop technologies and materials that can deal with organic pollution and crude oil pollution in water.

Currently, the most commonly used materials and technologies for oil spill remediation can be categorized into four different types: (i) chemical methods (dispersants, solidifiers), (ii) *in situ* burning, (iii) bioremediation, and (iv) mechanical recovery (booms, skimmers, sorbents). Among these methods and technologies, sorbent materials are more attractive for oil spill cleanup because of the possibility of collection and complete *in situ* removal of the oil from the water surface, while having no adverse effect on the environment. Therefore, physical adsorption is an effective and facile method to deal with such pollution. A good sorbent material should have hydrophobic and oleophilic properties, high oil sorption capacity, and low material cost. Among oil absorbents, the adsorption performance of graphene-based materials is particularly outstanding. Graphene is intrinsically hydrophobic and oleophilic, and graphene-based materials typically have extremely low densities. These features allow graphene

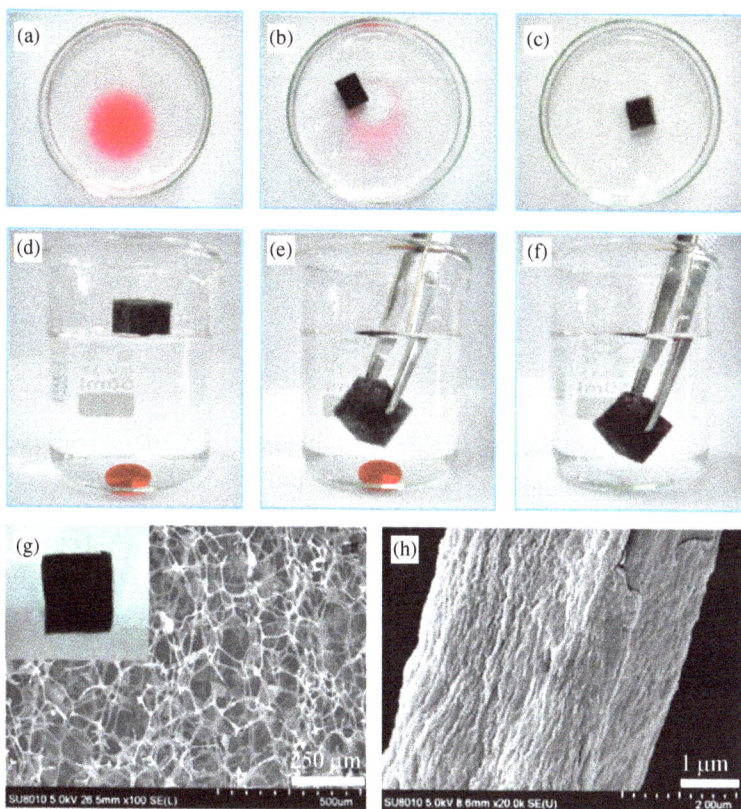

Figure 6.64. Selective adsorption of the rGO-modified melamine foam composite for soybean oil (dyed with Oil red O) (a–c) on water and chloroform (dyed with Oil red O) (d–f) in water. (g–h) SEM images of the rGO-modified melamine foam composite, insets are their corresponding digital photos.[73]

materials to exhibit high oil adsorption capability and outstanding adsorption capacity.

Optimizing the shape, internal pore structure, and mechanical properties of graphene-based adsorption materials can effectively enhance the adsorption performance of graphene materials. The rGO-modified melamine foam composite is superhydrophobic and superoleophilic and shows high capability for collecting various oils and organic solvents from water, such as soybean oil and chloroform (Fig. 6.64).[73] The maximum oil adsorption capacity is 112 times the weight of the initial composite, which is attributed to its porous structure (Fig. 6.64(g)) and the melamine

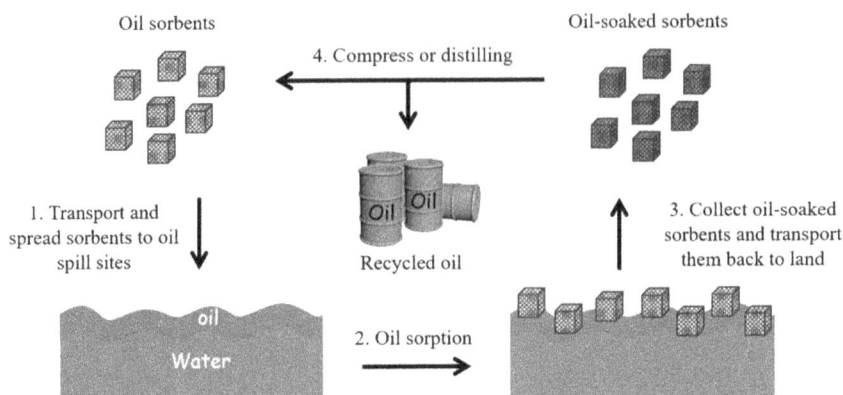

Figure 6.65.   Operation processes for the oil spill cleanup and recycling of oil absorbents.[74]

skeleton covered with complete layers of hydrophobic rGO. The adsorption capacity does not decay even after the foam is used 20 times, and it still exhibits excellent recyclability and stability against cavitation erosion and corrosion liquids.

After the absorbents are full of oils (such as crude oil), oil can be recovered from the absorbents by squeezing or distilling in order to reuse the absorbents (Fig. 6.65).[74] Oil-absorbent-based devices can continuously collect oils from the water surface under an external force and separate the adsorbed oils from the absorbents. This simultaneously achieves cleanup and recovery of oils.

It is worth noting that although most absorbents exhibit high adsorption capacity for low-viscosity oils, the adsorption speeds for high-viscosity oils (such as crude oil) are relatively low (crude oil has a high viscosity, which makes it penetrate the adsorbents at a lower velocity). In order to increase the adsorption rate of adsorbents for crude oil, some special adsorbent materials and devices have been developed. Joule-heated graphene-wrapped sponge composites are one such material that can efficiently remove crude oil.[75] In this material, the graphene coating uniformly wraps the skeletons of the sponge substrate, which endows the composites with hydrophobic and conductive properties. The high hydrophobicity and the porous structure of graphene enable the composite to adsorb crude oil from the water surface, but the oil sorption speed is low due to the high viscosity of crude oil [Fig. 6.66(a)]. After applying a voltage to the composite, current flows through the graphene coating, and

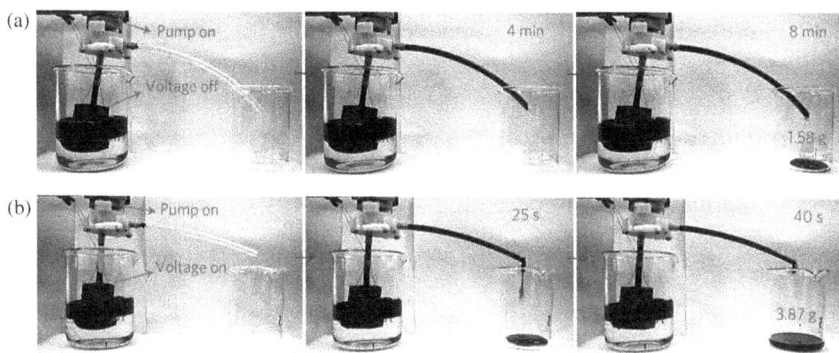

Figure 6.66. The process of collecting crude oil by graphene-wrapped sponge composites: (a) without heating and (b) with heating.[75]

Joule heat is generated, which quickly heats up the composite. Then, the hot composite can heat up the surrounding crude oil and thus the oil viscosity decreases, which then increases the diffusion coefficient of the crude oil into the composite. Therefore, the adsorption rate of the composite material to crude oil is increased (Fig. 6.66(b)). The oil adsorption time is reduced by 94.6% compared with that of a non-heated composite.

The oil/water separation efficiency of oil sorbents mainly depends on their surface wettability. Studies have shown that materials with a low surface energy and a rough surface exhibit excellent hydrophobicity and oleophilicity. In addition, excellent oil-absorbing materials should also have properties such as low density, good flexibility, high oil absorption capacity, chemical inertness, and high fire resistance. Oil-absorbent-based devices can not only effectively reduce the number of absorbent materials but also simplify and speed up the oil recovery processes. For oils with high viscosity such as crude oil, it is necessary to develop materials and technologies that can improve the adsorption and recovery efficiency. Besides, crude oil spills are often accompanied by complicated ocean weather; thus, oil absorbents should have the ability to overcome the destructive power of strong winds and big waves.

6.4.5 *Adsorption of toxic and harmful gases*

The emission of greenhouse gases (such as CO_2, CH_4, and Freon) is the cause of global warming, the release of Freon contributes to ozone depletion, and the resulting excessive exposure to UV radiation will threaten human health. In addition, toxic and harmful gases (such as NH_4, CO, H_2S, SO_2, NO_x, and N_2O_4) leaked into the environment during industrial production or chemical use will endanger the environment and human health. Therefore, it is necessary to take effective measures to deal with these harmful gases. Treating these harmful gases with graphene materials is convenient and reliable.

Graphene materials all have a porous structure and a high specific surface area, which gives them a high adsorption capacity for gases. Besides, graphene is easy to chemically functionalize. Functionalized graphene materials have a variety of functional groups, which can interact or react with adsorbed gas molecules, thereby improving the adsorption performance of graphene materials. For example, nitrogen, sulfur, and oxygen in functionalized graphene materials can interact with CO_2 molecules, and this additional force increases the absorption capacity of graphene materials for CO_2; polar groups such as $-COOH$, $-OH$, and NH_3 can interact with CO_2 or CH_4, thereby improving the adsorption capacity of graphene materials for CO_2 or CH_4.[76]

Highly active gases such as SO_2, H_2S, NO_2, and NH_3 can easily react with some composites to form stable covalent bonds. Using this principle, the functionalized graphene composites prepared by mixing graphene with active materials show a high adsorption capacity for these highly active gases. For example, hydroxyaluminium–zirconium polycation surfactants are attached onto the surface of GO to prevent the aggregation of GO sheets, resulting in space between graphene layers. This structure allows ammonia molecules to enter the space between the graphene layers and react with Al/Zr acidic centers to form stable chemical bonds. In addition, the water molecules accommodated in the interlayer space can also serve as binding sites to interact with ammonia molecules to enhance the adsorption of ammonia.

Optimizing the nanostructure of graphene materials can also improve the adsorption capacity for gas molecules. As mentioned earlier, GO intercalated with polycationic surfactants results in stable interlayer space. In addition, other adsorbents can be incorporated into graphene to form composites and the synergistic effect of graphene and other adsorbents is utilized to adsorb gaseous molecules, thus exhibiting higher adsorption capacity. These adsorbents include metal–organic frameworks (MOFs), covalent organic frameworks (COFs), and zeolite particles. A combination of multiple materials can provide more sites for gas adsorption and richer nanostructures, which allow the composite to have better and higher adsorption capacity.

6.4.6 *Summary*

In summary, in addition to the environmental applications mentioned earlier, functionalized graphene materials can also be used for the treatment of microorganisms and bacteria in wastewater and environmental monitoring. Although some developments have been made in the research of functionalized graphene materials in environmental protection, there are still some problems to be solved in this field and the following measures need to be taken: (i) further reducing the cost of graphene production and functionalization; (ii) understanding the detailed transformation mechanism of groups and structures during the preparation of functionalized graphene materials; (iii) inhibiting the aggregation of functionalized graphene materials during the application process to increase the specific surface area; (iv) improving the quality of rGO to improve the performance of corresponding composites; and (v) further investigating the toxicity of graphene-based materials to clarify whether such materials are harmful to the human body and environment. If these problems can be solved accordingly, graphene will have a greater application prospect in environmental protection.

6.5 Catalysis

At present, in addition to acid–base catalysis, homogeneous catalysis and heterogeneous catalysis have been widely used in industrial production. In particular, heterogeneous catalysis has been widely used in the fields of energy conversion, pharmaceuticals, materials science, and environmental protection in recent years. Metal catalysts such as platinum, palladium, ruthenium, cobalt, iron, and nickel are currently most widely used. Among them, noble metals such as platinum group metals are the dominant ones with high catalytic activity, but they are all rare metals with few earth reserves and high application costs. Therefore, it is necessary to develop new catalysts with abundant raw materials, high efficiency, and low cost. As a new type of 2D carbon material, graphene has great application potential in the field of heterogeneous catalysis. Graphene has a single-atom-thick 2D crystal structure, which provides two surfaces with

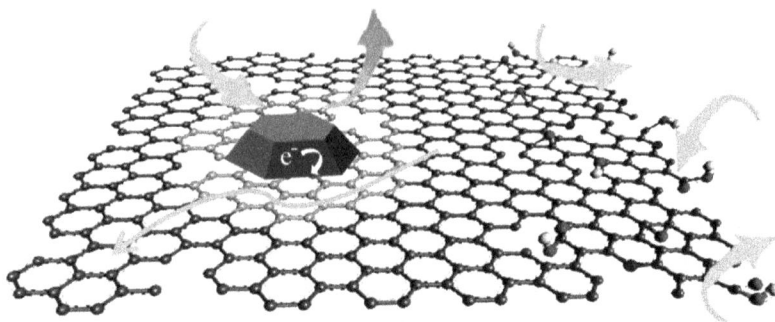

Figure 6.67. Illustration of the different roles of graphene in heterogeneous catalysis, i.e., its use as catalyst support and its intrinsic catalytic properties originating from the defects and heteroatom-containing functionalities.[77]

extremely small intervals to the environment, thus making it easy to interact with reactants and showing high surface activity. Graphene has a remarkably high electron mobility of 10,000 $cm^2/(V \cdot s)$ at room temperature, which is beneficial to electron transport in the catalytic process. Graphene has a theoretical specific surface area of 2630 m^2/g, which is able to adsorb more substrate molecules in the catalytic process. Graphene is easy to derivatize, and active sites for catalytic reactions can be easily introduced by doping and modification. The 2D plane of graphene provides a good support platform for other highly active catalysts and can be used as a carrier to prepare composite catalysts. In summary, graphene itself can be used as a highly active catalyst in catalytic reactions, it can be hybridized with other catalysts to prepare composite catalysts, and it can be used as a support for other catalysts. As shown in Fig. 6.67, graphene utilizes defects or heteroatoms as active centers to catalyze the reaction, and at the same time, it serves as a support material to load other active materials.[77]

6.5.1 *Graphene as a catalytically active material*

Graphene has a sp^2-hybridized carbon skeleton, so the carbon atoms at the edges (or defects) of graphene sheets contain dangling bonds. These dangling bonds are all in a high-energy state, and besides binding with hydrogen or other heteroatoms, they can also be used to catalyze reactions.

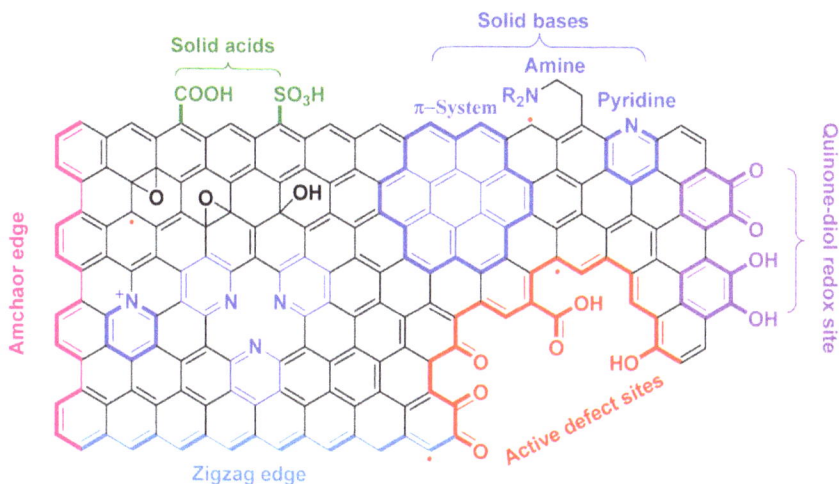

Figure 6.68. Schematic showing active catalytic sites in the graphene domain due to incorporation of heteroatoms.[78]

In addition, both functional groups and doping structures of graphene and its derivatives can alter the physical and chemical properties of graphene, making it exhibit high catalytic activity in several chemical reactions. Incorporating heteroatoms with electron-donating or electron-withdrawing behaviors, such as N, P, B, S, O, or F, into graphene will change the electronic properties of the graphene carbon framework, which can greatly enhance its electrochemical or chemical catalytic performance. As shown in Fig. 6.68, the main catalytic active sites of graphene and its derivatives are as follows: oxygen-containing groups (mainly including carbonyl, carboxyl, hydroxyl, epoxide, and benzyl alcohol groups), edge carbon (armchair and zigzag structures), and heteroatoms (N, P, B, S, O, and F).

6.5.1.1 *Oxygen-containing groups*

Oxygen-containing groups, such as carbonyl, carboxyl, hydroxyl, epoxide, and benzyl alcohol moieties, are distributed on GO and rGO. Because the oxygen atoms in these groups have unpaired electrons, these groups show nucleophilic activity in chemical reactions and attack the electron-deficient center to initiate the reaction. The nucleophilic activity of these groups depends on the electron density of the oxygen atom of the group. The higher

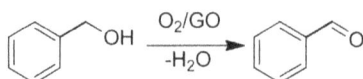

Figure 6.69.　GO as a catalyst for oxidation of benzyl alcohol to benzaldehyde.[81]

the electron density, the stronger the nucleophilic activity, and *vice versa*. Theoretical calculations show that compared with other oxygen-containing groups, the oxygen atom in the quinone group has the highest electron density, so its nucleophilic activity is higher than that of oxygen atoms in other groups. In addition, the nucleophilic activity of the oxygen-containing groups is also significantly influenced by the atomic configuration of graphene edges. For example, compared with the groups at the chair edge, the diketone groups at the zigzag edge present higher catalytic activity.

　　Due to their high catalytic activity, graphene and its derivatives with oxygen-containing groups have been widely used in a variety of catalytic reactions, such as C—H activations, redox reactions, ring-opening polymerizations, acetalization, and dehydration. For example, high-loading GO (20%~200%) can selectively catalyze the oxidation of benzyl alcohol to benzaldehyde (Fig. 6.69).[79] Theoretical calculations show that the epoxide groups on the GO surface provide active sites. However, it is found experimentally that this catalyst is deactivated after repeated use. The results of Fourier transform infrared spectroscopy, elemental analysis, and powder conductivity tests indicate that GO is gradually reduced to graphene during the catalytic reaction. Theoretical analysis reveals that the reaction occurs via the transfer of hydrogen atoms from the organic molecule to the epoxide groups on the basal plane of GO.[80] The epoxide groups are ring opened after receiving hydrogen atoms; this results in the formation of one equivalent of water and a double bond. The produced double bonds are re-oxidized in the presence of water. The decrease in catalytic activity indicates that the produced double bonds are not completely re-oxidized by oxygen molecules, which demonstrates that only if the re-oxidation of double bonds occurs will GO recover to its initial state and become usable as a catalyst again.

6.5.1.2　*N-doped graphene*

Doping graphene with heteroatoms has proven to be an effective method to modify the electronic properties of the carbon network, endowing the

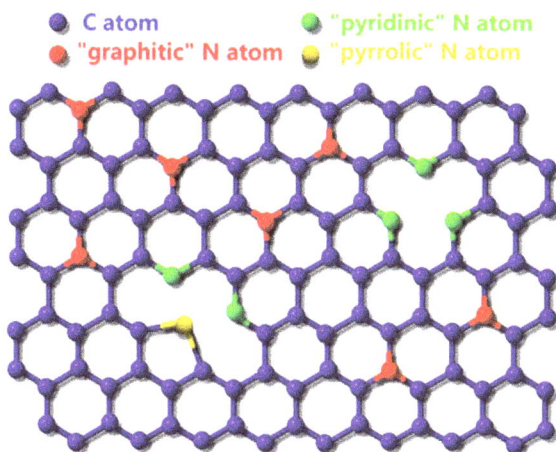

Figure 6.70. Schematic representation of the N-doped graphene.[82]

doped region with catalytic activity. Among all heteroatoms used for doping, nitrogen atoms are most commonly used. For example, *N*-doped graphene has been demonstrated to be a promising metal-free catalyst for the ORR, and the ORR is a key step in fuel cells and other renewable energy technologies, such as metal-air batteries and dye-sensitized solar cells. *N*-doped graphene exhibits high catalytic activity and durability, and it is an excellent substitute for noble metal catalysts such as platinum. *N*-doped graphene materials can be prepared by pyrolysis or bottom-up strategies. In *N*-doped graphene, nitrogen atoms have three forms: pyridinic-N, graphitic-N, and pyrrolic-N (Fig. 6.70).[82] Due to the different bonding characteristics inserted into the graphene network, the effects of the three forms of nitrogen atoms on the electronic structure of the graphene network are different. Hence, their roles in catalytic reactions are also different.

Experimental and theoretical analyses show that pyridinic-N is usually considered responsible for the catalytic activity of N-doped graphene. Based on local scanning tunneling microscopy/spectroscopy (STM-STS) measurements combined with DFT calculations, it has also been reported that carbon atoms adjacent to pyridinic-N possess a localized density of states in the occupied region near the Fermi level.[83] This suggests that the carbon atoms can behave as Lewis bases owing to the possibility of electron pair donation. The electronegativity of the nitrogen atom (3.04) is far larger than that of the carbon atom (2.55), so the carbon atom connected to the nitrogen atom is positively charged and serves as an active center to catalyze ORR. This catalytic process may proceed through two possible pathways: one is a four-electron mechanism (Eq. (8.16)) and the other is a two-electron process (Eq. (8.17)).

$$\text{Four-electron process: } O_2 + 4H^+ + 4e^- \rightarrow H_2O \qquad (8.16)$$

$$\text{Two-electron process: } O_2 + 2H^+ + 2e^- \rightarrow H_2O_2 \qquad (8.17)$$

The initial process of these two reaction pathways is the same; that is, the oxygen molecule is first adsorbed at the carbon atom next to pyridinic-N followed by protonation of the adsorbed O_2. In the four-electron mechanism, the other two protons attach to the two oxygen atoms, leading to breakage of the O—OH bond and formation of OH species. The additional proton then reacts with the adsorbed OH to form H_2O. In the two-electron pathway, H_2O_2 is formed by a reaction of the adsorbed O—OH species with another proton, followed by readsorption of H_2O_2 and its reduction by two protons to generate H_2O. The OH species may arise from the four-electron mechanism, but it is also possible that the OH species next to pyridinic-N may arise from a reaction with H_2O_2 in the two-electron mechanism. In either pathway, the carbon atoms next to pyridinic-N with Lewis basicity play an important role as the active sites at which oxygen molecules are adsorbed, forming the initial step of the ORR. From this analysis, it can be concluded that the carbon atom connected to pyridinic-N is a Lewis base, which serves as the active site to catalyze the ORR.

6.5.1.3 *B-doped graphene*

The electronegativity of boron atoms (2.04) is less than that of carbon atoms (2.55), so boron doping in the graphene framework leads to redistribution of electron density, where the electron-deficient boron sites provide enhanced binding capability. This electron-deficient center will adsorb the electron-rich reaction substrate in the catalytic reaction and catalyze the reaction as the active center. For instance, B dopants exhibit superior ORR catalytic performance because the electron-deficient B dopants can facilitate chemisorption of negatively polarized O atoms and promote O—O cleavage during ORR. In addition, the formation of *B*-doped graphene is also reported to present good OER activity in alkaline solution.

As shown in Fig. 6.71, *B*-doped graphene in the BC_3 structure can efficiently catalyze the electrochemical reduction of nitrogen to ammonia

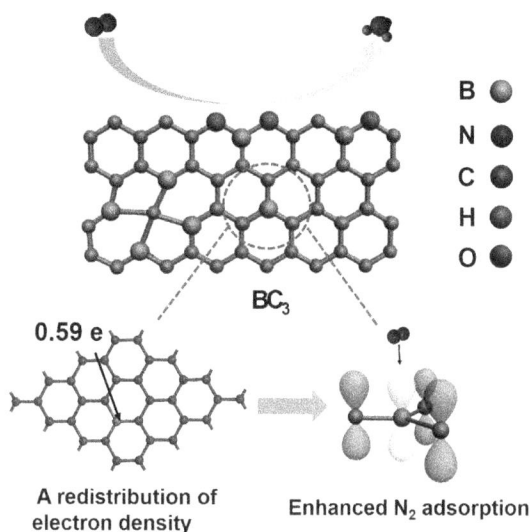

Figure 6.71. Schematic of the structure, energy level, and nitrogen adsorption process of B-doped graphene.[85]

under mild conditions.[85] The graphene framework doped with boron atoms can retain its original sp^2 hybridization and conjugated planar structure, but boron atoms become positively charged (+0.59 eV) due to the difference in electronegativity between boron and carbon atoms. The positively charged boron dopant is beneficial for adsorbing N_2 to lower the energy barrier for N_2 electroreduction to NH_3. In addition, these electron-deficient boron sites can also prohibit the binding of Lewis acid H^+ at these sites (under acidic conditions), which can promote the N_2 reduction reaction (NRR) and faradic efficiency (FE) as well as depress the HER. At a doping level of 6.2%, the boron-doped graphene achieves an NH_3 production rate of 9.8 $\mu g/(h \cdot cm^2)$ and one of the highest reported faradic efficiencies of 10.8% at −0.5 V versus reversible hydrogen electrodes in aqueous solutions at ambient conditions.

6.5.1.4 *P-doped graphene*

Similar to the boron atom, the electronegativity of the phosphorus atom (2.19) is also lower than that of the carbon atom (2.55) and has a higher electron-donating ability, so the phosphorus atom can also destroy the electrically neutral conjugated framework of graphene, producing positively charged active centers that can strongly adsorb oxygen molecules, thereby accelerating ORR. For instance, a thermal annealing approach has been used to fabricate *P*-doped graphene with triphenylphosphine as the P source.[86] This functionalized graphene material exhibits high electrocatalytic activity, long-term stability, and excellent tolerance to the crossover effects of methanol in ORR in an alkaline medium. The catalytically active center is the positively charged phosphorus atom.

6.5.1.5 *S-doped graphene*

From the abovementioned discussion, it is known that since the electronegativity of B or P is smaller than that of carbon atoms after graphene is doped with these atoms, positively charged electron-deficient centers will be produced at the heteroatom-doped sites. In ORR, these electron-deficient centers have a strong adsorption capacity for oxygen molecules and are the active sites for catalyzing ORR. Unlike B or P, although sulfur (2.58) has an electronegativity close to that of carbon (2.55), *S*-doped graphene also exhibits high catalytic activity in ORR. For example, *S*-doped graphene is obtained by directly annealing GO and benzyl disulfide (BDS) in argon (Fig. 6.72).[87] Structural analysis shows that sulfur atoms are at the edge or at the defect sites in two distinct forms: as sulfide groups (—C—S—C—) and as oxidized sulfur (—C—SO$_x$—C—, x = 2~4) such as sulfate or sulfonate. The electrocatalytic performances show that the *S*-doped graphene can exhibit excellent catalytic activity,

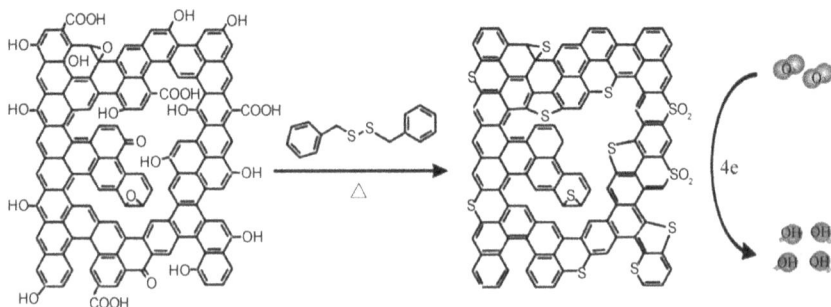

Figure 6.72. Schematic illustration of sulfur-doped graphene preparation and catalytic ORR process.[87]

long-term stability, and high methanol tolerance in alkaline media for ORRs. Analysis indicates that this *S*-doped graphene also catalyzes ORR in a four-electron mechanism. The active center of this catalytic reaction may be C—S bonds. However, considering that sulfur has an electroneutrality similar to that of carbon, the doping of sulfur does not significantly change the charge distribution of the doped graphene framework. Therefore, the mechanism of ORR catalyzed by *S*-doped graphene is different from that of *N*- or *B*-doped graphene. DFT calculation reveals that, compared with atomic charge density, the spin density is more important in determining the catalytic active sites. Therefore, the spin density of sulfur atoms on the *S*-doped graphene determines the catalytic activity of this material in ORR.

6.5.1.6 *Co-doped graphene*

In order to further improve the catalytic performance of heteroatom-doped graphene, co-doped graphene materials have been developed. In these materials, a variety of the abovementioned heteroatoms (N, P, B, S, O, or F) are spontaneously incorporated, giving rise to further improved catalytic performance under their synergistic effect. For example, *N*- and *B*-co-doped graphene shows superior catalytic activity over single-heteroatom-doped graphene.[88] The *N*-, *P*-, and *F*-co-doped graphene fabricated by thermal activation of polyaniline-coated GO with ammonium hexafluorophosphate exhibits excellent electrocatalytic activities for ORR, OER, and HER.[89] *N*- and *S*-co-doped graphene is prepared by thermally annealing graphene grown on nanoporous Ni with pyridine and thiophene, and its structure is shown in Fig. 6.73.[90] Owing to *N*- and

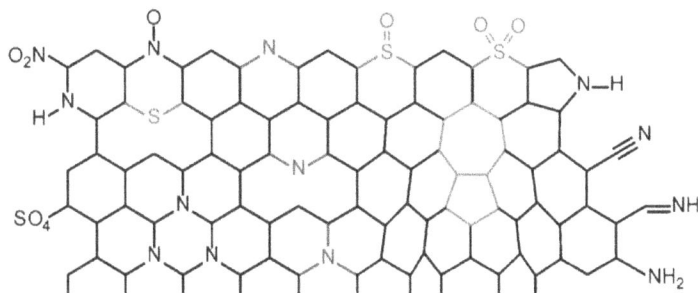

Figure 6.73. Potential defect structure in N- and S-co-doped nanoporous graphene.[90]

S-doping in various forms, this co-doped material exhibits extremely high catalytic activity in HER — showing a low reaction overpotential comparable to the best-performing metal-free catalyst (2D MoS$_2$).

6.5.1.7 *Defect/edge catalysis*

Defects are ubiquitous in graphene, but they are also very important. Defects can change the local electronic structure of graphene or inject charge into *sp*^{2}-conjugated frameworks and thus tailor the physical and chemical properties of graphene. Various defects in graphene are produced during synthesis or fabrication, or they can also be introduced later. Studies have shown that there are two types of defects in graphene: point defects and line defects.[91] Point defects include Stone–Wales defects (Fig. 6.74(b)), single vacancy (Fig. 6.74(c)), double vacancies (Fig. 6.74(d)), pentagon rings at the zigzag edge (Fig. 6.74(e)), octagon and fused pentagon carbon ring line defects with odd number of octagon rings (Fig. 6.74(f)), octagon and fused pentagon carbon ring line defects

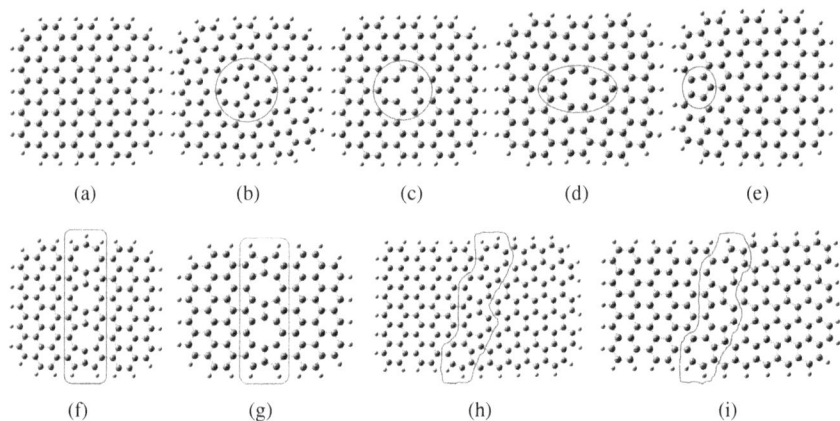

(a)   (b)   (c)   (d)   (e)

(f)   (g)   (h)   (i)

Figure 6.74. Perfect and defective graphene clusters[91]: (a) Perfect graphene cluster; (b) Stone–Wales defect; (c) single vacancy; (d) double vacancies; (e) edge defect with pentagon ring at the zigzag edge; octagon and fused pentagon carbon ring line defect with (f) odd number of octagon rings and (g) even number of octagon rings; and pentagon–heptagon pair line defects with (h) odd number of heptagon rings and (i) even number of heptagon rings. The larger gray and smaller white balls denote carbon and hydrogen atoms, respectively.

with even number of octagon rings (Fig. 6.74(g)), pentagon–heptagon pair line defects with odd number of heptagon rings (Fig. 6.74(h)), and pentagon–heptagon pair line defects with even number of heptagon rings (Fig. 6.74(i)). Due to the different structures and positions of these defect sites, their performance in catalytic reactions is also different. For example, theoretical calculations show that only the pentagon ring located at the zigzag edge structure shows catalytic activities for ORR, while other point defects, such as Stone–Wales defects, single vacancy, and double vacancies, exhibit no ORR catalytic activity. Among line defects, only the structure containing the odd number of heptagon or octagon carbon rings generates spin density and can catalyze ORR.

Due to its high catalytic activity, defective graphene has been widely used in various reactions instead of platinum. For instance, defective graphene (DG, Fig. 6.75(a)) prepared via the nitrogen removal procedure from nitrogen-doped graphene exhibits extremely high catalytic activity in three key electrochemical reactions: ORR, OER, and HER.[92] Studies have shown that the defect sites in this graphene material are mainly distributed at the edge of graphene and mainly contain three types: edge pentagon, 5-8-5 defect, and 7-5-5-7 defect (Fig. 6.75(b–d)). It is these defects that are fundamental in enhancing the electrocatalytic activity of the material, presumably due to the locally modulated electronic environment associated

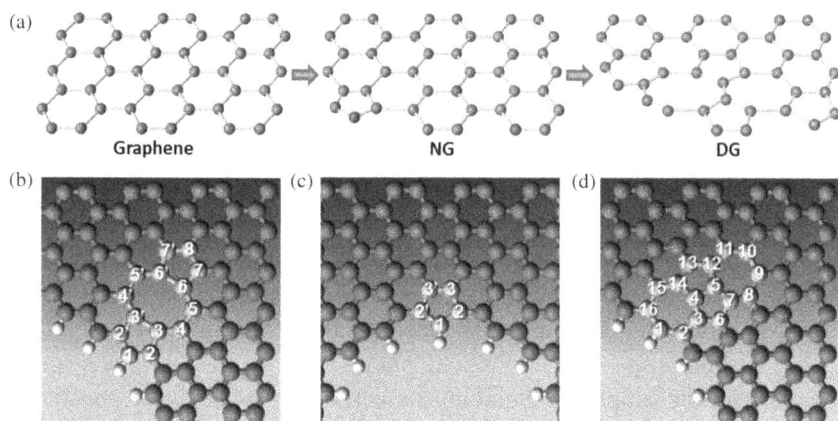

Figure 6.75. Example of graphene with defects.[92] (a) schematic of the formation of defective graphene and (b–d) schematic of defective graphene with three types of structures.

with the defects. It is also important that as these defects perturb the surface properties of the graphene, they induce other catalytic effects such as changes in specific surface area and surface hydrophobicity. These changes in surface properties are beneficial to mass transfer in aqueous electrolytes, thereby enhancing the catalytic activity of defective graphene.

6.5.2 *Graphene as a catalyst support*

Supported catalysts are the most important types of catalysts in heterogeneous catalysis. In these catalysts, the support plays a crucial role in the catalytic activity. Among various supports used in heterogeneous catalysis, carbon materials have been attracting growing interest owing to their large surface area and chemical stability, which facilitate high loading of active sites and contribute to their resistance to degradation in both acidic and basic media. In addition, simple combustion treatment is enough to recover the catalytic behavior of carbon supported metal catalysts, in particular, for the carbon supported noble metal catalysts. Compared with other carbon supports, graphene and its derivatives (e.g., GO) have unique properties, as mentioned earlier, that make them attractive for use in heterogeneous catalysis.

6.5.2.1 *Interaction with catalytic phases*

The intrinsic chemical activity, selectivity, and stability of a supported catalyst not only depend on its composition, structure, size, and shape but also on its interaction with the support. For supported metal catalysts, in particular, it is expected that the chemical bonding and associated charge transfer at the interface between the metal particles and support can be used to tune the electronic and chemical properties of the catalytically active sites to achieve higher catalytic activity and selectivity. Furthermore, good dispersion and improved stability of the supported nanoparticles are usually observed in catalysts featuring stronger support interactions. However, graphene is chemically inert because of the strong sp^2 bonding among carbon atoms in the graphene plane, which leads to weaker interactions with the supported metal clusters. To address this issue, mechanical strain, defects, and functional groups have been introduced to strengthen the interaction between the catalyst and graphene.

First-principles calculations reveal that the application of a moderate tensile strain to graphene greatly increases the adsorption energy of different types of metal clusters on graphene.[93] More specifically, in the case of supported Au clusters on graphene for CO oxidation, the strain in graphene reverses the charge transfer between the Au clusters and graphene support, resulting in a significant reduction in the reaction barrier.

Introducing structural defects is a more efficient way to increase the interaction between graphene and the supported catalytic species. As mentioned earlier, defects in graphene can change the local electronic structure of the place where they are located, inject charges into the sp^2 system, and change the surface properties of graphene. Therefore, defects are effective sites for graphene to hybridize active materials, especially metal nanoparticles. For example, platinum atoms can form strong Pt-C bonds with carbon atoms at defects, so that platinum metal nanoclusters are strongly adsorbed on defective graphene, leading to increased charge transfer between the catalyst system and the support.[94] Studies have indicated that the diffusion barrier of a metal atom on defective graphene is always considerably higher than that on pristine graphene.

Incorporating functional groups into graphene is another effective means to enhance its interaction with active materials. For example, GO has a large number of oxygen-containing functional groups and topological defects, which can serve as more favorable anchoring centers and nucleation sites for the active species or the precursors. Therefore, GO and its derivatives, rather than pristine graphene, have been used to prepare graphene-supported catalysts. Compared with graphene with a perfect structure, the advantages of using GO as a support or precursor can be summarized as follows: (i) GO is cheaper and more readily available. (ii) The amphiphilic nature and excellent dispersibility of GO in both aqueous and organic solvents render the surface more accessible to different precursors that can subsequently enhance the accessibility of the supported catalytically active species. (iii) The binding energy of the oxygen-doped surface of GO toward metal atoms is significantly higher than that of pristine graphene, which subsequently increases the stability of the supported nanoparticles and their resistance to sintering. Moreover, the associated charge transfer may significantly lower the reaction barrier without the need for introducing defects or strain. (iv) The covalent immobilization of catalytically active organic species or enzymes can be easily achieved by surface functionalization. (v) The redox reaction between GO and many metal ions may lead to spontaneous deposition of the metal nanoparticles on the basal plane of graphene.

It is worth noting that graphene, GO, and rGO supports may show significant differences in catalytic performance, especially for supported electrocatalysts. The advantages of GO as support have been discussed earlier, but its drawbacks and the advantages of the other two types of graphene should also be considered. The presence of functional groups and structural defects in GO will significantly reduce the electron mobility of pristine graphene, which will subsequently affect electron transfer of the catalytically active species. In particular, for the supported noble metal electrocatalysts, a crucial role of the support is to provide a network with high electron conductance to minimize ohmic losses associated with electron transport. Therefore, using graphene supports with low defect density and high electrical conductivity is more desirable for electrocatalysis. Therefore, compared to GO, highly conductive graphene, such as graphene prepared by the CVD method or exfoliated by solvent, is a better choice. In Section 6.4, semiconductor nanoparticle-loaded graphene is synthesized for the photodegradation of organic pollutants (Fig. 6.57), for which graphene serves as an acceptor for the photogenerated electrons, which can accept photoexcited electrons and leave behind holes in semiconductor nanoparticles.[65] This enhances the effective separation of excitons (holes and electrons) and inhibits their recombination, thereby enhancing the efficiency of the photodegradation reaction.

6.5.2.2 *Influence of mass transfer*

In heterogeneous catalysis, mass transfer is also one of the important factors affecting the catalytic activity. For conventional supported catalysts with a porous structure, the reaction only occurs when the reactant molecules come in contact with the supported active sites, which are usually located inside the pores. Typically, seven consecutive steps are involved in a classical heterogeneous catalytic reaction: (i) diffusion of the reactants from the bulk phase to the external surface of the catalyst (external diffusion); (ii) diffusion of the reactants inside the pores to the immediate vicinity of the active sites (internal diffusion); (iii) adsorption of the reactants onto the inner catalytic surface; (iv) reaction at specific active sites; (v) desorption of the products from the inner surface; (vi) diffusion of the products from the interior of the catalyst to its external surface (internal diffusion); and (vii) diffusion of the products from its external surface to the bulk fluid (external diffusion). Consequently, the observed reaction rate and selectivity of the targeted product also rely on the physical mass transfer/diffusion steps involved. Many heterogeneous catalytic reactions are diffusion controlled, especially for reactions in the liquid phase.

Therefore, in order to enhance the performance of supported catalysts, the selection of support materials and the optimization of their structures become crucial. It is expected that graphene support materials with a large surface area and interesting surface chemistry not only afford high loading and stronger interactions with the targeted species but also promote mass transfer during the reaction. In addition, the morphology of the graphene support is also crucial to enhance the catalytic activity. In contrast to 2D graphene supports, 3D graphene frameworks/networks may be a better choice for use in solid-state catalysts. The 3D structure not only provides high specific surface areas and high strength endurance for catalyst loading but also facilitates the rapid mass transfer of the reactants, products, and electrolytes.

6.5.2.3 *Single-atom catalysis*

Supported metal nanoparticles are the most widely used type of heterogeneous catalysts in industrial processes. Numerous studies have shown that the size of metal particles is a key factor in determining the performance of such catalysts. Compared with large particles, nanoparticles exhibit high specific surface areas and quantum size effects. When the volume of nanoparticles decreases, the exposed atoms on the surface gradually increase, and the surface atomic structure, electronic structure, and surface curvature of nanoparticles also change accordingly. In particular, because low-coordinated metal atoms often function as the catalytically active sites, the specific activity per metal atom usually increases with decreasing size of the metal particles (Fig. 6.76).[95] The ultimate small-size limit for metal particles is the single-atom catalyst (SAC), which contains isolated metal atoms singly dispersed on supports. Size reduction allows the increment of surface free energy of the metal components, and the metal sites become more and more active for chemical interactions with the support and adsorbates, which accounts for the size effects of metal nanocatalysts. In extreme case of SACs, because of the highly active valence electrons, the quantum confinement of electrons, and the sparse quantum level of metal atoms, the surface free energy of metal species reaches a maximum, which then leads to promoted chemical interactions with the support and

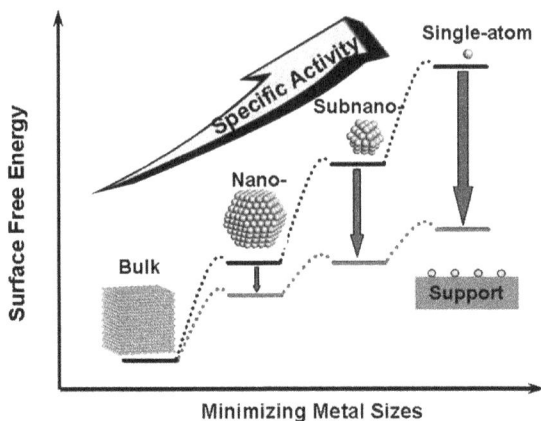

Figure 6.76.   Schematic illustrate the changes of surface free energy and specific activity per metal atom with metal particle size and the support effects on stabilizing single atoms.[95]

the unique chemical properties of SACs. Metals commonly used as SACs mainly include Ir, Ag, Cu, Fe, Rh, Ni, Pt, Pd, and Co.

However, the surface free energy of metals increases significantly with decreasing particle size, promoting aggregation of small clusters, and the isolated single metal atom is no longer retained. Therefore, the role of the support in the catalyst system becomes crucial. Using an appropriate support material that strongly interacts with the metal species prevents this aggregation, creating stable, finely dispersed metal clusters with a high catalytic activity. In addition to stabilizing single-metal atoms, charge transfer between metal atoms and the support occurs, thereby affecting the performance of the catalyst system. Therefore, the surface atoms, electronic structure, and morphology of the support can be optimized, which effectively improve the interaction between the support and single-metal atoms and ultimately improve the activity and selectivity of heterogeneous catalysts.

As shown in Fig. 6.77, there are different types of SACs according to the chemical interactions between the mononuclear metal atom and supports, including single-metal atoms anchored to metal oxides (FeO_x, Al_2O_3, ZnO, and MoS_2), metal (Au, Cu) surfaces, and graphene and its analogs (such as C_3N_4).[95] Among them, graphene-based supports have a 2D planar structure consisting of sp^2-hybridized carbon atoms; so, they are increasingly used in SAC research. The corresponding catalysts have

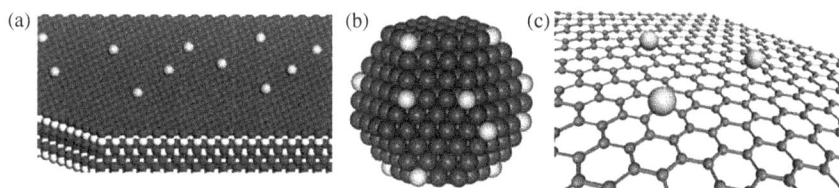

Figure 6.77. Schematic diagrams illustrating different types of SACs: Metal single atoms anchored to (a) metal oxide; (b) metal surfaces; and (c) graphene.[95]

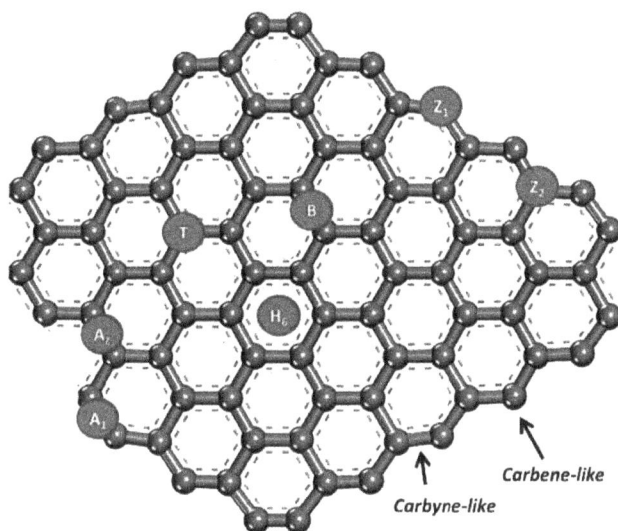

Figure 6.78. Adsorption sites on sp^2 carbon materials: Hollow H, bridge B, top T, A armchair, and Z zigzag sites.[96]

been widely used in redox reactions, catalytic hydrogenation, photocatalysis, and electrochemical reactions.

Defect-free graphene can adsorb metal atoms at edges and sp^2-hybridized 2D conjugated planes (Fig. 6.78).[96] The edges of graphene layers are the most active sites, owing to high densities of unpaired electrons, which are capable of binding metal atoms. On the basis of configurations of the edge carbon atoms, the zigzag sites are carbene-like, and the armchair sites are carbyne-like. For defect-free materials, the different possible adsorption sites above graphene are top, T- (top of a carbon atom); bridge, B-(above a C-C bond); and hollow, H6- (above the center of hexagons) sites.

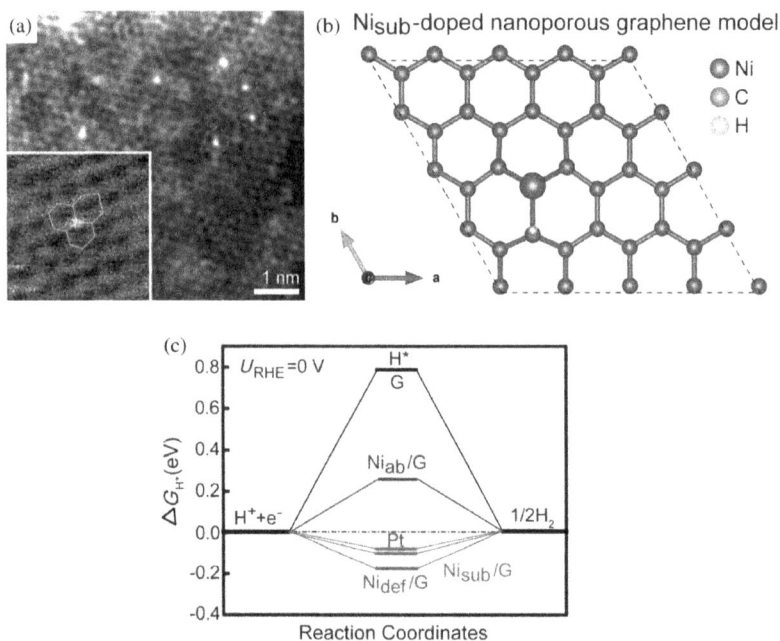

Figure 6.79. Example of functionalized graphene as support for single-metal atoms[97]: (a) HAADF-STEM image of Ni-doped graphene; (b) hydrogen adsorption sites and configuration of the model; and (c) calculated Gibbs free energy diagram of HER at equilibrium potential for a Pt catalyst and Ni-doped graphene.

However, in practical applications, graphene supports usually contain defect sites. Therefore, the edges and active surfaces of graphene interact with metal atoms and defect sites, which are used for catalytically active sites; for example, graphene-supported single-atom nickel can be used as a catalyst for HER.[97]

The preparation process for this graphene-supported single-atom nickel catalyst is relatively simple: Graphene is grown on nickel foam by the CVD method, and then the nickel template is etched with hydrochloric acid to obtain nickel-doped graphene (Fig. 6.79(a–b)).[97] Theoretically, nickel dopants can present as (i) interstitial atoms in the hollow centers of the benzene rings (Ni_{ab}), (ii) substitutional dopants occupying C sites in the graphene lattice (Ni_{sub}), or (iii) anchoring atoms on defect sites (Ni_{def}). DFT calculations reveal that the binding energies of these three configurations are −0.94 eV, −7.54 eV, and −8.97 eV, respectively. Apparently, substitutional and anchored Ni dopants are energetically favorable because the binding energies are even lower than the cohesive energy of bulk Ni (−4.44 eV). High-angle annular dark-field scanning transmission electron microscopy (STEM) images show that the monoatomic Ni species occupy carbon sites in the graphene lattices. Theoretical calculations show that in HER reaction, the Gibbs free energy profiles of intermediate H* in three configurations of Ni loading for graphene catalysts are different. Among them, in the reaction catalyzed by Ni_{sub}/G, the Gibbs free energy of intermediate H*, $|\Delta G_{H*}|$, is 0.10 eV, which is closest to zero (Fig. 6.79(c)). These results demonstrate that single Ni atoms occupying carbon sites of graphene are the most efficient HER catalysts and therefore play a critical role in the outstanding HER activities.

The defects, vacancies, doped heteroatoms, and surface functional groups on the carbon framework of functionalized graphene can be utilized to immobilize single-metal atoms through covalent and coordination bonds. Point defects and line defects on graphene can change the electronic structure of the local sp^2 carbon framework, thereby modifying the adsorption properties of graphene on metal atoms. As mentioned in Section 6.5.1, heteroatom doping can also tailor the electronic properties of the graphene carbon framework, thereby affecting its ability to adsorb metal atoms. In addition, nitrogen atoms contain unpaired electrons and are considered to be a Lewis base. Therefore, *N*-doped graphene can strongly adsorb metal atoms, and nitrogen atoms can also change the electronic structure of adsorbed metal atoms to further optimize the performance of the catalyst. Theoretical calculations reveal that *N*-doped graphene can adsorb metal atoms in various forms (Fig. 6.80), among which a 4N center structure, with the adsorbed metal atoms and four nitrogen atoms, exhibits the highest binding force, and its binding energy is higher than 7 eV.[98]

N-doped graphene as a support has been widely used in the field of SACs, for example, single-atom Co supported on *N*-doped graphene (Co-NG).[99] GO and $CoCl_2 \cdot 6H_2O$ are mixed uniformly in water (mass ratio GO/Co=135:1) and the solution is then freeze-dried and calcined in an ammonia atmosphere to obtain catalyst Co-NG (Fig. 6.81(a)). Defects and a large structural disorder are observed on the *N*-doped graphene framework using a scanning transmission electron microscope (STEM). HAADF STEM analysis shows that Co atoms are well dispersed in the carbon matrix (Figs. 6.81(c) and 6.81(d)). The size of these bright spots, corresponding to Co atoms, is in the range of 2 ~ 3 Å, indicating that each bright dot corresponds to one individual Co atom; an enlarged view of the

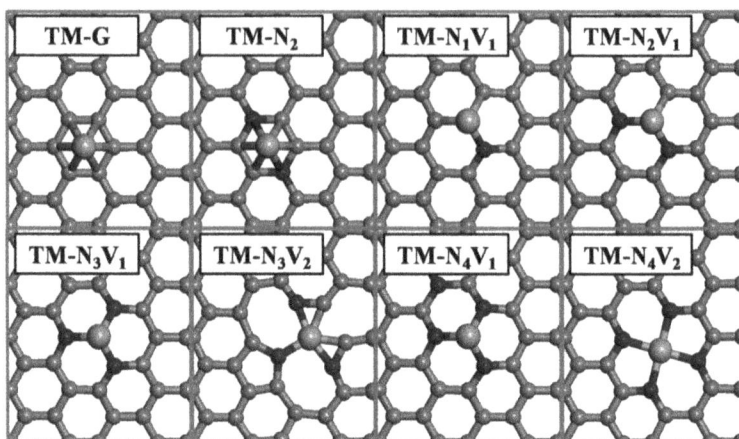

Figure 6.80. Optimized configurations for metal atoms embedded into different types of N-doped graphene.[98]

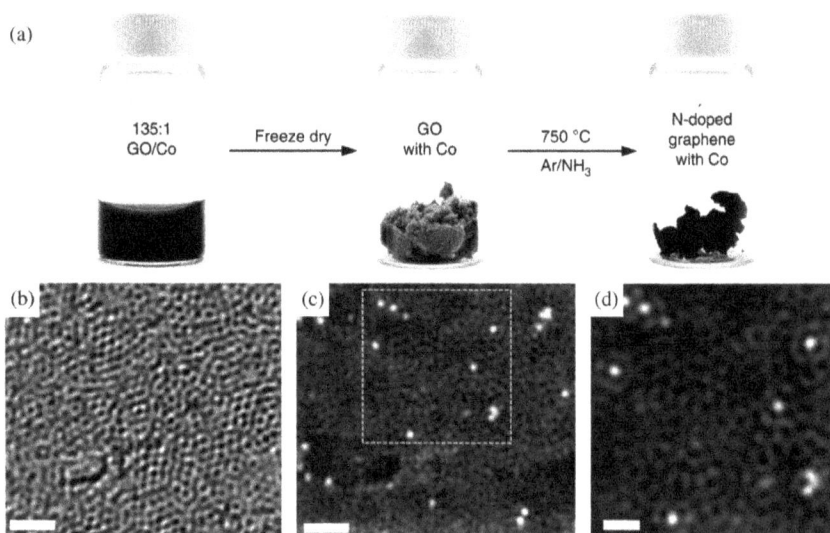

Figure 6.81. Preparation and morphology characterizations of single-atom Co supported on N-doped graphene (Co-NG)[99]: (a) schematic illustration of the synthetic procedure of the Co-NG catalyst; (b) bright-field aberration-corrected STEM image of the Co-NG; (c) HAADF-STEM image of the Co-NG; and (d) the enlarged view of the selected area in (c).

selected region reveals that each Co atom is centered by the light elements (C, N, and/or O). Analyses by various means indicate that in the catalyst Co-NG, the Co is atomically dispersed in the *N*-doped graphene matrix and it is in the ionic state with nitrogen atoms in the cobalt's first coordination sphere. This SAC functions as a highly active and robust HER catalyst in both acid and base media.

6.5.3 *Summary*

In summary, due to their unique properties, functionalized graphene materials can be used as support materials for heterogeneous catalysts as well as supported catalysts, especially SACs. The catalytic activity and loading capacity of functionalized graphene materials originate from their edges/ defects, functional groups, and doping structures. Although graphene-based catalysts have been applied in a variety of catalytic reactions and have shown excellent performance, this field is still in its infancy, and there are still many problems to be solved in order to realize practical applications. For instance, compared with commercial catalysts, the performance and stability of graphene-based catalysts still need to be improved. Due to the complexity of the structure of graphene-based catalysts, the actual catalytic sites are still difficult to identify and further studies are required. In addition, the catalytic mechanism and influencing factors of such catalysts are still not very clear and further research is needed.

6.6 EMI Shielding

Electronic and information technology in today's world is developing rapidly. All kinds of electronic devices have penetrated every part of our lives, such as family life, medical treatment, agricultural production, manufacturing, aerospace, and communication. Various electronic and electrical equipment can be found around us. This has caused electromagnetic waves to be present everywhere in our environment, which have two components: an oscillating electric field and a perpendicular, comoving magnetic field that oscillates at the same frequency, but with a phase shift of 90°. Electromagnetic waves are electromagnetic fields that propagate in the form of waves and they exhibit wave–particle duality. The ubiquitous electromagnetic waves may damage some high-precision electronic instruments and equipment and even have adverse effects on the living environment of human beings. In 1821, Faraday discovered for the first time that an enclosure made of a conducting material shields the inside from an external electric charge or field, or shields the outside from an internal electric charge or field. This is known as the "Faraday cage." In order to eliminate the effects of electromagnetic waves, various materials for electromagnetic shielding based on this principle have emerged. One can use electromagnetic shielding materials to surround instruments and equipment to prevent interference from external electromagnetic fields or suppress their ability to cause interference to other areas. Electromagnetic interference shielding effectiveness (EMISE) is defined as the logarithm of the ratio, expressed in decibels, of the incident power (P_i) to the transmitted power (P_o). The attenuation of an electromagnetic wave occurs by three mechanisms: reflection (SER), absorption (SEA), and multiple reflection (SEM). In order to achieve EMI reflection, the shielding material must have mobile charge carriers. This means the material used for shielding must be conductive. Absorption of electromagnetic waves is caused by interaction with electrical (or magnetic) dipoles of the shielding material. Therefore, the material with better electrical conductivity and magnetic permeability presents higher shielding effectiveness. The adsorption and multiple reflections of electromagnetic waves arise from scattering centers and interfaces or defect sites within the shielding materials. Thus, shielding effectiveness is the sum of all the three terms:

$$SE_T = SE_R + SE_A + SE_M = 10\log_{10}\left(\frac{P_i}{P_o}\right) \tag{8.18}$$

The greater the electromagnetic shielding efficiency value, the weaker the electromagnetic wave passing through shielding material, that is, the stronger the electromagnetic interference shielding (EMI) effect. That is to say, most of the electromagnetic radiation is absorbed or reflected by the shielding material, and only a small amount of energy passes through the shielding material. An ideal EMI shielding material is of a light weight and can achieve high EMI shielding efficiency (SE) with strong absorption and weak secondary reflection. Commonly used electromagnetic shielding materials are metals such as copper and aluminum.

6.6.1 *Graphene as EMI shielding material*

Graphene is a new type of electromagnetic shielding material. As mentioned earlier, graphene has a very high specific surface area, high electrical conductivity, and excellent thermal conductivity. These properties are beneficial to improving its electromagnetic shielding performance. In addition, compared with conventional metal shielding materials, graphene-based materials exhibit the advantages of low density, good flexibility, and corrosion resistance. Currently, graphene-based electromagnetic shielding materials have been widely reported.

Incorporation of functional groups into graphene, through doping, leads to the preparation of graphene electromagnetic shielding materials. Doping can improve the conductivity of graphene to enhance the reflection and absorption of electromagnetic waves and ultimately increase the EMI SE of graphene materials. At the same time, structure optimization of graphene shielding materials can also improve the EMI SE of the material. A laminated structure of sulfur-doped rGO (SrGO) is an excellent example.[100] SrGO is prepared through a reaction between GO and H_2S gas at elevated temperatures (Fig. 6.82). Because of the *n*-type doping

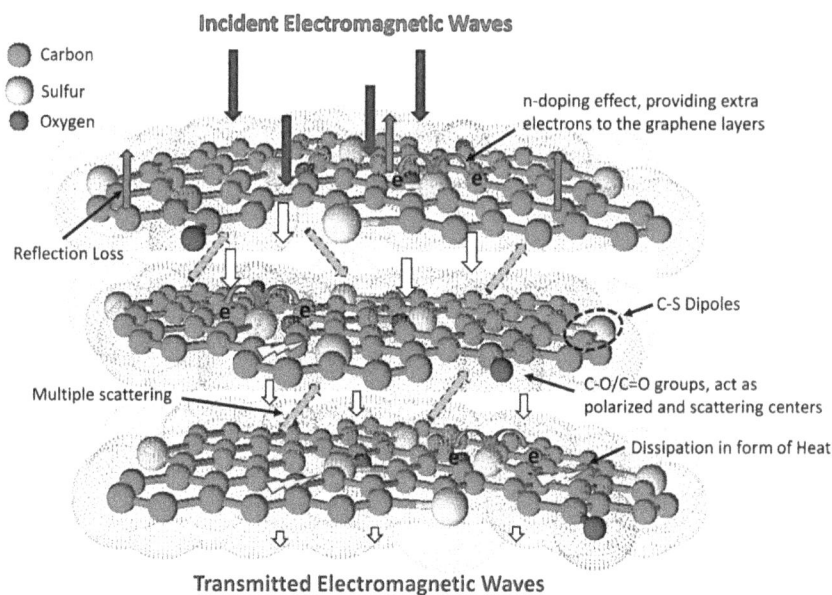

Figure 6.82. Proposed mechanism for EMI shielding with the effect of sulfur doping.[100]

contribution of the S atom to the doped graphene, the SrGO laminate reveals a 47% greater electrical conductivity than undoped rGO laminate. The improvement of electrical conductivity can significantly enhance the reflection and absorption of graphene materials on electromagnetic waves. Meanwhile, the laminate structure of this graphene functional material can effectively enhance the multiple internal scattering of electromagnetic waves. Moreover, the residual C—O and C=O bonds present in GO also serve as polarization centers. Under these multiple effects, the EMI shielding effectiveness of the *S*-doped rGO material is further increased.

Incorporating other materials with graphene to make functionalized graphene composites is also one of the effective means to prepare graphene electromagnetic shielding materials. Polymer materials are often used to make functionalized graphene composites due to their advantages of light weight, excellent plasticity, and easy adjustment of their electrical conductivity, for instance, polyetherimide (PEI)/graphene composite foams.[101] As shown in Fig. 6.83, lightweight microcellular polyetherimide (PEI)/graphene nanocomposite foams are fabricated, whose unique structure is believed to attenuate the incident electromagnetic microwaves by reflecting and scattering. Studies have shown that the specific EMI SE of PEI/graphene foam is 44.1 db/(g/cm^3) for 10 wt.% loading at the X-band.

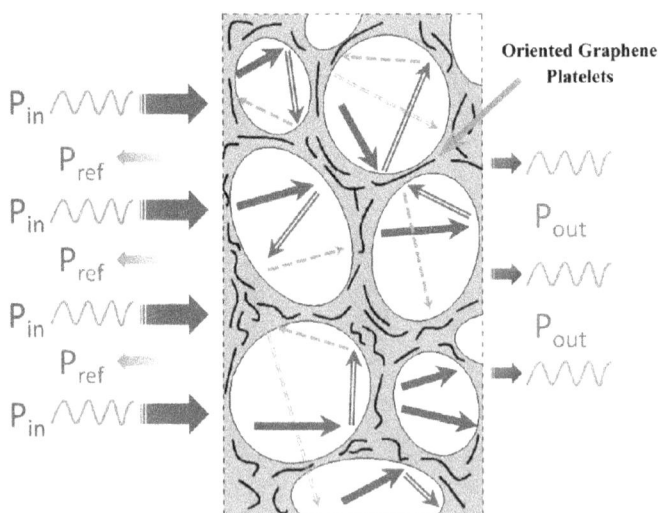

Figure 6.83.   Schematic description of the microwave transfer across PEI/graphene nanocomposite foam.[101]

The decreased reflection of microwaves is due to the composition and microcellular cell structure of the polymer/graphene composite. PEI, a kind of high-performance polymer, possesses excellent flame retardancy, low smoke generation, and good mechanical properties, due to which it can be used as a matrix for composites. Graphene is used as a filler in the composite to enhance its electrical conductivity. The enrichment and orientation of graphene during cell growth further improves the electrical conductivity of the composite, which allows the composite to adsorb and reflect electromagnetic waves. The spherical microscale air bubbles in the foams could attenuate the incident electromagnetic microwaves by reflecting and scattering between the cell wall and nanofillers, and the microwaves cannot escape without difficulty from the sample before being absorbed and transferred to heat. In addition, the foam structure of the composite makes it exhibit high mechanical strength (Young's modulus: 180~290 MPa) with a low density (0.3 g/cm^3).

The selection of different polymers and the optimization of the structure of the composite can further improve the EMI SE. As shown in Fig. 6.84, poly(dimethyl siloxane) (PDMS) is coated on the graphene grown on a nickel foam, and the nickel foam is then etched away using HCl solution to obtain the graphene/PDMS foam composite.[102] Its shielding effectiveness is as high as 30 dB in the 30 MHz–1.5 GHz frequency range and 20 dB in the X-band frequency range for a very low graphene loading of < 0.8 wt.%. This composite has a density of 0.06 g/cm^3, with a specific EMI shielding effectiveness of up to 500 dB/(g/cm^3). In addition, incorporating graphene with high-conductivity polymer materials, such as polyaniline (PANI), to fabricate graphene composite with better

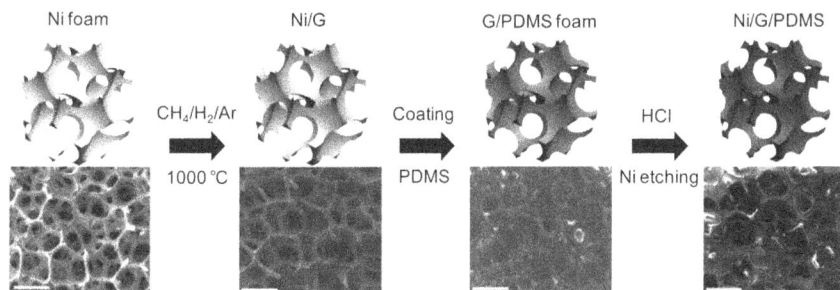

Figure 6.84. Schematic of the procedure for fabricating graphene/PDMS foam composites.[102]

conductivity can improve the reflection and absorption of electromagnetic waves, thereby enhancing EMI SE.[103]

As mentioned earlier, improving the magnetic permeability of materials can effectively enhance the EMI SE of graphene composites. Decorating graphene with magnetic materials is one of the effective means of improving the electromagnetic wave absorption performance of composites, including Fe_3O_4 or Fe_2O_3.[104] However, the EMI shielding materials are sometimes utilized to protect high-precision electronic instruments, where the EMI shield should be non-magnetic as the interference of magnetic materials also influences the instruments. Nanocomposites fabricated by graphene nanoribbons and non-magnetic MnO_2 is an excellent example for satisfying such requiement.[105] The presence of MnO_2 on graphene nanoribbons enhances the interfacial polarization, multiple scattering, natural resonances and the effective anisotropy energy, which leads to absorption dominated high shielding effectiveness of −57 dB (blocking >99.9999% radiation) by a 3 mm thick sample.

6.6.2 *Summary*

In summary, although advances in graphene-based EMI shielding materials are transpiring, the research on such materials is still in its infancy and great effort is still required to further the development in this field. First, one must fabricate and optimize graphene-based EMI shielding materials with various architectures, such as lamellar, paper-like, foam or porous structure. Second, one should use graphene-based EMI shielding materials that have a light weight, broadband, high efficiency, and stability at elevated temperatures. Third, researchers should further discuss the mechanism of EMI shielding, matching EM parameters, and synergizing both reflection and absorption at suitable penetration depths. In addition, green application is also an important factor to be considered, which needs strong absorption toward penetrating EM waves and weak secondary reflection from the surface.

6.7 Seawater Desalination

Seawater desalination refers to the removal of a large amount of salt and various impurities from seawater, after which seawater is turned into freshwater. A seawater desalination membrane is an essential component of the desalination process. Separation membranes can be fabricated using a variety of materials, such as polymers, zeolites, organic silicon, carbon nanotubes, and graphene.

6.7.1 *GO as desalination material*

Among many materials, GO can be more easily fabricated and engineered due to its size, surface chemistry, and structure. In addition, its manufacturing consumes considerably less energy, thus making it economically competitive (e.g., 500–1000 MJ/kg for solvent exfoliation of graphite or chemical reduction of GO, compared to 100,000 MJ/kg for CNTs).[106]

The large number of deprotonated carboxyl groups on GO gives rise to electrostatic repulsion between the sheets. Due to the electrostatic repulsion between the sheets, GO is water dispersible and the sheets do not aggregate. This feature allows for convenient processing of graphene materials as water can be used in lieu of organic solvents. For technical applications, oxygen-based functionality allows for enhanced surface hydrophilicity and aqueous stability. Further, GO can undergo various physical transformations. For example, 2D GO is structurally engineered to have different crumpled morphologies to give specific properties (e.g., aggregation resistance). These morphologies include paper ball-like spheres and corrugated surfaces (Fig. 6.85), among others. Another attractive feature of graphene is its intrinsic antimicrobial properties, which have led to applications in antimicrobial coatings and antifouling membranes. Proposed mechanisms of bacterial inactivation/cell membrane damage include physical disruption, oxidative stress, and extraction of phospholipids from cell membranes.

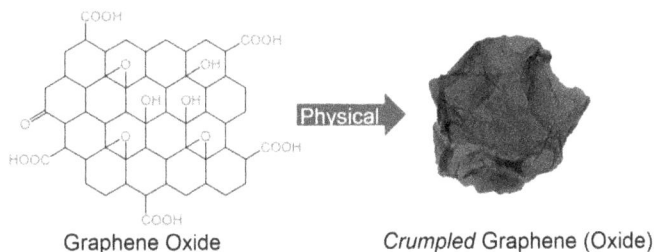

Graphene Oxide   *Crumpled* Graphene (Oxide)

Figure 6.85. Illustration of synthesis of crumpled graphene (oxide) from a 2D GO material.[106]

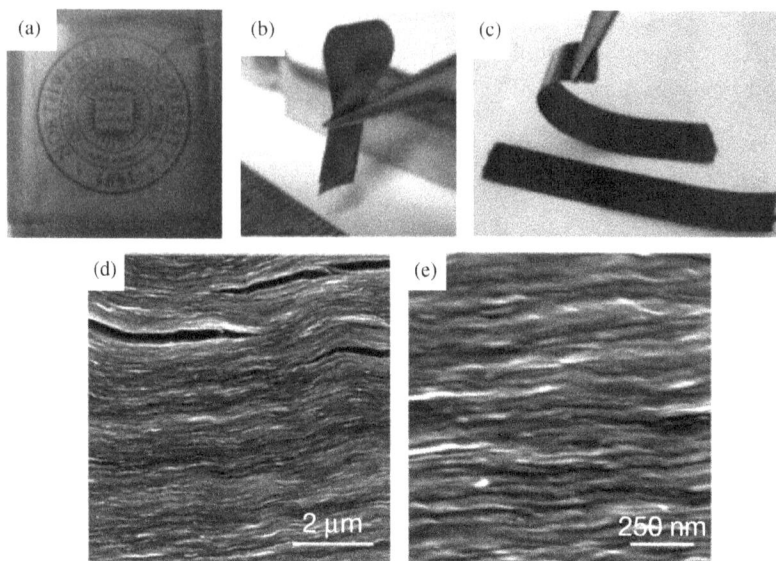

Figure 6.86.   Morphology and structure of GO paper.[107] (a) ~1-mm-thick film; (b) folded ~5-mm-thick semitransparent film; (c) folded ~25-mm-thick strip; (d) middle-resolution SEM image; and (e) high-resolution SEM image.

GO paper (free-standing GO without support) is an important GO-based separation membrane (Fig. 6.86). GO paper laminates are a surface-deposited collection of micron-sized GO crystallites forming an interlocked layered structure. GO layer-to-layer distance (d-spacing) has been estimated to be about 0.83 nm from X-ray diffraction experiments. Such spacing is hypothesized to allow for low-friction flow of water, while being able to reject other gases, including helium (i.e., H_2O vapor permeates through the membranes at least 10^{10} times faster than He). While such layers are vacuum-tight in a dry state, if immersed in water, they can act as molecular sieves, blocking all solutes with hydrated radii larger than 4.5 angstroms. Interestingly, smaller ions permeate through the membranes at rates thousands of times faster than what is expected for simple diffusion, which is attributed to capillary-like high pressures.

Another example is a kind of functionalized GO membranes, which are GO paper-like selective and/or functional layers on top of porous supports (e.g., polymeric polysulfone (Psf), polyethersulfone (PES) membranes) (Fig. 6.87). Conceptually, such membranes are made by deposition

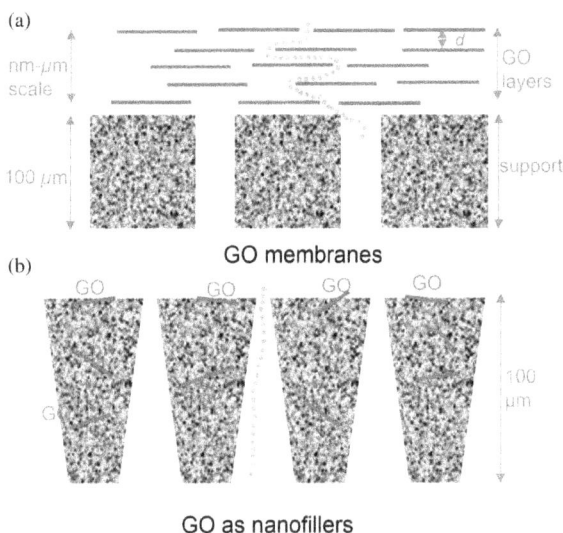

Figure 6.87.   GO/polymer composites for seawater desalination[106]: (a) schematic diagram of GO membrane structure on top of porous supports for water purification and (b) schematic diagram of GO as nanofillers in polymeric membranes.

of a thin layer of GO or GO nanocomposites (a few nm to μm) onto a relatively thick support membrane (usually >100 μm) via various techniques, such as vacuum filtration and layer-by-layer deposition. GO layers deposited onto polymeric supports are typically thinner than the free-standing GO papers. The deposited layers are hypothesized to form nanochannels, which facilitate fast water transport, while also achieving selectivity.

In addition, small amounts of GO (usually 0.1–2 wt% with respect to polymer) can be incorporated into conventional polymer structures, including PSF, PES, polyvinylidene fluoride (PVDF), and polyamide RO membranes (Fig. 6.87(b) and Fig. 6.88). GO is hypothesized to migrate to the top (membrane) surface during phase inversion. The water contacting angle measurements find an average decrease of ca. 20° after the introduction of GO, making it more hydrophilic. In addition to an increase in surface hydrophilicity, overall porosity also increases, and as a result, 2–20-fold enhancement in water flux is observed (due to GO additions). Rejection improvement can vary from a few percent to almost 3 times, depending on the polymers, GO percentage, and test foulants. Generally, an optimal GO percentage balances water permeability and rejection

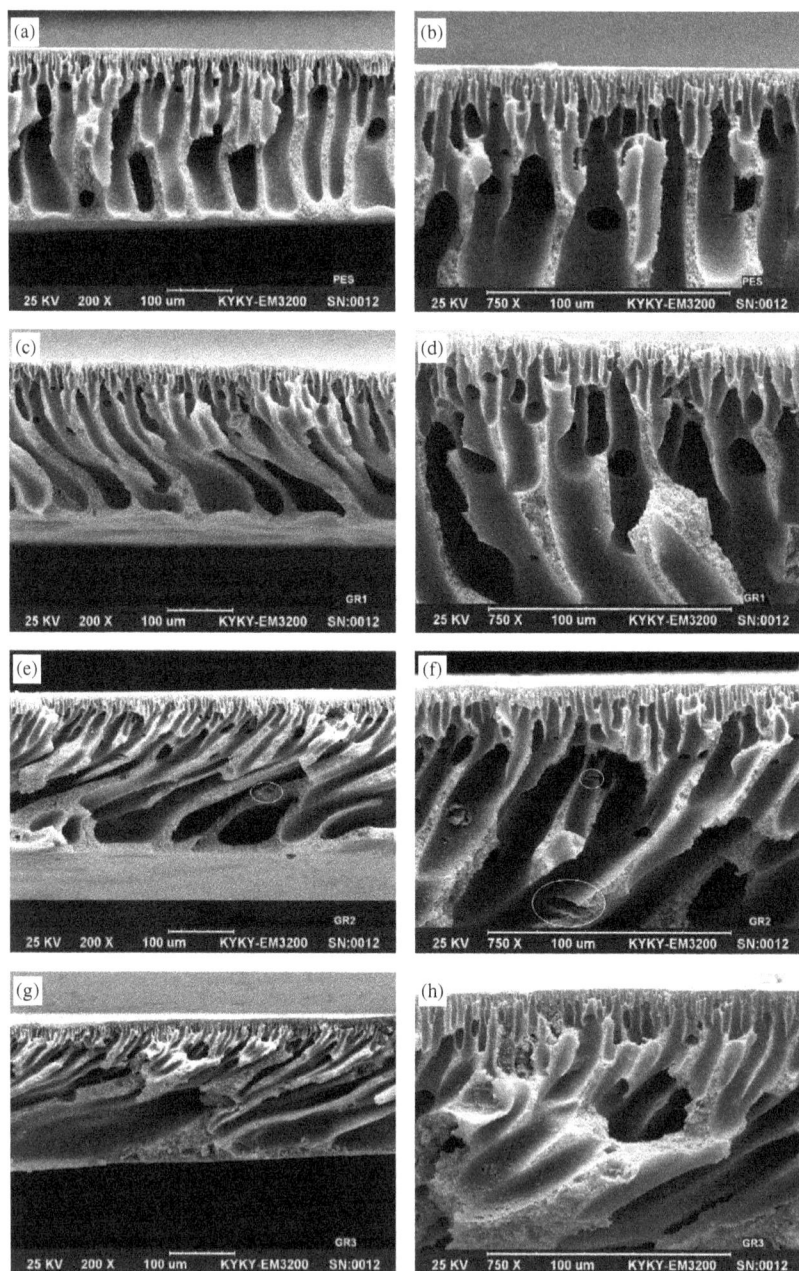

Figure 6.88. Cross-section SEM images of the prepared GO nanofiller for seawater desalination[108]: (a–b) unfilled polymer; (c–d) GO 0.1 wt.%; (e–f) GO 0.5 wt.%; and (g–h) GO 1wt.%.

rates, which conforms to the classical trade-off between permeability and selectivity associated with nano- and ultrafiltration membranes. With regard to RO technologies, size-fractionated GO (10–200 nm) was dispersed in the aqueous solution of *m*-phenylenediamine (MPD) before interfacial polymerization to make GO-embedded polyamide layers. Water permeability and the anti-biofouling property are found to have increased by approximately 80% and 98% (based on the biovolume), respectively, without loss of salt rejection.

6.7.2 *Summary*

The most distinctive advantage of GO as a nanofiller is the ease with which it can be readily integrated into current state-of-art technologies for membrane fabrication, including phase inversion or interfacial polymerization. With regard to GO, fundamental questions remain, including the relationships of GO properties (size, surface chemistry, etc.) with fabrication processes, the performance of GO, and better dispersion approaches of GO or GO nanocomposites in polymeric solutions. In general, a detailed understanding of the effects of GO addition on the thermodynamic and kinetic aspects of the phase inversion and interfacial polymerization processes is necessary.

References

1. Terrones, M., Martín, O., González, M., Pozuelo, J., Serrano, B., Cabanelas, J. C., Vega-Díaz, S. M., and Baselga, J. (2011). Interphases in graphene polymer-based nanocomposites: Achievements and challenges, *Adv. Mater.*, 23(44), 5302–5310.
2. Zhang, Y., Gong, S., Zhang, Q., Ming, P., Wan, S., Peng, J., Jiang, L., and Cheng, Q. (2016). Graphene-based artificial nacre nanocomposites, *Chem. Soc. Rev.*, 45(9), 2378–2395.
3. Xie, G., Zhang, K., Guo, B., Liu, Q., Fang, L., and Gong, J. R. (2013). Graphene-based materials for hydrogen generation from light-driven water splitting, *Adv. Mater.*, 25(28), 3820–3839.
4. Meng, L., Zhang, Y., Wan, X., Li, C., Zhang, X., Wang, Y., Ke, X., Xiao, Z., Ding, L., Xia, R., Yip, H.-L., Cao, Y., and Chen, Y. (2018). Organic and solution-processed tandem solar cells with 17.3% efficiency, *Science*, 361(6407), 1094–1098.
5. Hsu, C.-L., Lin, C.-T., Huang, J.-H., Chu, C.-W., Wei, K.-H., and Li, L.-J. (2012). Layer-by-layer graphene/TCNQ stacked films as conducting anodes for organic solar cells, *ACS Nano*, 6(6), 5031–5039.

6. Zhang, D., Xie, F., Lin, P., and Choy, W. C. H. (2013). Al-TiO_2 composite-modified single-layer graphene as an efficient transparent cathode for organic solar cells, *ACS Nano*, 7(2), 1740–1747.

7. Liu, Z., Liu, Q., Huang, Y., Ma, Y., Yin, S., Zhang, X., Sun, W., and Chen, Y. (2008). Organic photovoltaic devices based on a novel acceptor material: Graphene, *Adv. Mater.*, 20(20), 3924–3930.

8. Bonaccorso, F., Balis, N., Stylianakis, M. M., Savarese, M., Adamo, C., Gemmi, M., Pellegrini, V., Stratakis, E., and Kymakis, E. (2015). Functionalized graphene as an electron-cascade acceptor for air-processed organic ternary solar cells, *Adv. Funct. Mater.*, 25(25), 3870–3880.

9. Yun, J.-M., Yeo, J.-S., Kim, J., Jeong, H.-G., Kim, D.-Y., Noh, Y.-J., Kim, S.-S., Ku, B.-C., and Na, S.-I. (2011). Solution-processable reduced graphene oxide as a novel alternative to PEDOT:PSS hole transport layers for highly efficient and stable polymer solar cells, *Adv. Mater.*, 23(42), 4923–4928.

10. Yang, D., Zhou, L., Chen, L., Zhao, B., Zhang, J., and Li, C. (2012). Chemically modified graphene oxides as a hole transport layer in organic solar cells, *Chem. Commun.*, 48(65), 8078–8080.

11. Stratakis, E., Savva, K., Konios, D., Petridis, C., and Kymakis, E. (2014). Improving the efficiency of organic photovoltaics by tuning the work function of graphene oxide hole transporting layers, *Nanoscale*, 6(12), 6925–6931.

12. Beliatis, M. J., Gandhi, K. K., Rozanski, L. J., Rhodes, R., McCafferty, L., Alenezi, M. R., Alshammari, A. S., Mills, C. A., Jayawardena, K. D. G. I., Henley, S. J., and Silva, S. R. P. (2014). Hybrid graphene-metal oxide solution processed electron transport layers for large area high-performance organic photovoltaics, *Adv. Mater.*, 26(13), 2078–2083.

13. Low, F. W. and Lai, C. W. (2018). Recent developments of graphene-TiO_2 composite nanomaterials as efficient photoelectrodes in dye-sensitized solar cells: A review, *Renew. Sust. Energ. Rev.*, 82, 103–125.

14. Yang, N., Zhai, J., Wang, D., Chen, Y., and Jiang, L. (2010). Two-dimensional graphene bridges enhanced photoinduced charge transport in dye-sensitized solar cells, *ACS Nano*, 4(2), 887–894.

15. Akilimali, R., Selopal, G. S., Benetti, D., Serrano-Esparza, I., Algarabel, P. A., De Teresa, J. M., Wang, Z. M., Stansfield, B., Zhao, H., and Rosei, F. (2018). Hybrid TiO_2-Graphene nanoribbon photoanodes to improve the photoconversion efficiency of dye sensitized solar cells, *J. Power Sources*, 396, 566–573.

16. Ju, M. J., Kim, J. C., Choi, H.-J., Choi, I. T., Kim, S. G., Lim, K., Ko, J., Lee, J.-J., Jeon, I.-Y., Baek, J.-B., and Kim, H. K. (2013). N-Doped graphene nanoplatelets as superior metal-free counter electrodes for organic dye-sensitized solar cells, *ACS Nano*, 7(6), 5243–5250.

17. Guarracino, P., Gatti, T., Canever, N., Abdu-Aguye, M., Loi, M. A., Menna, E., and Franco, L. (2017). Probing photoinduced electron-transfer in graphene–dye hybrid materials for DSSC, *PCCP*, 19(40), 27716–27724.

18. Tsai, C.-H., Chuang, P.-Y., and Hsu, H.-L. (2018). Adding graphene nanosheets in liquid electrolytes to improve the efficiency of dye-sensitized solar cells, *Mater. Chem. Phys.*, 207, 154–160.

19. Yang, T., Gao, L., Lu, J., Ma, C., Du, Y., Wang, P., Ding, Z., Wang, S., Xu, P., Liu, D., Li, H., Chang, X., Fang, J., Tian, W., Yang, Y., Liu, S., and Zhao, K. (2023). One-stone-for-two-birds strategy to attain beyond 25% perovskite solar cells, *Nat. Commun.*, 14(1), 839.

20. Kakavelakis, G., Maksudov, T., Konios, D., Paradisanos, I., Kioseoglou, G., Stratakis, E., and Kymakis, E. (2017). Efficient and highly air stable planar inverted perovskite solar cells with reduced graphene oxide doped PCBM electron transporting layer, *Adv. Energy Mater.*, 7(7), 1602120.

21. Xie, J., Huang, K., Yu, X., Yang, Z., Xiao, K., Qiang, Y., Zhu, X., Xu, L., Wang, P., Cui, C., and Yang, D. (2017). Enhanced electronic properties of SnO_2 via electron transfer from graphene quantum dots for efficient perovskite solar cells, *ACS Nano*, 11(9), 9176–9182.

22. Jokar, E., Huang, Z. Y., Narra, S., Wang, C.-Y., Kattoor, V., Chung, C.-C., and Diau, E. W.-G. (2018). Anomalous charge-extraction behavior for graphene-oxide (GO) and reduced graphene-oxide (rGO) films as efficient p-contact layers for high-performance perovskite solar cells, *Adv. Energy Mater.*, 8(3), 1701640.

23. Heo, J. H., Shin, D. H., Kim, S., Jang, M. H., Lee, M. H., Seo, S. W., Choi, S.-H., and Im, S. H. (2017). Highly efficient $CH_3NH_3PbI_3$ perovskite solar cells prepared by $AuCl_3$-doped graphene transparent conducting electrodes, *Chem. Eng. J.*, 323, 153–159.

24. Kim, G.-H., Jang, H., Yoon, Y. J., Jeong, J., Park, S. Y., Walker, B., Jeon, I.-Y., Jo, Y., Yoon, H., Kim, M., Baek, J.-B., Kim, D. S., and Kim, J. Y. (2017). Fluorine functionalized graphene nano platelets for highly stable inverted perovskite solar cells, *Nano Lett.*, 17(10), 6385–6390.

25. Yang, M.-Q., Zhang, N., Pagliaro, M., and Xu, Y.-J. (2014). Artificial photosynthesis over graphene–semiconductor composites. Are we getting better? *Chem. Soc. Rev.*, 43(24), 8240–8254.

26. Pan, H., Zhu, S., Lou, X., Mao, L., Lin, J., Tian, F., and Zhang, D. (2015). Graphene-based photocatalysts for oxygen evolution from water, *RSC Adv.*, 5(9), 6543–6552.

27. Li, K., An, X., Park, K. H., Khraisheh, M., and Tang, J. (2014). A critical review of CO_2 photoconversion: Catalysts and reactors, *Catal. Today*, 224, 3–12.

28. Bell, N. J., Ng, Y. H., Du, A., Coster, H., Smith, S. C., and Amal, R. (2011). Understanding the enhancement in photoelectrochemical properties of

photocatalytically prepared TiO_2-reduced graphene oxide composite, *J. Phys. Chem. C*, 115(13), 6004–6009.

29. Kumar, D., Lee, A., Lee, T., Lim, M., and Lim, D.-K. (2016). Ultrafast and efficient transport of hot plasmonic electrons by graphene for Pt free, highly efficient visible-light responsive photocatalyst, *Nano Lett.*, 16(3), 1760–1767.

30. Zhong, C., Deng, Y., Hu, W., Qiao, J., Zhang, L., and Zhang, J. (2015). A review of electrolyte materials and compositions for electrochemical supercapacitors, *Chem. Soc. Rev.*, 44(21), 7484–7539.

31. El-Kady, M. F., Strong, V., Dubin, S., and Kaner, R. B. (2012). Laser scribing of high-performance and flexible graphene-based electrochemical capacitors, *Science*, 335(6074), 1326–1330.

32. Cao, X., Shi, Y., Shi, W., Lu, G., Huang, X., Yan, Q., Zhang, Q., and Zhang, H. (2011). Preparation of novel 3D graphene networks for supercapacitor applications, *Small*, 7(22), 3163–3168.

33. Xie, Y., Liu, Y., Zhao, Y., Tsang, Y. H., Lau, S. P., Huang, H., and Chai, Y. (2014). Stretchable all-solid-state supercapacitor with wavy shaped polyaniline/graphene electrode, *J. Mater. Chem. A*, 2(24), 9142–9149.

34. El-Kady, M. F. and Kaner, R. B. (2013). Scalable fabrication of high-power graphene micro-supercapacitors for flexible and on-chip energy storage, *Nat. Commun.*, 4(1), 1475.

35. Gao, W., Singh, N., Song, L., Liu, Z., Reddy, A. L. M., Ci, L., Vajtai, R., Zhang, Q., Wei, B., and Ajayan, P. M. (2011). Direct laser writing of micro-supercapacitors on hydrated graphite oxide films, *Nat. Nanotechnol.*, 6(8), 496–500.

36. Liang, S., Yan, W., Wu, X., Zhang, Y., Zhu, Y., Wang, H., and Wu, Y. (2018). Gel polymer electrolytes for lithium ion batteries: Fabrication, characterization and performance, *Solid State Ionics*, 318, 2–18.

37. Kim, H., Park, K.-Y., Hong, J., and Kang, K. (2014). All-graphene-battery: Bridging the gap between supercapacitors and lithium ion batteries, *Sci. Rep.*, 4(1), 5278.

38. Yan, Y., Yin, Y.-X., Xin, S., Guo, Y.-G., and Wan, L.-J. (2012). Ionothermal synthesis of sulfur-doped porous carbons hybridized with graphene as superior anode materials for lithium-ion batteries, *Chem. Commun.*, 48(86), 10663–10665.

39. Jiang, T., Bu, F., Feng, X., Shakir, I., Hao, G., and Xu, Y. (2017). Porous Fe_2O_3 nanoframeworks encapsulated within three-dimensional graphene as high-performance flexible anode for lithium-ion battery, *ACS Nano*, 11(5), 5140–5147.

40. Suresh, S., Wu, Z. P., Bartolucci, S. F., Basu, S., Mukherjee, R., Gupta, T., Hundekar, P., Shi, Y., Lu, T.-M., and Koratkar, N. (2017). Protecting silicon film anodes in lithium-ion batteries using an atomically thin graphene drape, *ACS Nano*, 11(5), 5051–5061.

41. Chen, K.-S., Xu, R., Luu, N. S., Secor, E. B., Hamamoto, K., Li, Q., Kim, S., Sangwan, V. K., Balla, I., Guiney, L. M., Seo, J.-W. T., Yu, X., Liu, W., Wu, J., Wolverton, C., Dravid, V. P., Barnett, S. A., Lu, J., Amine, K., and Hersam, M. C. (2017). Comprehensive enhancement of nanostructured lithium-ion battery cathode materials via conformal graphene dispersion, *Nano Lett.*, 17(4), 2539–2546.
42. Zhou, G. (2017). Design, fabrication and electrochemical performance of nanostructured carbon based materials for high-energy lithium–sulfur batteries: Next-generation high performance lithium–sulfur batteries, in *Graphene–Pure Sulfur Sandwich Structure for Ultrafast, Long-Life Lithium-Sulfur Batteries*, pp. 75–94 (Springer, Singapore).
43. Song, J., Yu, Z., Gordin, M. L., and Wang, D. (2016). Advanced sulfur cathode enabled by highly crumpled nitrogen-doped graphene sheets for high-energy-density lithium–sulfur batteries, *Nano Lett.*, 16(2), 864–870.
44. Fang, R., Zhao, S., Pei, S., Qian, X., Hou, P.-X., Cheng, H.-M., Liu, C., and Li, F. (2016). Toward more reliable lithium–sulfur batteries: An all-graphene cathode structure, *ACS Nano*, 10(9), 8676–8682.
45. Xing, L.-B., Xi, K., Li, Q., Su, Z., Lai, C., Zhao, X., and Kumar, R. V. (2016). Nitrogen, sulfur-codoped graphene sponge as electroactive carbon interlayer for high-energy and -power lithium–sulfur batteries, *J. Power Sources*, 303, 22–28.
46. Aurbach, D., McCloskey, B. D., Nazar, L. F., and Bruce, P. G. (2016). Advances in understanding mechanisms underpinning lithium–air batteries, *Nat. Energy*, 1(9), 16128.
47. Jeong, Y. S., Park, J.-B., Jung, H.-G., Kim, J., Luo, X., Lu, J., Curtiss, L., Amine, K., Sun, Y.-K., Scrosati, B., and Lee, Y. J. (2015). Study on the catalytic activity of noble metal nanoparticles on reduced graphene oxide for oxygen evolution reactions in lithium–air batteries, *Nano Lett.*, 15(7), 4261–4268.
48. Zhu, T., Li, X., Zhang, Y., Yuan, M., Sun, Z., Ma, S., Li, H., and Sun, G. (2018). Three-dimensional reticular material NiO/Ni-graphene foam as cathode catalyst for high capacity lithium-oxygen battery, *J. Electroanal. Chem.*, 823, 73–79.
49. Xiao, J., Mei, D., Li, X., Xu, W., Wang, D., Graff, G. L., Bennett, W. D., Nie, Z., Saraf, L. V., Aksay, I. A., Liu, J., and Zhang, J.-G. (2011). Hierarchically porous graphene as a lithium–air battery electrode, *Nano Lett.*, 11(11), 5071–5078.
50. Reina, G., González-Domínguez, J. M., Criado, A., Vázquez, E., Bianco, A., and Prato, M. (2017). Promises, facts and challenges for graphene in biomedical applications, *Chem. Soc. Rev.*, 46(15), 4400–4416.
51. Masoudipour, E., Kashanian, S., and Maleki, N. (2017). A targeted drug delivery system based on dopamine functionalized nano graphene oxide, *Chem. Phys. Lett.*, 668, 56–63.

52. Ren, L., Zhang, Y., Cui, C., Bi, Y., and Ge, X. (2017). Functionalized graphene oxide for anti-VEGF siRNA delivery: Preparation, characterization and evaluation in vitro and in vivo, *RSC Adv.*, 7(33), 20553–20566.

53. Ryoo, S.-R., Lee, J., Yeo, J., Na, H.-K., Kim, Y.-K., Jang, H., Lee, J. H., Han, S. W., Lee, Y., Kim, V. N., and Min, D.-H. (2013). Quantitative and multiplexed microRNA densing in living cells based on peptide nucleic acid and nano graphene oxide (PANGO), *ACS Nano*, 7(7), 5882–5891.

54. He, Q., Wu, S., Yin, Z., and Zhang, H. (2012). Graphene-based electronic sensors, *Chem. Sci.*, 3(6), 1764–1772.

55. Gao, Z., Xia, H., Zauberman, J., Tomaiuolo, M., Ping, J., Zhang, Q., Ducos, P., Ye, H., Wang, S., Yang, X., Lubna, F., Luo, Z., Ren, L., and Johnson, A. T. C. (2018). Detection of sub-fM DNA with target recycling and self-assembly amplification on graphene field-effect biosensors, *Nano Lett.*, 18(6), 3509–3515.

56. Pang, Y., Zhang, K., Yang, Z., Jiang, S., Ju, Z., Li, Y., Wang, X., Wang, D., Jian, M., Zhang, Y., Liang, R., Tian, H., Yang, Y., and Ren, T.-L. (2018). Epidermis microstructure inspired graphene pressure sensor with random distributed spinosum for high sensitivity and large linearity, *ACS Nano*, 12(3), 2346–2354.

57. Yang, K., Zhang, S., Zhang, G., Sun, X., Lee, S.-T., and Liu, Z. (2010). Graphene in mice: Ultrahigh in vivo tumor uptake and efficient photothermal therapy, *Nano Lett.*, 10(9), 3318–3323.

58. Ma, X., Qu, Q., Zhao, Y., Luo, Z., Zhao, Y., Ng, K. W., and Zhao, Y. (2013). Graphene oxide wrapped gold nanoparticles for intracellular Raman imaging and drug delivery, *J. Mater. Chem. B*, 1(47), 6495–6500.

59. Wang, Y., Wang, H., Liu, D., Song, S., Wang, X., and Zhang, H. (2013). Graphene oxide covalently grafted upconversion nanoparticles for combined NIR mediated imaging and photothermal/photodynamic cancer therapy, *Biomaterials*, 34(31), 7715–7724.

60. Perreault, F., Fonseca de Faria, A., and Elimelech, M. (2015). Environmental applications of graphene-based nanomaterials, *Chem. Soc. Rev.*, 44(16), 5861–5896.

61. Boruah, P. K., Borthakur, P., Darabdhara, G., Kamaja, C. K., Karbhal, I., Shelke, M. V., Phukan, P., Saikia, D., and Das, M. R. (2016). Sunlight assisted degradation of dye molecules and reduction of toxic Cr(vi) in aqueous medium using magnetically recoverable Fe3O4/reduced graphene oxide nanocomposite, *RSC Adv.*, 6(13), 11049–11063.

62. Li, J., Shao, Z., Chen, C., and Wang, X. (2014). Hierarchical GOs/Fe_3O_4/ PANI magnetic composites as adsorbent for ionic dye pollution treatment, *RSC Adv.*, 4(72), 38192–38198.

63. Chen, Y., Chen, L., Bai, H., and Li, L. (2013). Graphene oxide–chitosan composite hydrogels as broad-spectrum adsorbents for water purification, *J. Mater. Chem. A*, 1(6), 1992–2001.

64. Kemp, K. C., Seema, H., Saleh, M., Le, N. H., Mahesh, K., Chandra, V., and Kim, K. S. (2013). Environmental applications using graphene composites water remediation and gas adsorption, *Nanoscale*, 5, 23.

65. Jaihindh, D. P., Chen, C.-C., and Fu, Y.-P. (2018). Reduced graphene oxide-supported Ag-loaded Fe-doped TiO_2 for the degradation mechanism of methylene blue and its electrochemical properties, *RSC Adv.*, 8(12), 6488–6501.

66. Yusuf, M., Khan, M. A., Abdullah, E. C., Elfghi, M., Hosomi, M., Terada, A., Riya, S., and Ahmad, A. (2016). Dodecyl sulfate chain anchored mesoporous graphene: Synthesis and application to sequester heavy metal ions from aqueous phase, *Chem. Eng. J.*, 304, 431–439.

67. Luo, S., Xu, X., Zhou, G., Liu, C., Tang, Y., and Liu, Y. (2014). Amino siloxane oligomer-linked graphene oxide as an efficient adsorbent for removal of Pb(II) from wastewater, *J. Hazard. Mater.*, 274, 145–155.

68. Venkateswarlu, S., Lee, D., and Yoon, M. (2016). Bioinspired 2D-carbon flakes and Fe_3O_4 nanoparticles composite for arsenite removal, *ACS Appl. Mater. Interfaces*, 8(36), 23876–23885.

69. Li, L., Luo, C., Li, X., Duan, H., and Wang, X. (2014). Preparation of magnetic ionic liquid/chitosan/graphene oxide composite and application for water treatment, *Int. J. Biol. Macromol.*, 66, 172–178.

70. Salam, M. A., Fageeh, O., Al-Thabaiti, S. A., and Obaid, A. Y. (2015). Removal of nitrate ions from aqueous solution using zero-valent iron nanoparticles supported on high surface area nanographenes, *J. Mol. Liq.*, 212, 708–715.

71. Fei, Y., Li, Y., Han, S., and Ma, J. (2016). Adsorptive removal of ciprofloxacin by sodium alginate/graphene oxide composite beads from aqueous solution, *J. Colloid Interface Sci.*, 484, 196–204.

72. Tang, Y., Guo, H., Xiao, L., Yu, S., Gao, N., and Wang, Y. (2013). Synthesis of reduced graphene oxide/magnetite composites and investigation of their adsorption performance of fluoroquinolone antibiotics, *Colloids Surf. A Physicochem. Eng. Asp.*, 424, 74–80.

73. Song, S., Yang, H., Su, C., Jiang, Z., and Lu, Z. (2016). Ultrasonic-microwave assisted synthesis of stable reduced graphene oxide modified melamine foam with superhydrophobicity and high oil adsorption capacities, *Chem. Eng. J.*, 306, 504–511.

74. Ge, J., Zhao, H.-Y., Zhu, H.-W., Huang, J., Shi, L.-A., and Yu, S.-H. (2016). Advanced sorbents for oil-spill cleanup: Recent advances and future perspectives, *Adv. Mater.*, 28(47), 10459–10490.

75. Ge, J., Shi, L.-A., Wang, Y.-C., Zhao, H.-Y., Yao, H.-B., Zhu, Y.-B., Zhang, Y., Zhu, H.-W., Wu, H.-A., and Yu, S.-H. (2017). Joule-heated graphene-wrapped sponge enables fast clean-up of viscous crude-oil spill, *Nat. Nanotechnol.*, 12(5), 434–440.

76. Seredych, M. and Bandosz, T. J. (2009). Graphite oxide/AlZr polycation composites: Surface characterization and performance as adsorbents of ammonia, *Mater. Chem. Phys.*, 117(1), 99–106.
77. Fan, X., Zhang, G., and Zhang, F. (2015). Multiple roles of graphene in heterogeneous catalysis, *Chem. Soc. Rev.*, 44(10), 3023–3035.
78. Su, C. and Loh, K. P. (2013). Carbocatalysts: Graphene oxide and its derivatives, *Acc. Chem. Res.*, 46(10), 2275–2285.
79. Dreyer, D. R., Jia, H.-P., and Bielawski, C. W. (2010). Graphene oxide: A convenient carbocatalyst for facilitating oxidation and hydration reactions, *Angew. Chem. Int. Ed.*, 49(38), 6813–6816.
80. Boukhvalov, D. W., Dreyer, D. R., Bielawski, C. W., and Son, Y.-W. (2012). A computational investigation of the catalytic properties of graphene oxide: Exploring mechanisms by using DFT methods, *ChemCatChem*, 4(11), 1844–1849.
81. Navalon, S., Dhakshinamoorthy, A., Alvaro, M., Antonietti, M., and García, H. (2017). Active sites on graphene-based materials as metal-free catalysts, *Chem. Soc. Rev.*, 46(15), 4501–4529.
82. Wei, D., Liu, Y., Wang, Y., Zhang, H., Huang, L., and Yu, G. (2009). Synthesis of N-doped graphene by chemical vapor deposition and its electrical properties, *Nano Lett.*, 9(5), 1752–1758.
83. Kondo, T., Casolo, S., Suzuki, T., Shikano, T., Sakurai, M., Harada, Y., Saito, M., Oshima, M., Trioni, M. I., Tantardini, G. F., and Nakamura, J. (2012). Atomic-scale characterization of nitrogen-doped graphite: Effects of dopant nitrogen on the local electronic structure of the surrounding carbon atoms, *Phys. Rev. B*, 86(3), 035436.
84. Guo, D., Shibuya, R., Akiba, C., Saji, S., Kondo, T., and Nakamura, J. (2016). Active sites of nitrogen-doped carbon materials for oxygen reduction reaction clarified using model catalysts, *Science*, 351(6271), 361–365.
85. Yu, X., Han, P., Wei, Z., Huang, L., Gu, Z., Peng, S., Ma, J., and Zheng, G. (2018). Boron-doped graphene for electrocatalytic N_2 reduction, *Joule*, 2(8), 1610–1622.
86. Zhang, C., Mahmood, N., Yin, H., Liu, F., and Hou, Y. (2013). Synthesis of phosphorus-doped graphene and its multifunctional applications for oxygen reduction reaction and lithium ion batteries, *Adv. Mater.*, 25(35), 4932–4937.
87. Yang, Z., Yao, Z., Li, G., Fang, G., Nie, H., Liu, Z., Zhou, X., Chen, X. A., and Huang, S. (2012). Sulfur-doped graphene as an efficient metal-free cathode catalyst for oxygen reduction, *ACS Nano*, 6(1), 205–211.
88. Zheng, Y., Jiao, Y., Ge, L., Jaroniec, M., and Qiao, S. Z. (2013). Two-step boron and nitrogen doping in graphene for enhanced synergistic catalysis, *Angew. Chem.*, 52(11), 3110–3116.

89. Zhang, J. and Dai, L. (2016). Nitrogen, phosphorus, and fluorine tri-doped graphene as a multifunctional catalyst for self-powered electrochemical water splitting, *Angew. Chem. Int. Ed.*, 55(42), 13296–13300.
90. Ito, Y., Cong, W., Fujita, T., Tang, Z., and Chen, M. (2015). High catalytic activity of nitrogen and sulfur Co-doped nanoporous graphene in the hydrogen evolution reaction, *Angew. Chem. Int. Ed.*, 54(7), 2131–2136.
91. Zhang, L., Xu, Q., Niu, J., and Xia, Z. (2015). Role of lattice defects in catalytic activities of graphene clusters for fuel cells, *PCCP*, 17(26), 16733–16743.
92. Jia, Y., Zhang, L., Du, A., Gao, G., Chen, J., Yan, X., Brown, C. L., and Yao, X. (2016). Defect graphene as a trifunctional catalyst for electrochemical reactions, *Adv. Mater.*, 28(43), 9532–9538.
93. Zhou, M., Zhang, A., Dai, Z., Feng, Y. P., and Zhang, C. (2010). Strain-enhanced stabilization and catalytic activity of metal nanoclusters on graphene, *J. Phys. Chem. C*, 114(39), 16541–16546.
94. Fampiou, I. and Ramasubramaniam, A. (2012). Binding of Pt nanoclusters to point defects in graphene: Adsorption, morphology, and electronic structure, *J. Phys. Chem. C*, 116(11), 6543–6555.
95. Yang, X.-F., Wang, A., Qiao, B., Li, J., Liu, J., and Zhang, T. (2013). Single-atom catalysts: A new frontier in heterogeneous catalysis, *Acc. Chem. Res.*, 46(8), 1740–1748.
96. Axet, M. R., Durand, J., Gouygou, M., and Serp, P. (2019). Chapter Two — Surface coordination chemistry on graphene and two-dimensional carbon materials for well-defined single atom supported catalysts, *Adv. Organomet. Chem.*, 71, 53–174.
97. Qiu, H.-J., Ito, Y., Cong, W., Tan, Y., Liu, P., Hirata, A., Fujita, T., Tang, Z., and Chen, M. (2015). Nanoporous graphene with single-atom nickel dopants: An efficient and stable catalyst for electrochemical hydrogen production, *Angew. Chem. Int. Ed.*, 54(47), 14031–14035.
98. Yang, M., Wang, L., Li, M., Hou, T., and Li, Y. (2015). Structural stability and O_2 dissociation on nitrogen-doped graphene with transition metal atoms embedded: A first-principles study, *AIP Adv.*, 5(6), 4922841.
99. Fei, H., Dong, J., Arellano-Jiménez, M. J., Ye, G., Dong Kim, N., Samuel, E. L. G., Peng, Z., Zhu, Z., Qin, F., Bao, J., Yacaman, M. J., Ajayan, P. M., Chen, D., and Tour, J. M. (2015). Atomic cobalt on nitrogen-doped graphene for hydrogen generation, *Nat. Commun.*, 6(1), 8668.
100. Shahzad, F., Kumar, P., Yu, S., Lee, S., Kim, Y.-H., Hong, S. M., and Koo, C. M. (2015). Sulfur-doped graphene laminates for EMI shielding applications, *J. Mater. Chem. C*, 3(38), 9802–9810.
101. Ling, J., Zhai, W., Feng, W., Shen, B., Zhang, J., and Zheng, W. G. (2013). Facile preparation of lightweight microcellular polyetherimide/graphene

composite foams for electromagnetic interference shielding, *ACS Appl. Mater. Interfaces*, 5(7), 2677–2684.

102. Chen, Z., Xu, C., Ma, C., Ren, W., and Cheng, H.-M. (2013). Lightweight and flexible graphene foam composites for high-performance electromagnetic interference shielding, *Adv. Mater.*, 25(9), 1296–1300.

103. Singh, A. P., Mishra, M., Sambyal, P., Gupta, B. K., Singh, B. P., Chandra, A., and Dhawan, S. K. (2014). Encapsulation of γ-Fe_2O_3 decorated reduced graphene oxide in polyaniline core–shell tubes as an exceptional tracker for electromagnetic environmental pollution, *J. Mater. Chem. A*, 2(10), 3581–3593.

104. Cao, M.-S., Wang, X.-X., Cao, W.-Q., and Yuan, J. (2015). Ultrathin graphene: electrical properties and highly efficient electromagnetic interference shielding, *J. Mater. Chem. C*, 3(26), 6589–6599.

105. Gupta, T. K., Singh, B. P., Singh, V. N., Teotia, S., Singh, A. P., Elizabeth, I., Dhakate, S. R., Dhawan, S. K., and Mathur, R. B. (2014). MnO_2 decorated graphene nanoribbons with superior permittivity and excellent microwave shielding properties, *J. Mater. Chem. A*, 2(12), 4256–4263.

106. Jiang, Y., Biswas, P., and Fortner, J. D. (2016). A review of recent developments in graphene-enabled membranes for water treatment, *Environ. Sci. Water Res. Technol.*, 2(6), 915–922.

107. Dikin, D. A., Stankovich, S., Zimney, E. J., Piner, R. D., Dommett, G. H. B., Evmenenko, G., Nguyen, S. T., and Ruoff, R. S. (2007). Preparation and characterization of graphene oxide paper, *Nature*, 448(7152), 457–460.

108. Zinadini, S., Zinatizadeh, A. A., Rahimi, M., Vatanpour, V., and Zangeneh, H. (2014). Preparation of a novel antifouling mixed matrix PES membrane by embedding graphene oxide nanoplates, *J. Membr. Sci.*, 453, 292–301.

109. Xia, J., Chen, F., Li, J., Tao, N. (2009) Measurement of the quantum capacitance of graphene, *Nature Nanotech.*, 4, pp. 505–509.

Chapter 7

Summary and Outlook

Since its discovery in 2004, graphene has achieved remarkable progress in research and application within two decades, maintaining a strong momentum of development and yielding a continuous stream of break-throughs. However, graphene's poor dispersibility in conventional solvents poses challenges for direct processing and application, limiting its ability to meet diverse practical needs. As a result, functionalizing graphene has become essential to enhance its applicability and versatility. Notably, a majority of graphene materials reported to date are functionalized, highlighting the necessity and significance of this process.

This book integrates the author's research experience and insights in the field of functionalized graphene materials, providing a systematic review and classification of recent advancements. It establishes a clear definition and categorization of functionalized graphene materials, organizes various types of such materials, and explores their applications across multiple critical fields.

Through this analysis, it is evident that functionalized graphene materials are still in their early stages of development. Key concepts remain ambiguous, many functionalization methods are underdeveloped, and some materials fall short of practical application requirements. To drive future progress, the following areas require focused attention:

Concept clarification: Despite the diversity of functionalized graphene materials, definitions remain unclear. For instance, what distinguishes graphene from functionalized graphene? How should functionalized graphene materials be classified? To advance the field, the industry must establish precise definitions and frameworks to eliminate ambiguity.

Standard unification: While numerous functionalized graphene materials exist, performance evaluation standards for similar materials or those used in the same field are inconsistent. For example, what are the essential properties of functionalized graphene materials for a specific application? What evaluation methods and benchmarks should be applied? The industry must develop unified standards based on theoretical research and practical needs to ensure consistent and reliable performance assessment.

Preparation standardization: Although a variety of preparation methods for functionalized graphene materials are available, their lack of standardization leads to inconsistent material quality and poor reproducibility. Establishing standardized protocols is crucial to produce materials with stable and predictable performance.

In-depth research: Current research often focuses on material preparation, method description, and basic performance demonstration, with insufficient exploration of underlying mechanisms. For example, what role does functionalized graphene play in a device? How does it function, and what factors influence its performance? Future studies must delve deeper into these questions to uncover the materials' full potential and optimize their applications.

Performance uniqueness: Some functionalized graphene materials lack distinct advantages in certain applications, as other materials can perform equally well or better. Many functionalized graphene materials are merely substitutes rather than leveraging their unique properties. Future research should emphasize uncovering and harnessing the unique characteristics of these materials to differentiate them from alternatives.

Goal-oriented research: Many studies on functionalized graphene materials lack clear objectives, often being exploratory rather than targeted. This limits their relevance to real-world applications and hinders in-depth research. Material research must be goal-driven, focusing on systematic and thorough investigations into the structure–performance relationship to address key scientific challenges and drive practical innovation.

Despite the challenges, the field of functionalized graphene materials has made significant strides. With increased collaboration between academia and industry, along with standardized research methods and strategies, we can unlock the hidden potential of graphene materials and showcase their extraordinary capabilities.

Functionalized graphene materials are both beautiful and fascinating — we have only begun to glimpse their true potential. Their future is bright, and with continued effort, their full promise will be realized.

Index

www.ingramcontent.com/pod-product-compliance
Lightning Source LLC
Chambersburg PA
CBHW050539190326
41458CB00007B/1847